MINNESOTA STUDIES IN THE PHILOSOPHY OF SCIENCE

Minnesota Studies in the PHILOSOPHY OF SCIENCE

RONALD N. GIERE, GENERAL EDITOR

HERBERT FEIGL, FOUNDING EDITOR

VOLUME XVI

Origins of Logical Empiricism

EDITED BY

RONALD N. GIERE

AND

ALAN W. RICHARDSON

MINNESOTA

University of Minnesota Press

Minneapolis

London

A shorter version of Peter Galison, "Constructing Modernism: The Cultural Location of *Aufbau,*" appeared in F. Stadler, ed., *Scientific Philosophy: Origins and Developments,* pp. 75–93. Copyright 1993 by Kluwer Academic Publishers. Reprinted by permission of the publisher.

Published by the University of Minnesota Press
111 Third Avenue South, Suite 290, Minneapolis, MN 55401–2520
Printed in the United States of America on acid-free paper

Library of Congress Cataloging-in-Publication Data

Origins of logical empiricism / edited by Ronald N. Giere and Alan W. Richardson
p. cm. — (Minnesota studies in the philosophy of science ; v. 16)
ISBN 0-8166-2834-3 (hc)
1. Logical positivism. I. Giere, Ronald N. II. Richardson, Alan W. III. Series.
Q175.M64 Vol. 16
[B824.5]
501 s–dc20
[146′.42] 96–19926

Contents

PART III
LOGIC, MATHEMATICS, AND PHILOSOPHY

PART IV
EXPERIENCE, EMPIRICAL KNOWLEDGE, AND EMPIRICISM

PART V
AFTERWORD

Preface

Publication of this volume on the origins of logical empiricism to some extent represents an effort on the part of the Minnesota Center for Philosophy of Science to recover its own historical origins. The Center's founder, Herbert Feigl, was a student of Moritz Schlick, the organizing force behind the Vienna Circle. Feigl was one of the signers of the Circle's 1929 manifesto "Wissenschaftliche Weltauffassung: Der Wiener Kreis." He was among the first to publicize the works of Circle members in the United States. In founding the Minnesota Center in 1953 he re-created something of both the spirit and the organization of the Vienna Circle in the American Midwest. And the early volumes of the Minnesota Studies in the Philosophy of Science, which Feigl founded and edited, significantly shaped the field of philosophy of science in the following years. It is thus more than appropriate that this volume should appear now in that same series.

As general editor of the series, and coeditor of this volume, I would particularly like to thank my coeditor, Alan Richardson, for his many contributions to this project. It was largely through his efforts that we were able to bring together the contributors in the first place. And when the time came, it was Alan who read all the final drafts and put together a meaningful introduction. In the end, of course, it is the contributors who deserve our greatest thanks.

The original workshop was supported by the National Science Foundation through its program, Studies in Science, Technology, and Society. Additional funding was provided by the College of Liberal Arts and the Graduate School of the University of Minnesota. Finally, we all thank the one truly indispensable person, the principal secretary of the Center, Steve Lelchuk. He handled all the correspondence, saw to all the physical arrangements for the workshop, collected the manuscripts, and prepared the final bibliography.

RONALD N. GIERE

Alan W. Richardson

Introduction: Origins of Logical Empiricism

This volume grew out of a workshop on the origins of logical empiricism held in October 1993 under the auspices of the Minnesota Center for Philosophy of Science at the University of Minnesota. This workshop was dedicated to the idea that the ongoing reappraisal of logical empiricism is an endeavor worthy of being broadened and deepened.[1] Thus, a number of leading workers in the history of analytic philosophy, philosophy of science, logic, and science were brought together to share their perspectives and concerns. This volume, therefore, presents something of the state of the art of thought on the origins of logical empiricism.

Why is there now a reappraisal of logical empiricism and its role in analytic philosophy and philosophy of science? There are, no doubt, many reasons, but a few stand out as the important ones. First, perhaps more than any other aspect of analytic philosophy, logical empiricism did dominate its field of endeavor — principally philosophy of science — for a very long period. Logical empiricism provided the working framework of most philosophers of science from roughly the 1930s to the 1960s. Moreover, the issues placed at the heart of philosophy of science by the logical empiricists — the analysis or explication of important scientific and metascientific terms (confirmation, explanation, and so on) — continue to play a major role in philosophy of science, even as criteria of analytic adequacy change. It would not be too great an exaggeration to claim that philosophy of science as a discipline distinct from epistemology would not exist without the impetus of logical empiricism.[2] For this reason alone, philosophers of science should have an interest in the project of logical empiricism — regardless of whether they see this project as, in the end, fatally flawed.

More important than the historical role of the project, however, is the continuing role that its *demise* plays in the motivation of more recent projects in philosophy of science (and interdisciplinary science studies generally). Seemingly everyone has some story to tell about how their own project has overcome a difficulty inherent to logical empiricism. This is true, for example, of the adherents of the semantic approach to theories, realists, the historiographical schools following Thomas Kuhn, and naturalists.[3] In large measure such projects are defined against aspects of the logical empiricist project; for example, the easiest way to motivate the semantic approach to theories, given its myriad formulations, is via the problems it raises for the syntactic approach. Logical empiricism has been dead since the 1950s or the 1960s depending on whether W. V. O. Quine or Kuhn is taken to have delivered the death blow. Nonetheless, even into the 1990s it is not uncommon to see philosophers of science eagerly engaging in overcoming logical empiricism when explaining their views.[4]

This curious circumstance indicates that, at least in part, the attractiveness of these contemporary projects depends on their having exposed real weaknesses in logical empiricism. Thus, if philosophy of science is seeking its way from the old received view (logical empiricism) to a new received view (naturalism, realism, or what have you), part of the struggle for dominance is through continued concern with logical empiricism. Thus, an account of logical empiricism and its weaknesses is crucial to the contemporary battle for the hearts and minds of philosophers of science. Informed philosophers of science, groping toward the future with unsteady steps, therefore, only stand to gain by studying the history of their discipline, especially the role of logical empiricism in it.

Beyond this, it has become increasingly evident that certain of the tales told about logical empiricism are deficient in important ways. Some of the interpretative stories — for example, those that see logical empiricism as "first philosophy," as engaged in "foundationalist epistemology," or as espousing a view of logic as a "normative" discipline and *thus* distinct from psychology — impose upon logical empiricism philosophical perspectives quite at odds with the logical empiricists' own favored ways of expressing the project. We should, therefore, be very careful about explaining the project in these terms. Conversely, some of the favored vocabulary of the logical empiricists themselves is frequently ignored. Whereas, for example, the early work of the logical empiricists was frequently dominated by questions of the objectivity of knowledge, they are retrospectively seen as concerned only with justification or verification. These issues are not obviously identical, and any simple slide from one to the other threatens to misrepresent the epistemological project of early logical empiricism. Similarly, the rhetoric of a scientific philosophy that is ubiquitous in the early motivational writings of the logical empiricists (and should be, one might think, a matter of interpretative interest to naturalistic epistemologists and philosophers of science) is usually simply forgotten.

This problem is due in part to a rather naive historiography of philosophy. Analytic philosophers are not renowned for their attention to nuance in their reading of historical texts or their breadth of knowledge of the context in which those texts were written. Whether this is a failing of analytic philosophy or analytic philosophers need not concern us here. For the logical empiricists the problem is exacerbated by at least two additional factors.[5] First, as the logical empiricists were analytic philosophers themselves, any qualms an analytic philosopher may feel about imposing analytic frameworks on historical figures are ameliorated — Rudolf Carnap, Moritz Schlick, and the others surely share this framework, it is thought. But this is not so. For, on the one hand, the logical empiricists cannot even be said to share among themselves more than a very general scientific orientation (in their own words, *eine wissenschaftliche Weltauffassung*) in their philosophy. On the other hand, precisely the history of logical empiricism is as responsible as any other development for the founding of contemporary analytic perspectives. But this is something the logical empiricists were groping toward and surely cannot be simply imputed from the word "go" without thereby rendering the importance of the historical development empty. Moreover, it would be hard to see why contemporary analytic philosophers might feel warranted in dismissing certain central features of logical empiricism as utterly mistaken while feeling this theoretical kinship and sense of interpretative transparency with respect to it. The second problem is that in the course of development and particularly with the movement of logical empiricism out of Europe, the terms in which the project was motivated were significantly altered. The logical empiricists were masters of the misleading expression of their philosophical points of view as a way to let others into the project. As Carnap, Hans Reichenbach, and others moved to the United States, they took up native concerns and philosophical vocabularies. In this way the roots of the project were covered over, motivations obscured, and historical links severed.

For reasons such as these, it was decided in the planning and execution of our workshop and this volume that we should concentrate on logical empiricism in its beginnings — the time of the early writings of the future founders of logical empiricism (around 1910 for Schlick and Otto Neurath and other older members, later in the 1910s and into the early 1920s for the younger members such as Carnap and Reichenbach) to the diaspora in the mid-1930s. The mid-1930s were both politically and philosophically crucial — logical empiricism dispersed into the world at a point in which early debates were shutting down and mature projects (for example, Carnap's *Wissenschaftslogik* and Neurath's encyclopedic vision of unified science) were being promulgated. By placing logical empiricism firmly into the philosophical, scientific, and cultural contexts of its genesis, we felt we could explore features of the project that are little known but should be genuinely interesting to many contemporary philosophers. Further, as an exercise in the broadening of intellectual perspective, it was felt that if current attitudes are that the only good logical empiricist is a dead one, we could at least show that many of the dead logical empiricists were in fact good ones.

Upon reviewing the contributions to this volume, several features of both logical empiricism and the historical approaches taken to it seem notable. These features are interwoven and support the value of a more broadly construed historical approach to philosophy as well as of the philosophy that is here so scrutinized. These features include:[6]

- an accounting of the broader context of the philosophy of the logical empiricists. This includes taking into account the broader philosophical background to logical empiricism (including phenomenology, Marxism, and neo-Kantianism), but is not limited to this. Methodological features of the contemporary sciences, cultural features of German and Austrian society, and the political context and motivations of the logical empiricists are examined. We achieve a picture less of an isolated movement in philosophy reaching back to eighteenth-century Britain to rehabilitate old-fashioned empiricism than of a modernist movement in philosophy trying to bring the forward-looking technical and problem-solving features of science into philosophy itself.
- a correlative illumination of distinctions among the logical empiricists through consideration of the very distinct scientific and methodological problem situations in which they matured. For example, the broad differences between Neurath's naturalism and Carnap's logical philosophy are fairly well known. A consideration of Neurath's place in political and social-scientific controversies in Austria and Bavaria and Carnap's place in conventionalist methodological controversies in physics can lead to a better understanding of both the differences in the philosophical concerns of these philosophers and their mutual feeling of philosophical kinship. Moreover, as we both see and begin to understand the differences as well as the similarities among the logical empiricists, we also see the way that certain major themes

dominated the concerns of individual members of the movement throughout their careers.

- a concern to consider the important terms with which the logical empiricists themselves expressed their points of view, motivated their philosophical perspectives, and connected their work to or severed it from other intellectual movements. Thus, in these pages, one sees meditations on terms not considered very important in contemporary analytic philosophy, for example, *Eindeutigkeit* (univocality or uniqueness), or whose importance has changed due in part to changes of cultural significance, for example, "construction" in the particular sense of an *Aufbau.* Similarly, we see how *within this particular period of intellectual history* logic could have been conceived as a methodological and technical tool that at once severs philosophy's tie to metaphysics and connects it fully to science.[7]
- an emphasis — based on taking the terms of the art of the logical empiricists seriously in this way — that certain debates long thought finished can be reconsidered from a different perspective. For example, if the point of the analytic/synthetic distinction within Carnap's own philosophy has been misunderstood, then we can go back again to his texts to see whether his definitions can serve *his* purposes. Similarly, a more patient examination of the protocol-sentence debate may at once reveal lessons of value for current debates over the observational basis of science and uncover different perspectives regarding the point of such debates. None of this should be seen as an illicit attempt to revive the dead. The point is rather to maximize philosophical perspectives at a time when philosophy of science can little afford to luxuriate in close-mindedness.[8] Also, this can free philosophers of science of a felt need for refutations of logical empiricism and allow them to rest content with rejections. Refuting a philosophy that makes much of its lack of theses and concentrates rather on goals and methods can be fruitless business — far better simply to offer new goals and methods.[9]

In order to facilitate the reader who is interested in particular issues in the interpretation of logical empiricism, the editors have divided the essays into thematic groups. Inasmuch as no grouping of the essays would be wholly "natural," given that the themes have important interconnections, we suggest the reader view the thematic divisions as a way of highlighting important issues and not as erecting hard and fast interpretative categories. The first group of essays concentrates on aspects of the cultural and philosophical context in which logical empiricism arose in the 1920s and 1930s. The second group focuses on questions of science and scientific methodology in this same period. The third group considers the question of what logic was and what it was for in the work of the logical empiricists. The final group considers the question of experience and the empirical basis of science in the work of the logical empiricists.

1. The Cultural and Philosophical Background

Michael Friedman provides a detailed and nuanced rendering of the *Auseinandersetzung* between Carnap and Martin Heidegger on the philosophical primacy of logic and metaphysics. He presents the shared background of neo-Kantian epistemology motivating both Carnap and Heidegger in the late 1920s and early 1930s and, indeed, the extent to which Heidegger and Carnap agreed on the nature of their philosophical disagreement. By tracing the main lines of philosophical influence of Carnap and Heidegger to competing projects in neo-Kantianism, respectively the Marburg and the Southwest schools, Friedman exposes how these thinkers can be seen as pushing opposing tendencies in the thought of the neo-Kantians. The major problem for the neo-Kantians was how to continue to find an a priori transcendental grounding for objective knowledge of spatio-temporal objects after the rejection of a separate faculty of sensibility and forms of pure intuition. The question becomes one of whether formal logic can play this role, assigned by Kant to *transcendental* logic. The Marburg school claims that it can, whereas the Southwest school looks for another basis for the transcendental function. Friedman then presents Carnap as radicalizing the Marburg proposal and Heidegger as doing the same for the Southwest proposal.

Friedman notes the extent to which the fundamental cleavage in twentieth-century philosophy between analytic and Continental philosophy derives from just this dispute over the priority of logic or fundamental philosophical insight in grounding objective knowledge. It is remarkable that what began as a local concern internal to a neo-Kantian project could ramify to such an extent and can, subsequently, be read back into the entire nineteenth century. We have here an example of how detailed examination of the historical record can lead to fundamental changes in our view of the structure of the history of modern philosophy. It can also lead to meditations on the current state of philosophy of science and science studies, although Friedman does not do this. It is clear that a number of the sociologists of science working today use methods derived from the *geisteswissenschaftlicher* point of view so thoroughly rejected by the logical empiricists. If sociology of science uses methods derived from ethnomethodology and ethnomethodology derives from hermeneutic and phenomenological philosophical perspectives, then we may well be seeing in current debates the faint afterimage of the very disputes to which Friedman draws our attention.

Peter Galison's essay draws out another unexpected aspect of early logical empiricism. Galison investigates the cultural significance attached to the word *Aufbau* — a word known to contemporary philosophers mainly through the title of Carnap's first book, *Der logische Aufbau der Welt* (Carnap 1928a). Galison's research uncovers a political significance to this term in Germany between the world wars. Within the project of logical empiricism, use of this term, therefore, serves to connect it with other intellectual, artistic, and political movements of the time. In this period the term referred primarily to a left-wing, technocratic, modernist rebuilding of society. The old political order having been overthrown by war and/or revolution, the primary concern was to rebuild or reconstruct society using the rational methods and resources of the sciences. After discussing the early political aims of the logical empiricists, Galison notes how those aims were covered over in their movement to the United States.

Nancy Cartwright and Jordi Cat expose a different type of connection between logical empiricism, politics, and political philosophy. They seek to explain Neurath's use of the curious notion of conceptual *Ballungen* (congestions, concentrations) in the protocol-sentence debate by connecting it to a crucial use of similar notions in a much earlier debate, which had renewed currency at the time of the protocol-sentence debate. This earlier argument, which raged from the 1890s to the 1910s, was a methodological debate among Marxists about the status of their conception of history. At stake was the scientific status of materialist history and the extent to which it could accommodate the causal claims of religious, artistic, philosophical, or other aspects of history. Cartwright and Cat argue that the notion of complex, or *complexus,* found in the work of Marxist theoreticians Antonio Labriola and Georgy Plekhanov formed the basis for Neurath's notion of a *Ballung.* They argue further that this notion was crucial to Neurath's antimethodism as it emerged in his work in the protocol-sentence debate.

Thomas Uebel explores the overall development of Neurath's philosophy and the interplay between its focus on social science and its political motivations. He first reconstructs the connections and then provides a local historical context within which Neurath came to this view. In this historical account we see the influence of Ernst Mach and the French conventionalists, but also other, less well-known influences — Austrian and German social theorists and activists such as Ferdinand Tönnies, Josef Popper-Lynkeus, Otto Bauer, and, ultimately, Karl Marx. We are asked to see Neurath as engaged in a self-consciously renewed Enlightenment project of clarified and rationally controlled science serving social ends. Neurath's social naturalism appears then as a robust theory of science that differs from now current social alternatives mainly in its optimistic stance and explicit political mission.

These essays individually and as a group indicate how wide-ranging the influences on logical empiricism were. Moreover, we see again and again how engaged the logical empiricists were with their political and intellectual contexts. Their projects were far from being *merely* technical; indeed, the very technical nature of the projects was to be their source of genuine engagement with the empirical realm and, hence, with the social order. A hopeful modernist reliance on science and technology to resolve tremendous social and economic difficulties is everywhere in evidence. This may seem naive to contemporary thinkers, but perhaps the difference between modernist hope and postmodern irony is ultimately the difference between critical engagement and cynical theorizing.

2. Science, Philosophy, and Scientific Philosophy

A more well-known set of influences on logical empiricism came from the mature sciences, especially physics and mathematics. A major impetus to the reappraisal of logical empiricism came early on from highly technical work in philosophy of physics that led to interpretative concerns about how well logical empiricist accounts of the theory of relativity accorded with the content and methods of that theory.[10] While the consensus seems to be that, in fact, the early conventionalism of Reichenbach, in particular, fails to respect the theory it was designed for, still the depth of the technical engagement with relativity was a watershed for the new logical empiricist philosophy of science. The next two essays in our volume speak to the continuing relevance for the contemporary philosopher of science of the early methodological and conceptual debates surrounding relativistic physics.

Thomas Ryckman provides a rich discussion of early attempts by Hermann Weyl and others to extend the general theory of relativity to a unified theory of all known forces. At issue was whether and how the geometrization of forces could be extended beyond Einstein's account of gravitation. Ryckman contends that Reichenbach's metric conventionalism, especially as systematically carried through in his early *Axiomatik der Relativistischen Raum-Zeit-Lehre* (1924), is directed against Weyl's "unified theory" — in particular, against Weyl's foundational criticism of accepting rods and clocks as primitive concepts in general relativistic theories of the space-time metric. Among the issues Ryckman discusses in recounting the history of Weyl's project and objections to it from Einstein, Wolfgang Pauli, Reichenbach and others is, of course, the question of mathematics and its relation to empirical science. Ryckman finds, ultimately, that remnants of neo-Kantian epistemology color Reichenbach's vision of the methodological situation here — especially in his understanding of the relation between metrical concepts, nonmetrical concepts, and the behavior of physical objects. Despite Ryckman's negative assessment of Reichenbach's metrical views, the depth of engagement between logical empiricism and the furthest reaches of theoretical physics of the time is evident, as is the fruitfulness of continued examination of this engagement.

Don Howard provides an object lesson in the multiple connections among epistemological, scientific, and logical points of view in the period of the development of logical empiricism. His topic is the issue of *Eindeutigkeit* (univocality) as it appears in various contexts in that time period. It is an important notion in the foundations of physics, especially for questions regarding the structure of space-time and Einstein's search for the system of generally covariant field equations. It also occurs as a term of art in various epistemological writings, such as those of the "relativistic positivist" Joseph Petzoldt. Howard's main interest is the connection that arose between, on the one hand, the physical and the epistemological questions and, on the other hand, the technical questions of uniqueness up to isomorphism of models of axiom systems in Carnap's work in the mid- to late 1920s. In Howard's view, we have here an early, little explored attempt by Carnap to convert a methodological or epistemological concern into a question of logic. In this way, the episode gives a snapshot of the early Carnap's approach to philosophical problems.

3. Logic, Mathematics and Philosophy

Logical empiricism would have been impossible without the advent of the new logic of Gottlob Frege and Bertrand Russell. But no univocal view of formal logic nor its role in philosophy can be found among the logical empiricists. Moreover, of course, the myriad advances in formal logic did not leave the way in which logic was to be used in philosophy unchanged. The work of Carnap, in particular, was punctuated by changes in what counted as logic, as well as in technical and philosophical resources brought to bear on logical issues. Moreover, Carnap is famous for apparently failing to achieve a central goal of his logical work — a workable definition of analyticity. Perhaps no questions are more fundamental to the interpretation of logical empiricism than what the philosophical significance of logic was meant to be and whether this significance was fulfilled by the formal logics promulgated.

Warren Goldfarb provides a map of the philosophical landscape of the logical empiricists on just these issues up to Carnap's *Logical Syntax of Language* (1934; trans. 1937). He shows how confused the early logical empiricists were about the nature and extent of logic throughout the 1920s and into the 1930s. Perhaps it is a measure of how fundamental logic was to their philosophical project that they could not even formulate much of a coherent perspective for talking about it until Carnap took over David Hilbert's metalogical perspective in the 1930s. Goldfarb contrasts Carnap's view of what is at stake in foundations of mathematics and the role of logic therein with the robustly realist views of Kurt Gödel. He makes the telling point that an account of the foundations of mathematics will be compelling or not based primarily on what one takes the foundational issues to be. The way Carnap seeks to replace certain philosophical questions with technical ones means that he is under no obligation to answer those philosophical questions.

Thomas Ricketts surveys the development of Carnap's use of analyticity from *Logical Syntax* to *Introduction to Semantics* (Carnap 1942). He argues that Carnap's early enthusiastic adoption of Alfred Tarski's technique for defining truth makes no sharp break with the project of *Logical Syntax.* In *Logical Syntax,* Carnap, on semitechnical grounds, rejects truth as both semantically intractable and otiose for the logic of science. Tarski's semantic approach showed Carnap how to overcome these difficulties. Ricketts argues that this technical development brings great conceptual difficulty, however; the move to semantics blunts the force of Carnap's diagnosis of metaphysics as confusion engendered by lack of appreciation for the distinction between the formal and material modes of speech. Richard Creath points to a different difficulty in the relation of analyticity to logical truth in *Logical Syntax.* He notes that Carnap's definition proceeds from a distinction between logical and descriptive vocabulary to a definition of logical truth to a definition of analyticity. But Creath argues that the definition can lead to there being languages with no logic — no class of logical vocabulary and no logical truths. He argues that this may be a principled failure. While this may seem more bad news for analyticity, Creath proposes a change in view — that rather than following Carnap in explaining analyticity in terms of logical truth, we should explain logical truth in terms of analyticity. In this way, we can, according to Creath, keep what is epistemologically right about the notion of analyticity without engaging in a hopeless explanatory exercise with respect to it.

4. Experience, Empirical Knowledge, and Empiricism

It seems the most natural thing in the world to say that the logical empiricists were empiricists. Indeed, this claim is largely true. But the precise way in which they were empiricists is hard to capture. This is because in large measure their concern in empirical knowledge was not so much verification and evidence as objectivity, and there they took a generally formalist line that was very hard to square with empiricism of any kind, let alone hard-core verificationism. In this way, too, the logical empiricists were more subtle (and, perhaps, more confused) than rigid common stories about their philosophies allow. Moreover, the shape of the protocol-sentence debate indicates there were deep divisions among the logical empiricists about how to characterize empiricism and what a commitment to empiricism amounts to. The final group of essays explores some aspects of this situation.

Joia Lewis Turner's essay is a meditation on Schlick's conceptual difficulties with maintaining his views of the conceptual nature of scientific knowledge while sustaining a connection between knowledge and experience. She notes that Schlick was adamant in denying that pure sensory experience — or, to use his term, "intuition" — was knowledge at all. This raises a completely general question of how to connect the conceptual knowledge of science to such experience. Turner claims that Schlick's epistemological precursors in this view were others trained in physics rather than philosophy, principally Herman Helmholtz and Max Planck. She then turns her attention to three ways in which Schlick attempted to explicate the difference between sensory intuition and conceptual knowledge and explain their connection. Turner argues that none of these attempts was particularly successful. She ends by noting Schlick's movement toward thinking that his difficulty was an *empirical* one — that is, that his analysis would only have to account for a de facto distinction between knowledge of a shared empirical world and private sensation. He would not have to give general philosophical grounds for the possibility of such shared knowledge. In this spirit Turner offers some more up-to-date theorizing from cognitive science that bears on the issue that concerned Schlick.

Thomas Oberdan offers another consideration of Schlick, concentrating on Schlick's point of view in the protocol-sentence debate. Oberdan contends that questions about the role of *Konstatierungen* (affirmations) in conferring certainty to the basis of empirical knowledge misrepresent the main import of Schlick's mature view. Oberdan argues that, in the end, *Konstatierungen* were meant mainly to guarantee the connection between language and experience. This was crucial to Schlick in his attempt to give a semantic account of language that countered Carnap's syntacticism. Oberdan claims that the main difference between Carnap and Schlick was over the question of the relation of linguistic conventionalism and empiricism — contact with the empirical world is constitutive of genuine language for Schlick but not so for Carnap. Thus, a semantic empiricism was more central to Schlick's philosophy than to Carnap's.

Finally, my contribution is meant to illuminate the development of Carnap's philosophy of empirical knowledge in the 1930s. It is perhaps too infrequently noted that Carnap argued against the whole project of epistemology during these years and offered logical empiricist philosophy of science as a replacement for epistemology. I argue that he was led to this conclusion by tensions in his own account of epistemology in the *Aufbau.* Once again the vexed issue of the relation of Carnap's empiricism to his logical work is raised. I claim that until his paper "Testability and Meaning" (Carnap 1936–37), Carnap had no coherent way to be an empiricist; in that paper he gives an account of empiricism as commitment to a restricted range of the logically admissible languages.

Notes

1. Within the Anglo-American context, this reappraisal began in earnest early in the 1980s, particularly in writings of Alberto Coffa and a number of authors whose interest in the history of analytic philosophy was stimulated by Burton Dreben (including Michael Friedman, Warren Goldfarb, Thomas Ricketts, Peter Hylton, Bill Hart, and Richard Creath). This period also saw the opening of the major collections in the Archive for Scientific Philosophy in the Twentieth Century at the University of Pittsburgh, which has greatly facilitated research into the founding figures of logical empiricism. In recent years, the work has increased in pace with the publication of major books by Joelle Proust (1986), Alberto Coffa (1991), Thomas Uebel (1992b), and Thomas Oberdan (1993) as well as important edited volumes (such as Rescher 1985; Uebel, ed., 1991; Bell and Vossenkuhl 1992; Haller and Stadler 1993; Stadler 1993) and dedicated journal issues (*Noûs* [December 1987]; *Synthese* [October/November 1992]; *Erkenntnis* [July 1992]). In part this is due to happenstance, especially the Schlick-Neurath centennial of 1982 and the

Carnap-Reichenbach-Zilsel centennial of 1991, but it also reflects a genuinely emerging reawakening of philosophical interest in the writing of the logical empiricists.

2. This dominance of logical empiricism was greatly facilitated by the early volumes of Minnesota Studies in Philosophy of Science. Thus, it has been a matter of considerable poignancy for the contributors to this volume that it appears in the very same series.

3. Particularly interesting examples of this can be found in the first chapters of Suppe 1989 and Giere 1988.

4. In this they resemble nothing so much as the early broadcasts of *Saturday Night Live*. Just as Chevy Chase continually informed his audience that Franco remained dead, so too do today's philosophers of science seek to soothe us with the thought that logical empiricism remains moribund. Thus, logical empiricism remains alive in philosophical memory chiefly by the significance of its death. (I take the death of logical empiricism to be a social fact of analytic philosophy — few, if any, consider themselves to be continuing the project. It should be obvious that I do not believe that philosophical death in this sense entails lack of intellectual vitality; neither does the entailment go in the other direction.)

5. Perhaps we should add a third: the logical empiricists in general evinced little interest in history of philosophy and notably little interest in history of science. Thus, they played a role in creating the interpretative and hermeneutic shallowness of analytic philosophers, almost guaranteeing their own very subtle perspectives would be given very unsubtle treatment. But, one might argue that more has been made of logical empiricism's allegedly antihistorical approach than is strictly warranted; Zilsel and Neurath, especially, had highly developed interests in history of science. Moreover, early work connecting logical empiricism with the history of modern physics by Frank and Reichenbach was arguably extremely important for facilitating the reception of logical empiricism in North America. Such work may misconceive the "role for history" in philosophy of science, but it does not ignore it.

6. Conjointly these features might be called "the riches of historicism" and indicate the introduction of somewhat more nuanced historiography into the history of analytic philosophy.

7. This connects to the strangely little-understood lesson of the history of rejections of metaphysics that one must pay close attention to what is being rejected and what the argument for the rejection draws upon. This can lead us finally away from the twin errors of thinking that all rejections of metaphysics are ultimately identical and ultimately "just more metaphysics." Very few antimetaphysicians are such bad philosophers that they can be legitimately accused of relying in their arguments against metaphysics on the very metaphysics they are arguing against.

8. Thus, this historical work can serve as part of an attempt to revitalize what we dearly hope is yet living.

9. Of course, one can refute any particular version of logical empiricism by showing that its goals are unattainable by its methods. This has been attempted with more than a small measure of success, but usually is done best by those who have first made a substantial interpretative effort to uncover these goals and how the methods are meant to fulfill them. In giving examples of this strategy I would point, for example, to Ricketts 1982 and Friedman 1992a. Some of the essays in this volume also exhibit this strategy — especially the essays by Ricketts and by Ryckman.

10. The most important example of this work is Friedman 1983b.

PART I

THE CULTURAL AND PHILOSOPHICAL CONTEXT

Peter Galison

Constructing Modernism: The Cultural Location of Aufbau

1. The Cultural Meaning of *Aufbau*

Between the end of World War I and the immediate post–World War II period, almost a hundred journals and multiauthored volumes appeared in the German-speaking world with the word *Aufbau* in their titles. Practically none existed before the end of World War I, and only a handful remained after 1947. Put into a histogram, the journals fall into three spikes: the largest burst between 1919 and 1927, a middle peak between 1934 and 1937, and a final surge between 1945 and 1947.

These three extraordinary spikes, and the absence of any equivalent spike before 1918 or after the late 1940s, are significant. The first peak corresponds to a staggering proliferation of journals polemically announcing the rise of a new world from the chaos and rubble of the Great War and more specifically from the ignominy and devastation of the German defeat. Out of the rubble, a new structure would emerge; translating *Aufbau* as "rebuilding" or "reconstruction" fails to capture the novelty of what these many authors hoped would come to pass. On new, and for the first time firm, foundations, they would erect a political, philosophical, and aesthetic world separate from everything that had come before. It would (in most instances) be socialist, internationalist, practical, and deeply scientific and technological. As a shorthand designation, I will refer to this cluster of usages as the left-technocratic period of *Aufbau.*

The second peak corresponds to the rise of Nazism, and the *Aufbau* journals and volumes come from both right and left — from the right, organicist volumes espousing work and the empowering of the state; from the left, antifascist journals rising to oppose the destruction already wrought and planned by the dictatorship. Finally, the third *Aufbau* spike was essentially a literature of rebuilding after World War II, a literature of liberal democracy and cultural renovation that would reintegrate West Germany into the world community; in East Germany, the Aufbau-Verlag espoused the building of the new communist society.

What all three moments have in common is this: a rupture with the past, a construction not based on precedent, and a profound conviction that the *Aufbau* process could not be superficial. It had to embody not just the trappings of political change — it had to transform culture, education, architecture, and the modes of reasoning that guide us through the world. Precisely because of its scope, the meaning of *Aufbau* was contested. Right and left both sought to appropriate the impulse to sweep out an old order and to build anew.

This essay focuses on the first spate of *Aufbau* usages. More particularly, I want to exploit *Aufbau* as a keyword (in Raymond Williams's sense [Williams 1983]), a word that embodies a nodal point between culture and society. Even more specifically, the word can help us understand something about the links between cultural domains in the interwar period, for this lexical element was indeed emblematic of a moment at which art, architecture, politics, education, and philosophy could appear to many as participating in a common, quintessentially *modernist* endeavor. Just because the term stood where the tracks of so many lines of culture crossed, groups across the cultural-political spectrum struggled over its appropriation; like modernism itself (for which it is often a stand-in), everyone wanted an *Aufbau* of his or her own design: Nazis, liberals, socialists, and communists.

Though I will make reference to the Nazi and liberal democratic periods of *Aufbau* usage (and to the struggle over its meaning), the spotlight here will be on the left-technocratic period, when the term (principally) connoted the modernist project of building fresh without reference to history, the superficial, or the ornamental. I trace the uses of *Aufbau* in the hope that exploring this particular strand of modernism, located temporally in the interwar years and culturally in the German-speaking lands, will shed some further light on the Vienna Circle's often-repeated self-conception of being at one with the spirit of modernism.

Consider, for example, *Der Aufbau: Flugblätter an Jugend* (Der Afbau: Leaflets for youth), published in 1921, which included an article by Hans Reichenbach on socialism (see fig. 1). The introductory essay in the series pursued the relationship between *Aufbau* and revolution:

> What is *Aufbau?* Is *Aufbau* "changing the world"? Evidently not! For what changes the world more than revolution? And yet we say: a revolution creates only the conditions for an *Aufbau.* Change is attached to the given, to the substantial; it is a redirection, a shifting of forces;

Der Aufbau
Flugblätter an Jugend

5.

Student und Sozialismus

von

Hans Reichenbach

Mit einem Anhang:
Programm der Sozialistischen Studentenpartei

Herausgegeben im Auftrage der Freien Hochschul-
gemeinde, Berlin: Hermann Schüller

Figure 1. Cover page from volume 5 of Der Aufbau: Flugblätter an Jugend *(1919), featuring work by Hans Reichenbach. From Harvard University, Widener Library, Cambridge, Massachusetts.*

> it is the technical. (Schüller 1921, 1:5; see also Reichenbach 1919, 5:3–9)

Even this brief excerpt captures themes that draw together many of these manifestos. For in this postwar time of revolutionary governments, from Russia to Bavaria, there was a sense that politics alone failed to plumb the depth of the transformation necessary in society. Means of transport, modalities of production and distribution, the architecture of buildings, and the planning of cities — these material and technical realms held out a real and symbolic importance that for many lay deeper than politics alone. At the same time, in many of the *Aufbau* journals, this emphasis on the need for change at the deepest level, below politics, so to speak, was coupled with an almost mandatory reference to the destructiveness of war that had made it necessary. This was explicitly the case in the journal *Umsturz und Aufbau* (Destruction and construction), which began publishing in 1919, as the following excerpt indicates:

> We have lived through the most unheard of catastrophe that a people has ever encountered. In need and death, in blood and tears there has been only one asset — but a powerful one — left to us: the spirit. To make the spirit fruitful for the renewal of our people is the goal of our series *Umsturz und Aufbau.* Here we want to assemble documents.... [W]e want to remove the rubble of the European killing fields and with heart and head to prepare for the erection of a new humanity. (*Umsturz und Aufbau* 1919 [1974], 22–23)

Digging out from the catastrophe was a trope repeated over and again, as Europeans — especially in the German-speaking world — contemplated their present and their future. The left-leaning Swiss periodical *Der Aufbau: Sozialistische Wochenzeitung* concurred that the war years had shed light on the "catastrophe" but saw in socialism an exit from the destruction (see fig. 2). In the first edition (December 1919) of the journal, the editors pronounced that their weekly publication would show itself to be

> born out of the need of our time,... aimed at those who, like us, feel that need.... In these war years, they have experienced the last horrible but inevitable consequences of one of the innermost laws of life of a fallen world. They know that if they do not want to be destroyed, then a new existence [*ein Neues*] is necessary. A renewal, with its roots in the deepest essence of humanity, but which for that reason also is expressed in the outer structures of life, must transform them. *Der Aufbau* must proceed on new fundamentals and according to new principles. (*Der Aufbau: Sozialistische Wochenzeitung* 1919, 1)

DER AUFBAU
SOZIALISTISCHE WOCHENZEITUNG

Erscheint jeden Freitag — No. 1 — Zürich, 5. Dezember 1919

Abonnementspreis: Jährlich Fr. 6.50; ½-jährlich Fr. 3.50; ¼-jährlich Fr. 2.—. Ausland: jeweiliger Portozuschlag. Preis der Einzelnummer 20 Cts.

Herausgeber: Genossenschaft zur Herausgabe einer sozialistischen Zeitung. Redaktion: MAX GERWIG, MAX GERBER. Druck, Administration und Expedition: Pfingstweidstrasse 57, Zürich 5.

Insertionspreis: Die einspaltige Petitzeile 30 Cts. Ausländische Annoncen 50 % Zuschlag.

Was wir wollen.

I.

Was dieses neue Wochenblatt will, wird es selbst zeigen müssen. Geboren aus der Not unserer Zeit, wendet es sich an alle, die mit uns diese Not empfinden. Wir wissen, dass ihrer auch in unserem Volke nicht wenige sind. Sie haben in den vergangenen Kriegsjahren die letzten schauerlichen, aber naturnotwendigen Konsequenzen einer von den innersten Lebensgesetzen abgefallenen Welt miterlebt und wissen, dass, wenn sie nicht zugrunde gehen wollen, ein Neues nötig ist. Eine Erneuerung, die ihre Wurzeln freilich im tiefsten Wesen des Menschen selbst haben, aber gerade deshalb auch in den äussern Lebensordnungen zum Ausdruck kommen, sie umwandeln muss. Der Aufbau muss auf neuen Fundamenten und nach neuen Prinzipien erfolgen. Noch nie wohl war deshalb die Zeit dem sozialistischen Ideal so günstig als jetzt, wo die bisherige Wirtschaftsordnung im Lichte katastrophaler Ereignisse und im Lichte ihrer «unlösbaren Fragen» gerichtet erscheint. Aber wie jedem Ideal, so drohen nun auch dem Sozialismus desto grössere Gefahren, je näher er seiner Verwirklichung reift. Da ist einmal die Bedrohung von aussen. Seit in der russischen Revolution das sozialistische Ideal greifbare Wirklichkeit anzunehmen schien, seit es klar wird, dass der Kampf nicht um ein paar Reformen mehr oder weniger, sondern um eine grundsätzliche Neugestaltung der Wirtschaft geht, haben sich alle Mächte der Reaktion zu einem entscheidenden Verzweiflungskampfe aufgerafft und suchen mit allen Mitteln die Entwicklung zum Sozialismus zu hindern. Und doch liegt hier die kleinere Gefahr, denn der Sozialismus kann nicht von aussen, durch die absterbenden Mächte einer vergangenen Epoche überwunden werden. Der Sieg in d i e s e m Kampfe ist ihm gewiss, wenn er nur sich selbst, seinem innersten Wesen treu bleibt. «Aber die ungeheure Gefahr ist, dass Schlendrian und Nachahmung sich auch der Revolutionäre bemächtigen und sie zu Philistern des Radikalismus, des tönenden Wortes und der Gewaltgebärde machen, dass sie nicht wissen, und nicht wissen wollen: d i e U m w a n d l u n g d e r G e s e l l s c h a f t k a n n n u r i n L i e b e, i n A r b e i t, i n S t i l l e k o m m e n.» (Gustav Landauer.)

Wir begreifen nun freilich, wie berauschend auf die in Not und Elend versunkenen Massen der bolschewistische Gedanke, durch Gewalt und vorübergehende Diktatur in kurzer Zeit das Alte zu stürzen und eine neue Ordnung sozialer Gerechtigkeit und Freiheit zu errichten, wirken muss. Wir wissen, wie viel Idealismus und verhüllter Glaube in den Z i e l e n des Bolschewismus steckt. Wir wissen auch, dass auch seine Methoden nur zum kleinsten Teil aus niedrigen Gewaltinstinkten und Leidenschaften, vielmehr aus grosser Hoffnung und grosser Enttäuschung geboren sind. Die grossen Massen können den Gedanken nicht ertragen, dass das gewaltige Völkerringen mit seinen Verwüstungen zu Ende gehen soll, ohne dass der Grund gelegt werde zu einer andern, solidarischen und brüderlichen Welt. Sie haben, zu oft schon getäuscht und betrogen, kein Vertrauen zu den Versprechungen bürgerlicher Machthaber und Parlamente, und sehen nur noch den Ausweg gewaltsamer Revolution. Schliesslich betonen wir, dass gerade die Vertreter des Bestehenden am wenigsten Grund haben, sich über den Bolschewismus zu entrüsten, denn der in ihm steckende Gewaltgeist ist der Geist der bestehenden Ordnungen und Mächte; der Bolschewismus will nichts als die Probleme mit denselben, vielleicht etwas vergröberten Methoden lösen, die ihm Geschichte und Gegenwartspolitik gelehrt haben. Aber gerade hierin liegt nun auch sein Grundirrtum und sein Verhängnis. Denn es ist eine tiefe, mit der Sicherheit eines Naturgesetzes wirkende Wahrheit, dass das Gute seine eigenen Kampfmethoden hat und dass eine Sache nur durch diejenigen Mittel verwirklicht werden kann, die ihrem innersten Wesen entsprechen. Ein Reich der Gewalt kann mit Gewalt errichtet werden, die Freiheit kann nur realisieren, wer an die Freiheit glaubt und es mit ihr wagen will. Wo der Sozialismus mit ihm wesensfremden Methoden zum Ziel gelangen will, da gibt er etwas von seinem innersten Wesen, seiner dem Bestehenden überlegenen Kraft preis und verrät dadurch sich selbst. Gerade das ist eine Not und Gefahr der Gegenwart, dass auch diejenigen, die etwas Neues möchten, in so hohem Grade selbst dem «Zeitgeist» zum Opfer gefallen sind, dass auch das sozialistische Ideal vielfach verfälscht wird gerade von den Mächten aus, die es überwinden möchte; dass wir auch in der Sozialdemokratie, in deren politischen Zielen doch eine neue Wahrheit aufleuchtet, so oft dieselbe geistige Einstellung finden, die das Verderbliche unserer Kultur ausmacht: den Glauben an das Quantum, an die Zahl (z. B. die Mitgliederzahl), den Glauben an die Gewalt, den Aberglauben an das Wissen (z. B. in der Bildungsarbeit), die Ueberschätzung der Wissenschaft, die unsere heutige Wirtschaftsordnung tragenden Gedanken, dass nur der Appell an das Interesse, an den Egoismus den sozialen Bau aufrechterhalte usw. Und doch sind wir fest überzeugt, dass durch nichts jetzt so sehr der Geist der alten Ordnung und damit diese selbst gestärkt werden könnte und das Leben der heutigen Gesellschafts- und Wirtschaftsformen künstlich verlängert als durch einen Sozialismus, der stekken bleibt im bürgerlichen Glauben an die Gewalt, in der lähmenden Skepsis gegenüber dem Geist und den ideellen Triebkräften; der Gewaltgeist würde sich nur allzu bald wieder gegen den Sozialismus selbst kehren und ihm den Todesstoss versetzen, denn er ist von Natur aus der beste Verbündete derer, denen Gewaltherrschaft und Unterdrückung nicht nur Uebergangsepoche sondern auch Endziel sind. Jedenfalls ist ein solcher Ausgang sehr viel wahrscheinlicher als die aller historischen und psychologischen Erfahrung widersprechende Utopie, man könne die bösen Geister, die man gerufen, nach vollbrachtem Dienst nach Belieben wieder beurlauben, und so auf dem Wege vorübergehender Despotie in das Reich der Freiheit gelangen.

Es liegt aber im Wunderglauben, das Alte auf einen Schlag stürzen und auf seinen Trümmern das Neue erbauen zu können, noch eine weitere Gefahr. Er lenkt ab von den drängenden und bitter notwendigen Vorbereitungsaufgaben und den gegenwärtig schon vorhandenen Möglichkeiten. Das Interesse für sie erlahmt, weil sie im Lichte einer angeblich nahen sozialistischen Zukunft überflüssig scheinen. Die neue Wirtschaft ist aber kein Mechanismus, den man «nach Eroberung der Macht» konstruieren könnte wie eine neue Maschine, sondern ein Organismus, der aus tiefen Wurzeln wachsen und reifen muss. Wer u. a. im Hinblick auf das russische Beispiel die ungeheure Schwere der Aufgabe empfindet, nach dem Zusammenbruch des Kapitalismus die bisher im Prinzip durch den Anreiz des Profites getriebene Wirtschaft in neuen Formen und mit ganz andern Triebkräften aufrecht zu erhalten; wer weiss, dass der Kapitalismus «nicht das Wunder tun wird, in seinem platzenden Zusammenbruch den Sozialismus zu gebären», der wird schon jetzt nach den neuen wirtschaftlichen Formen suchen und sie nach Möglichkeit verwirklichen, um so wenigstens das Bett zu graben, in das sich einmal der Strom neuen Lebens ergiessen kann. Es gilt, mitten im bestehenden Wirtschaftssystem von u n t e n her nach neuen Prinzipien aufzubauen. In der Rede, dass

Figure 2. Title page from the first issue of the Zurich-based Der Aufbau: Sozialistische Wochenzeitung *(December 1919). From Harvard University, Andover-Harvard Theological Library, Cambridge, Massachusetts.*

Socialism was this transforming force and contributed the principles upon which these inner and outer mutations would build. As in so many of the journals, the authors of this one went on to elaborate their vision of a mutation in society that would not end with the "merely" political:

> We do not want to stop at politics in the narrow sense, for the new reality that we hope for is not simply a new political or economic system, but a renewed world from the depths on up. [This renewal] must come to a breakthrough in all other domains of life: public education, comradeship, the essence of child rearing, science, art, and religion. (ibid., 1)

Although in the period 1919–26 the *Aufbau* journals and other volumes tended to be left-technocratic, others already explored aspects of nationalism and Groß-Deutschland. From what we know of the contested history of modernism in the Nazi period (see, for example, Herf 1984), it comes as no surprise that *Aufbau* too was seized upon by the national socialists. *Aufbau und Erbe,* a multiauthored volume, was just one of the *Aufbau* texts sporting a swastika on its cover. Its content, ranging over music, poetry, and prose, returned to the Nazi themes of land, weapons, and handwork; *Aufbau* in this context meant building up the new fascist society, including the new fascist man. Indeed, the scope of the fascists' project of building up is no less thoroughgoing than that of the left. Verses tell of building weapons, of defending the fatherland, of physical strength, and of work. It is an *Aufbau* altogether different from the transparent construction that characterized the Dessau Bauhaus or the Vienna Circle.[1]

Precisely to oppose everything the fascist buildup envisioned, especially after the Nazi *Machtergreifung* (seizure of power), a spate of anti-Nazi *Aufbaus* emerged. One, founded on 1 December 1934 in New York, grew into a highly influential journal published in German primarily by and for German Jews. They announced in their *Aufbau Almanac:*

> Its name [*Aufbau*] is its program. Over there, in Europe, every pinnacle of achievement that the German-speaking Jewish people have erected has been leveled. In a few years, centuries were annulled. In the monstrous waves of hate that flooded a flourishing cultural landscape, that which in a thousand ways had organically grown and often had matured to the most beautiful bloom was scattered to the winds.... Resurrection on a new Earth, on new roots, the focusing of all creative elements to this purpose — that is the first task that the *Aufbau* serves. (*Aufbau Almanac* 1941, 6)

Between 1937 and 1945 the number of new *Aufbau* journals declined. But following the end of World War II they reemerged in large numbers, founded by German speakers flung from Österreich to Ohio. In Vienna in July 1946, the *Stadtbauamt* (office of city planning) began a monthly *Aufbau* dedicated to *Wiederaufbau.* Bundespräsident Karl Renner penned a dedication lamenting the terrible destruction wrought by the modern day "Caesars" and extolling the Viennese to follow the special course of Austrian art and architecture in the rebuilding of the city. Others reiterated the necessity of rebuilding but at the same time mourned the destruction that had been deeper than merely material, calling for a reconstruction that must be so as well:

> The old world is destroyed.
>
> Before its ruins stand an old and a young generation, both suffering greatly from the unhealthy consequences of a false politics, a poorly applied technology, and a failed organization.
>
> Now we, the people, must take a new path. Each person must find his own spiritual way, but in everyday life, in the practical, we must go together. . . .
>
> Reconstruction is preponderantly a technical problem. The organizer, the engineer, the architect will carry it out. . . .
>
> *Der Aufbau* is oriented towards all technical people, but beyond that to everyone struggling for the new form of society. (Körner 1946, 2)

Even in the far-flung POW camps in the United States, German prisoners gathered to create a journal entitled *Der Aufbau: Lagerzeitung für Camp Perry und Nebenlager* (Der Afbau: Prison newspaper for Camp Perry and neighboring prison camps). The typescript paper, with its symbol of a rising sun, announced in one of its first numbers that the history of Weimar had to be rewritten not as one of destructive chaos, but as the attempt at a new beginning. Linking that postwar time with their own (October 1945), the prisoners looked to a future that would, once again, build democracy out of the rubble (*Der Aufbau: Lagerzeitung für Camp Perry* 1945).

This is, of course, just a sampling of the literature of *Aufbau.* But with this plethora of *Aufbaus* in mind, we can begin to piece together what the cultural resonance of the term would have been in the post–World War I years. At the first, most direct level of interpretation, the term suggested a quite literal construction. Architecture played a crucial role because of the great need for housing necessitated by the massive urbanization that followed the Great War. In Vienna alone, between 1922 and 1925 some 20,600 dwelling units were added to the city's stock (Schuster 1926, 8). Second, as is well known, the new buildings, especially the *Siedlungen* (mass housing), became a principal focus of the modernist architecture in both Austria and Germany.[2] Political and architectural progressives clustered around this chance to shape the daily lives of workers. Finally, in each of the periods, left-technocratic (1918–26), Nazi/anti-Nazi (1933–37), and liberal-democratic (1945–48), the notion of *Aufbau* was taken far beyond the exigencies of mortar, steel, and glass. From the merely material, the notion of construction was elevated to a program for a way of life.

Otto Neurath bridged the two conceptions of *Aufbau:* the material building up and the construction of an ordered, rational *Lebensform.* Both, Neurath contended, were of the utmost importance following the terrible destruction of the Great War. On one side, he reacted to this devastation like so many others, with horror. Yet curiously he found a constructive aspect to the war economy; he saw it function with extraordinary results, drawing resources from a centralized agency and deploying them to maximum effect. In April 1919, Neurath, then president of the Central Office of Economics for the Munich Workers' Council, had responsibility for the socialization of Bavaria. Having passed from critic to actor, he now set about enforcing the peacetime analogue of wartime centralization (Neurath 1973, 124).[3]

With the war over, part of the challenge of centralization and planning was the immense problem of coping with the huge migration of Austrians from the countryside to Vienna. Neurath followed the new movements in housing with great care.[4] He organized propaganda for the *Siedlungen;* he attended lectures by contemporary architects such as Walter Gropius; and he worked in close collaboration with one of Austria's leading architects and architectural theorists, Josef Frank (Philipp Frank's brother). Collaborating on a range of projects, from museums to the *Isotype* system of universal signage, the sociologist and the architect came to share a vision of modernity that joined aspects of reformist Viennese architecture to elements of German architecture with its "scientific" approach to a myriad of social problems. Neurath and Frank were sufficiently in sympathy that Josef Frank was invited to give the first in a series of lectures to the Verein Ernst Mach in 1929. Not long afterward, Neurath, Carnap, Philipp Frank, and Hans Reichenbach all lectured at the Dessau Bauhaus, then the citadel of modernist architecture.[5]

Hannes Meyer, the Swiss architect, theorist, and Marxist who succeeded Gropius as director of the Dessau Bauhaus in 1924, invited all of these philosophers to give the lectures. Meyer's celebration of the technical was boundless. In one of his most famous articles, "The

New World," he began with a staccato hymn to the machine world: "Flash! The *"Norge"* has circled the North Pole! Completed! The Zeiss Planetarium at Jena. To be tested today: Flettner's rotorship! These were, in 1926, the most recent exploits in the mechanization of our globe" (Meyer 1926). Airplanes, ships, parachutes, gliders, microphones, gramophones—these and many other devices were (by Hannes Meyer's lights) changing the nature of space and time, and at the same time destroying patriotism and inculcating internationalism. Art as well as politics yielded to the new complex of gears and lenses: "The artist's studio becomes a scientific-technical laboratory, and his works are the fruits of analysis and invention" (ibid.). Neurath had nothing but appreciation for Meyer precisely because Meyer (unlike Gropius) had explicitly put engineering front and center. But the technical in Meyer (as in Neurath) was more than the physical instantiation of instrumental reason: the technical stood for and created a way of life, an orientation toward the world that would cleanse it of the full spectrum of irrational threats: above all clericalism, nationalism, and social divisiveness (see Galison 1990, 709–52).

At the time Meyer wrote his "New World," Neurath himself was deeply engaged with the direction and writing of a journal entitled (not surprisingly) *Der Aufbau* (1926) (see fig. 3). Sharing the masthead with him were the Austrian architects Franz Schuster and Franz Schacherl. Schuster, collaborating with Adolf Loos, Schacherl, and others, had participated in the design of some 1,589 apartments by 1926. Heinrich Tessenow, also on the directorship of the journal, was the head of the architecture school of the Dresden Academy of Art and had been professor in the Vienna *Kunstgewerbeschule* since 1919. Like Schuster and Schacherl, Tessenow too had participated in the Red Vienna architecture, contributing (for example) to the Kolonie Rannersdorf outside Vienna.[6] Martin Wagner, the Berlin architect, was *Stadtbaurat.* But probably the best known of the journal's leaders was another Berlin architect, Bruno Taut, creator of the famous Glass Pavilion for the Cologne Werkbund exhibition of 1914. In 1919, Taut and Gropius, two leading members of the Berlin Arbeitsrat für Kunst (Working council for arts), had drawn up a program similar to the one Gropius was setting out at just this time for the Bauhaus.[7]

The Viennese *Der Aufbau* specified its orientation in its subtitle, *Österreichische Monatshefte für Siedlung und Städtebau* (Austrian monthly for mass housing and city building). But its pages contained far more than plans for walls of reinforced concrete and comfortable rooms for the working class. Their founders' hope, expressed over and again in philosophical, methodological, architectural, and sociological tracts,

DER AUFBAU

ÖSTERREICHISCHE MONATSHEFTE FÜR SIEDLUNG UND STÄDTEBAU

HERAUSGEBER: ARCHITEKTEN **FRANZ SCHUSTER - FRANZ SCHACHERL**

MITARBEITER: REGIERUNGSRAT A. D. DR. **HANS KAMPFFMEYER** (WIEN), GENERALDIREKTOR DR. ING. **HERMANN NEUBACHER** (WIEN), DR. **OTTO NEURATH** (WIEN), STADTBAURAT A. D. ARCH. **BRUNO TAUT** (BERLIN), PROFESSOR DR. **HEINRICH TESSENOW** (DRESDEN), STADTBAURAT A. D. DR. ING. **MARTIN WAGNER** (BERLIN).

NUMMER 1 — WIEN, IM FEBRUAR 1926 — JAHRGANG I

UNSER WOLLEN.

Wir wollen mitwirken am kulturellen, sozialen und wirtschaftlichen Aufbau der Gesellschaft, die aus der gegenwärtigen Unklarheit, Unordnung und chaotischen Verworrenheit, nach Klärung und Ordnung strebt.

Unsere Zeit fordert von jedem Einzelnen und von der Gemeinschaft äußerste Anspannung und ernsteste Besinnung, um einer neuen Entwicklung eine richtige und bedeutende Bahn zu geben.

Jede Zeit hatte ihre Verantwortung vor der Zukunft, aber keine seit langem hatte sie so sehr wie unsere. In rascher und fast beängstigender Aufeinanderfolge hat das 19. Jahrhundert Erfindungen und Entdeckungen gezeitigt, die Jahrhunderte alte Träume und Märchen in besonderer Weise wahr gemacht haben. Geniale Maschinen und unheimliche Instrumente stehen uns zur Verfügung, um die weitverzweigten Bedürfnisse des Lebens in einer Art und Weise zu erfüllen, wie dies keine Zeit vor uns gekannt hat; den großen wissenschaftlichen und technischen Umwälzungen folgten politische und soziale. Altes steht unvermittelt neben Neuem, Chaos, Unklarheit, Unordnung, ist unser Erbe nach dem vergangenen Jahrhundert und unsere Städte, unsere Häuser und unsere Wohnungen sind ein getreues Abbild dieser Zerrissenheit und Haltlosigkeit! Unser Jahrhundert scheint berufen, wieder zu ordnen und zu scheiden, dem Alten seinen ihm zukommenden Platz zu geben und dem Neuem sein Recht!

Es soll unsere Aufgabe sein jene Probleme zu erörtern, die im Bauen auf Umwandlungen, Umstellungen und neue Wege hinweisen.

„Bauen" ist für uns nicht eine Angelegenheit des Einzelnen, des Bauherrn, des Architekten oder des Baumeisters, sondern der Gesamtheit.

Wir verurteilen daher die Einstellung, daß jeder Einzelne machen könne was ihm behage und wie er es nach Laune und persönlichem Dafürhalten wünsche; wir wissen, daß dies im Gegensatz steht zur individualistischen Auffassung breiter Schichten, aber wir wissen auch, daß nur größte Rücksicht auf die Gesamtheit, daß nur ein ernstes Eingehen auf die Forderungen und Notwendigkeiten, eines großen Gesamtwollens, den häßlichen und unhaltbaren Zustand unserer Städte überwinden hilft, der allen am Aufbau einer neuen Gesellschaft Arbeitenden in kultureller, sozialer und wirtschaftlicher Hinsicht die ernstesten Sorgen bereitet.

Der zerrissene, würdelose Großstadtrand mit seinen verwahrlosten Zinshausblöcken und Feuermauern dicht neben dem Ackerland und der alten Dorfstraße, ist uns ein Zeichen der Zerrissenheit und Würdelosigkeit unserer Zeit, deren Ueberwindung ein Ziel aller sein muß. Wir sind überzeugt, daß dies nicht geht ohne die breiten Massen, ohne daß die ganze Bevölkerung an diesen Fragen verstehenden Anteil nimmt. Aufklärung zu geben und den Kampf zu führen, soweit es im Rahmen und der Möglichkeit einer Zeitschrift liegt, soll Aufgabe des Aufbau sein.

Wir betrachten es als schwächlich und verlogen, den heutigen Zustand etwa mit einem Hinweis auf die „gute alte Zeit" zu einer freundlichen Veränderung bewegen zu wollen; wir glauben nicht an die Idylle der Postkutsche und die „Schönheit der gotischen Stadt", weil nicht umsonst die schöpferischen Köpfe der Menschheit aus der Postkutsche die Eisenbahn und das Auto und das Luftschiff gemacht haben und weil wir wissen, daß zum Beispiel das gotische Wien in allen Straßen so voll Schmutz war, daß jeder Bauer, der in die Stadt kam, eine Fuhr Unrat hinausführen mußte!

Figure 3. Title page of the first issue of Der Aufbau: Österreichische Monatshefte für Siedlung und Städtebau, *1/1 (Vienna, February 1926).*

was to create a new *Lebensgestaltung,* a new form of life. Housing pressure in Vienna began mounting even before World War I. Peasants were moving to the city from across the empire; roughly fourteen thousand new units were going up every year, and even that number could hardly keep pace with demand. During the war itself, construction essentially ceased (372 apartments in 1917), leaving a four-year gap in building. Following that lag in construction came a massive postwar influx of soldiers, the burst of new families, and droves of people displaced by the end of the Hapsburg regime. It is not hard, then, to see why politicians of every stripe put the housing situation front and center. Land on the outskirts of Vienna filled first with vegetable patches, then with animals and barns, and eventually with shanty-town dwellings. By 1920 some thirty thousand people who populated these "wild settlements" began organizing and launched two massive demonstrations that stunned Vienna. The newly elected socialists, under Otto Bauer, responded with donations of plumbing, windows, and doors — soon factories went up to provide these items in ever-increasing numbers. Appropriation of "underutilized" apartments began in earnest as did the seizure of unused lots, factories, and commercial space. Within a few years, the government distributed some ten thousand such units.[8]

But the most dramatic development was the beginning, in 1921, of the *Wohnbausteuer,* a highly progressive rent tax earmarked for the building of radical new *Siedlungen* — giant complexes of workers' apartments surrounded by stores and including kindergartens, laundries, and a collectively owned courtyard. These would be, by socialists' parlance, the "palaces" of the proletariat. From the beginning, these buildings, such as the Karl Marx Hof, were intended not merely to house but to transform the working class: to bring light and air into houses, to banish forever the backyard dark shacks that demeaned the workers. The new apartments would be space and work efficient, and so foster collective values of property; the new architecture would free women from domestic labor and instantiate the *factual* rationality that was the workers' birthright. This was a self-consciously politicized architecture that would exploit space, material, and design to combat the older bourgeois architecture. Older buildings divided families from each other and led people to cling to a dying past. Architects of Red Vienna thought of themselves as building — literally and figuratively — a new life (Marcuse 1986, 558–85). Perhaps as important as workers, the *Siedlungen* became the primary exhibition point for the host of new architects who were emerging from Austrian and Bauhaus movements. What was needed was an *Aufbau.*

The polemical opening of this particular *Aufbau* (the one that Neurath, Schuster, Schacherl, Wagner, Taut, and Tesserow directed) declaimed:

> We want to work together on the cultural, social, and economic *Aufbau* of society, which is striving from the present unclarity, disorder, and chaotic confusion toward clarity and order.

Not only in the manifesto, but in all subsequent issues, the journal made it plain that it despised the ornate in architecture; elimination of the "stylish" was essential in all aspects of life. Reforming the "interior" meant rationalizing personal life as well as public buildings. Together, this restructuring of the personal and the public would mesh people like precision gears into the technical-objective world. Modernity, in the form of this constructed world, opposed the romantic, the religious, the national, and the traditional. The *Aufbau* manifesto continued:

> We take it as weak and untruthful to wish to move the present situation to a "friendly" change with reference to the "good old times." We do not believe in the idyll of the postal carriage and the "beauty of the gothic state" because not for nothing did the creative minds of mankind make the train, the car, and the airship out of the postal coach; and because we know, for example, that gothic Vienna was so full of dirt in every street that each farmer who came to the city had to take out a cartful of filth! (*Der Aufbau: Österreichische Monatshefte* 1926, 1/1:1)

Modernity, for Neurath and his colleagues, was precisely the tearing down of this gothic Vienna and its replacement by the rational and factual (*sachlich*). In his first individual contribution to the journal in May 1926, "Rationalismus, Arbeiterschaft, und Baugestaltung," Neurath began by berating the lip service paid to modern architecture:

> The demand "construction projects should be handled objectively [*sachlich*]" is almost acknowledged by the architects of late capitalism. The demand of an architecture free from ornaments and decorations, which views buildings like a kind of machine, is self-evident. Nevertheless, it occurs so seldom despite the fact that it is spoken about so often.

Deriding the ornate façades of the Palace of Justice and the War Ministry in Vienna, Neurath contrasted approvingly the plan and appearance of a factory designed in part by Gropius as exemplary of the new rationalism (Neurath 1926b, 49).

Central to Neurath's analysis was his belief that an inexorable march of technology had transformed society, extending from social hygiene to the inner workings of the mass production factory. All those who held out against these developments (the train tracks, telephones, and telegraphs) were standing in "crass contradiction" to the rationalization that could be seen everywhere. The educated and upper classes — even artists and scientists who should know better — still resisted the physical instantiation of rationality into their homes. Nothing illustrated this recalcitrant attachment to traditional ways of life better, Neurath argued, than the ornate bronze lamps that took the style of gas lamps at a time when electrical bulbs had made their form utterly inappropriate (see fig. 4). This reluctance to shed the forms of a previous and now-outmoded life was particularly acute in Austria, Neurath contended, because the bourgeoisie lacked the factual (*sachlich*) tradition. In England, by contrast, the bourgeoisie had the independent existence needed to match inner and outer forms (ibid., 51).

All these vestiges of the past went together: the romantic way of thinking in private life, nationalism, clericalism, traditional furniture and clothing. Just because Neurath saw this full circle of cultural and political goals as intertwined into a *Lebensordnung* (order of life), *Lebensgestaltung* (form of life), or *Gesamtgestaltung* (total formation), he interwove political advocacy of a classless socialist society with a critical commentary on the constructivist paintings of Fernand Léger. Such art exemplified the superficial adherence to the hard-edged rationalism of modernism. To Neurath, the artist who appropriated the symbols of modern means of production without grasping production methods themselves resembled nothing so much as a child playing with retorts and test tubes without understanding the underlying chemical reactions. The reproduction of bits and pieces of machines was, in and of itself, meaningless. Outer forms counted for nothing. Architects would do far better to show the public blueprints than to show them façades. Is a house light and airy? That matters; stone cornices do not (ibid., 52). Real construction, an *Aufbau* based on simple structural forms and on rational principles, was needed.

For of all the elements of private life, it was in architecture and factory production that the rationality of the working class would be expressed in its objective-technical form. Rationalization penetrated everyday life not by the visible gears of machines, but by the thoroughgoing saturation of workers' lives through Taylorism in the giant factories of the late nineteenth and early twentieth centuries. Still, by Neurath's lights, it was by no means a foregone conclusion that the forces of *Sachlichkeit* would succeed, for arrayed against them was an Austrian "ruling class" that, far more than its British equivalent, was hostile to the new, more rational way of life (ibid., 51). To Neurath,

Figure 4. Electric lamp with the appearance of a gas lamp. From Der Aufbau: Österreichische Monatshefte für Siedlung und Städtebau, *1/1 (Vienna, February 1926).*

only the working class lay free of the claims of the past; only the working class saw the *Lebensordnung* of everyday life as a means to its happiness:

> To such a way of thinking [the proletariat's] architecture is a slice from the total form of life [*Gesamtgestaltung des Lebens*]! Its task is to grasp clearly this goal and, in its realm, to realize it. In architecture the workers' rationalism finds its new expression: in its factual-technical forms [*sachlich-technischen Formen*]. (ibid., 54)

Neurath and his colleagues on the board of *Der Aufbau* celebrated their ostensive break with the past. In one issue they contrasted a full-page collage of ancient monuments and statuary with icons of electricity, shipping, power, and factories (see fig. 5). Too many Sundays had been lost in the accumulation of antiquities, the first page proclaimed. Now the new world was at hand, and humankind's duty was to be worthy of it. "Art," classically conceived, was to be replaced by all that was rational and therefore international. The watchwords were *Welthandel, Weltindustrie, Weltverkehr,* and *Weltwirtschaft* (*Der Aufbau: Österreichische Monatshefte* 1926, 1/6:99). World structure and inner life were bound together: modifying even the vocabulary of expression became a way of modifying thinking.

Neurath's own writings on the necessity of bringing interior and exterior life into line with the time were thus part and parcel of a journal project within the larger literature and politics of post–World War I *Aufbau.* And it was within this particular circle of *Aufbau* connotations that Neurath's Vienna Circle pronouncements on "inner and outer life," and on an embattled but inevitable *Sachlichkeit,* must be understood.[9]

Perhaps most significantly, Neurath identified contemporary architecture with the creation of new ways of thinking, acting, and feeling, and he wrote article after article about substituting the hard-edged, hard-boiled rationalism of technical objectivity for false fronts and façades. The reformation (or perhaps creation *de novo* of the "modern man") was explicitly tied to the establishment of new "tools for living" associated with an age of ordered, structured reasoning and mechanical operations.

In December 1926, the year Neurath and his collaborators founded their *Aufbau* journal, the startling edifice of the Dessau Bauhaus opened. Perhaps inevitably, Neurath was there. The building's sharp-edged lack of ornamentation pleased him, though he reiterated his (Josef Frank–like) caution against elevating functionalism into a new stylistic fetish of its own. "When will the modern engineers run the Bauhaus?" he asked. For if the Bauhaus was truly to usher in the age of a new form of societal and personal life (*Neugestaltung des gesellschaftlichen und*

Figure 5. Detail from a modernist collage. From Der Aufbau: Österreichische Monatshefte für Siedlung und Städtebau, *1/1 (Vienna, February 1926).*

persönlichen Lebens), it could only be through the thoroughgoing adoption of the technical. Those who wanted to "enter the promised land," leaving the past behind, "will seize upon the formation of the new form of life [*Gestaltung des Lebens*] as a technical achievement. This is the thrust of the Bauhaus, unfettering the liveliest discussion and most vigorous efforts on all sides" (Neurath 1926a, 210–11).

In 1926, Neurath was thus deeply engaged in the definition and extension of a modernist world — at precisely the same time that he, Hans Hahn, and Philipp Frank joined Moritz Schlick in the formation of the "official" Vienna Circle. And it is this wider context of what one might call an Austrian and German "techno-modernism" that set the tone of the Vienna Circle "manifesto," the preface to Carnap's *Aufbau,* and much else in the work of the logical positivists of the mid-1920s.

The manifesto, "Wissenschaftliche Weltauffassung: Der Wiener Kreis," which Neurath, Hahn, and Carnap drafted during the summer of 1929, deliberately politicized their philosophical stance when they argued that their position emerged from "fierce social and economic struggles."[10] They continued: "[O]ne group of combatants, holding fast to traditional social forms, cultivates traditional attitudes of metaphysics and theology whose content has long since been superseded; while the other group...faces modern times, rejects these views and takes its stand on the ground of empirical science." Like so many of the architects writing in the various *Aufbau* journals, Neurath, Carnap, and their colleagues directly tied their enterprise to the "modern process of production, which is becoming ever more rigorously mechanised and leaves ever less room for metaphysical ideas" (Neurath 1973, 317). The echo of the *Aufbau* slogans of *Welthandel, Weltindustrie, Weltverkehr,* and *Weltwirtschaft* could not have been clearer. Or again, elsewhere in the same document: "We witness the spirit of the scientific world-conception penetrating in growing measure the forms of personal and public life, in education, upbringing, architecture, and the shaping of economic and social life according to rational principles" (ibid., 309). Rudolf Carnap shared some, but not all, of Neurath's ambitions to construct the world anew. As the preface to *Der logische Aufbau der Welt* indicated, he held high hopes for that rather technical work, hopes that echoed all the left-technocratic cultural meanings of *Aufbau* I have discussed. It was, he argued, part and parcel of a rejection of the past and a "faith that the future belongs to this attitude":

> We do not deceive ourselves about the fact that movements in metaphysical philosophy and religion which are critical of such [a scientific] orientation have again become very influential of late.

> Whence then our confidence that our call for clarity, for a science that is free from metaphysics, will be heard? It stems from the knowledge or, to put it somewhat more carefully, from the belief that these opposing powers belong to the past. We feel that there is an inner kinship between the attitude on which our philosophical work is founded and the intellectual attitude which presently manifests itself in entirely different walks of life; we feel this orientation in artistic movements, especially in architecture, and in movements which strive for meaningful forms [*Gestaltung des menschlichen Lebens*] of personal and collective life, of education and of external organization in general. We feel all around us the same basic orientation, the same style of thinking and doing. . . . Our work is carried on by the faith that this attitude will win the future. (Carnap 1967, xvii–xviii)

Imagine an American analytic philosopher of the 1950s resting his future reputation on the "inner kinship" between philosophy, collective life, and architecture. The thought-experiment fails. That Carnap's *Aufbau* could be introduced this way is testimony to the enormous amount of cultural work that had already prepared the ground for these affiliations between fields to make sense—perhaps even to seem natural.

I do not, however, mean to imply that the postwar usage of *Aufbau* determined the content of the philosophical projects of that title. We know by now far too much of the much older empiricist and the more recent neo-Kantian roots of many aspects of Carnap's project.[11] In the text of Carnap's *Aufbau* itself, the author is, above all (as Michael Friedman has stressed), seeking to define a system in which each concept would have a fully specified relational and therefore objective place (Friedman 1987). By building up from basic elements, Carnap hoped to exclude the superfluous, the mystical, and indeed all that was not rationally or empirically located. It was this eliminalist strategy, this systematic construction from clear starting points, that girded Carnap's *Aufbau,* broadly conceived, against reactionary assault.

As we now know, Carnap's *Logische(r) Aufbau* originally had a very different title. The first draft manuscript of 1924 was labeled *Vom Chaos zur Wirklichkeit* and soon took on the more neo-Kantian appellation *Konstitutionstheorie* (see Coffa 1991, 208). By 1928, however, the term *Aufbau* would have carried a much wider connotation and attracted considerably more attention as it would quite clearly have suggested to German speakers, and to the Vienna Circle in particular, a set of associations that would have resonated across architecture, philosophy, and politics. Not unlike the term "postmodernism" in the 1980s, *Auf-*

bau in the mid-1920s embedded a work into a chain of meanings — and a wider cultural struggle (Huyssen 1986; Jones 1994). Having read the work, and very much approved of it, Schlick was reluctant to see it issued under its old moniker. On 4 January 1928, he strongly urged Carnap to change it:

> As for the *title of the work,* I would like to say this: a book title not only has the task of correctly describing the work's content; it also must indicate the most important part of the content or the intent with suggestive force. This latter is the main thing. On these grounds I would decisively dissuade you from using the titles that you proposed at the close of your letter; they are too bland.... The fundamental thing in your book... [is] the general discussion of the principles of the constitution, to which the *Aufbau* of the system of knowledge is bound. For this foundation of principles, and more generally for the whole book, it seems to me that *Der logische Aufbau der Welt* is in fact the appropriate title. By these words it is, of course, to be understood that in the first instance the book has to do less with the actual and special development of a particular *Aufbau,* than with the principles of any such *Aufbau.*[12]

The idea of such a construction appealed to Schlick a great deal; as he had often stressed, the systematization of knowledge was just what was needed to root out bad reasoning, lies, and deception. As regards Schlick's suggestions, Carnap (never one to leap to a conclusion) hesitated, keeping his original title in the subtitle, but he eventually came to identify the work so closely with its new moniker that when the printers forgot to insert the *Konstitutionstheorie,* he didn't even notice. Increasingly, he had come to see this work in general, and the efforts of the Vienna Circle more generally, as part and parcel of a larger, critical cultural movement. Calling it *Der logische Aufbau der Welt* had come to be so natural, the old title had faded from view.

2. *Aufbau* in America

People move across oceans with relative ease; complexes of ideas do not. When László Moholy-Nagy became director of the New Bauhaus in Chicago, he recruited Carnap as an occasional lecturer and brought on board in a continuing fashion three members of the Unity of Science movement, including Charles Morris.[13] Throughout the 1930s, Morris had acted as an American clearinghouse for contact between Americans and the Vienna Circle, and by the time Carnap arrived in the United

States, Morris had become one of his closest philosophical contacts. In addition, for several years Morris had been the most active American in the ambitious series of conferences the Circle ran on the unity of science and was a coeditor (with Carnap and Neurath) of the successor journal to the Vienna Circle's *Erkenntnis: The International Encyclopedia for the Unity of Science.* Morris's philosophical program stressed the purification of language, the reduction of all talk to the simplest of starting points. Applying this very basic vision of the positivists, he tried his best to cast it as a variant of Dewey, presumably not to frighten his compatriots with a massive dose of philosophy with a German accent. In his courses at the New Bauhaus, he explicitly linked the goal of training in art to the reduction of language already imposed on science.

At the New Bauhaus and in his role as an editor of the *Encyclopedia of Unified Science,* Morris increasingly pushed his allies, especially Neurath, further from the political engagement that had marked the Circle's European incarnation. In a sense, the Americanization of the Unity of Science movement resulted from its now American funding and institutional setting, but the process was much more complex and interesting than that. In addition, Morris, Percy Bridgman, and other Americans infused pragmatic philosophy into the movement, while simultaneously the political pressure mounted to remove internationalism and Marxism from the movement. Partially a result of the generalized pressure to become scientifically "neutral," this depoliticization issued both from within the movement (Morris) and from without (Warren Weaver of the Rockefeller Foundation). By "depoliticization" I mean not just the deletion of Marxist or socialist ambitions, but a sharp narrowing of the broad cultural program of technological modernism that the left wing of the Vienna Circle had advocated. *Aufbau* had no significant role to play in the New Bauhaus.

Already in January 1935, when Morris first approached Weaver, he began to phrase the goals of the unified science program in terms that completely obscured these wider cultural ambitions: "I represent an international group of scientists and logicians interested in the unification of science[,] meaning by that term not a synthesis of factual results but the analysis and systematization of the language of science."[14] Weaver encouraged Morris and supported the movement with more than words: by 1941, the movement had a twenty thousand dollar grant.[15] Gone was the desire to meld a mode of art, architecture, and indeed a whole *Lebensgestaltung* into a language and demeanor. What was left was a tripartite program that included the coordination of scientists, the publication of an encyclopedia, and the establishment of a "unified center" that would stand against antiscience.

With Carnap, Reichenbach, Frank, and Neurath on the move, Morris's established position in Chicago and his growing links with Rockefeller's Weaver and the University of Chicago Press gave him vastly increased power. This power was reflected in the tenor of his letters. Morris's correspondence with the senior philosophers in Europe had been extremely deferential. This changed. For, grateful as they were to be out of the Nazis' grasp, members of the Circle were precariously perched in their adopted countries. Even Carnap had a hard time securing a position and was in very difficult and sometimes acrimonious discussion about the terms of his employment with the University of Chicago. And he was probably the most secure of all the refugee positivists. To Morris, Carnap reported on the anti-Semitic situation at Princeton, one that effectively barred Reichenbach or Feigl from the appointment that Carnap had declined, leaving both Feigl and Reichenbach somewhat vulnerable.[16]

Through the placement of Carnap at Chicago, through the management of the other logical positivists, and through his own growing reputation, Morris's position in the movement changed drastically. Leaving behind his tentative approaches to the Circle in the late 1920s, he now began to instruct his collaborators on how they must act in the United States. For example, Morris resolutely resisted Neurath's attempt to push Reichenbach out of the new Unity of Science movement, on the grounds that parochialism had to be combatted: "I am interested in the Unity of Science movement as such, and not in any special group such as the Wiener Kreis."[17]

Elsewhere, Morris deprecated Neurath's proposed title for the encyclopedia, which began with the word "international." Some people (Morris did not specify who) did not like the very word "international," presumably because it smacked (in 1937) of unpatriotic leftism.[18] Nor would Morris countenance Neurath's appellation of the contributors as "new encyclopedists" after the French *Encyclopédie*.[19] Writing to Neurath, Morris reminded his Austrian colleague of his own earlier comment that the journal would proceed "without entering into discussions of principle touching political ideologies." Morris interpreted the meaning of this phrase as opening the *Encyclopedia* to all comers, with no particular political ideology to be associated with the journal—it was quite evidently a warning to Neurath that his political views, such as those expressed in *Der Kampf,* would have no place.[20] Indeed (according to Morris) if the journal needed anything, it needed *more science* and *more scientists* (not more politics and philosophy) in order to facilitate its acceptance in the wider scientific community.

Neurath's early pleas for a new *Encyclopédie* met resistance from

Morris, partly because Neurath understood and wanted to foster the revolutionary associations of its eighteenth-century counterpart. Voided of this association, the historical antecedent was much less threatening, as was evident from Morris's enthusiasm over a 1938 evaluation of the first volume of the *Encyclopedia* that appeared in the *New York Times Book Review.* The review (written by Neurath's relative W. Kaempffert) put it this way:

> Indeed this work must serve a higher purpose than that of amalgamating the sciences for the benefit of specialists. The men who are making this encyclopedia have no desire to enter the political arena but every desire to influence intellectual leaders and to ensure freedom of scientific inquiry everywhere. The encyclopedists of the eighteenth century played their part in breaking the tyranny of their time without engaging in political struggles. So with this expression of empirical opinion.[21]

Here was a form of nonpolitical politics that Morris could endorse: tyranny would quake at the mere presence of scientific reasoning.

Neurath did not go gently into the Americanized, politically homogenized, and neutral Unity of Science movement. Repeatedly in his correspondence, Neurath returned to the old divisions between left and right, distinctions that for him still made a difference: "I remember in the WIENER KREIS as we read Wittgenstein, I again and again said: that is Metaphysics. It became dull and Hahn suggested I should speak of M. only to shorten the sounds[,] and since I too often said 'M,' he suggested I should only remark when I [was] satisfied by saying, 'NM.' "[22] Tainted by Wittgenstein, others on the right side of the Circle also fell from Neurath's favor: Schlick's "Wittgensteinization and Decarnapization has been the real 'tragedy' of his later arguing. Wittgenstein now is, where I expected him to be,... fighting scientific arguing and maintaining dull generalities on language, without any contact with empiricist science."[23]

But if Wittgenstein's metaphysical proclivities tended to divide the left from the right wings of the Circle, Neurath envisioned a second order of boundaries between the Vienna Circle and sympathetic movements outside: "James, who is to a certain extent somebody who helps us, is on the other hand full of metaphysical trends, liking Bergson etc[.] [T]he same is right of Peirce. Therefore we should avoid to be in too close contact with a name... which stresses the connection with these fine people, who nevertheless do not belong to our movement as such."[24]

Outside the universe of "fine people" who might disagree about tactics were the divisions that severed the Nazis from their enemies. Those philosophers who aligned themselves with the Nazi movement offended at an altogether different level:

> It is a sad thing, that now the German illness enters the States [. . .][,] but that is the same in many other cases [i]n history etc. The Nazidom of Heidegger did not sufficiently shock people on the other side of the Atlantic. I bought his Nazi declarations, they are TERRIBLE, of the worst Goebbels type, full of distorted details, freely told lies etc[.,] but you feel through all these sentences the HIGHER language of his philosophy.[25]

As with so many leftists during the 1940s, Neurath's opposition to the right did not imply an easy alliance with the specific structure of the Soviet Union. Ever since the mid-1930s, it had been clear that, for Moscow, the Vienna Circle amounted to a variant form of *Machismus,* hardly a term of recommendation after Lenin's bitter denunciation of Mach in *Empiriocriticism.*[26] Neurath found his new political niche in a delicately perched third position — that is, as a technocratic left democrat. And here, he believed, the "Wissenschaftliche Weltauffassung" would ply its true trade: "I think our LOGICAL EMPIRICISM will play a certain role in the democratization on Europe."[27]

The fourth and final division that Neurath drew during the war was, in some sense, against himself, against the earlier version of his own left wing. More precisely, politics had intervened. From a largely pro-Soviet Austromarxism, Neurath seems to have moved to a looser sort of decentralized and democratized socialism. His philosophy too turned away from even the hint of foundationalism that terms of reference seemed to suggest.

To a philosopher looking back today, it often appears self-evident that the project of an unmetaphysical philosophy was doomed to failure; there is, as Ernst Nagel has put it so well, no "view from nowhere." For the modern historian it seems equally apparent that the cry for an apolitical politics is similarly hopeless. Technocracy is not a day-pass from politics. One surviving visitor to the early Vienna Circle meetings (Peter Hempel) once told me that the Circle had an agreement to leave politics at the door. Our general view today is that politics, like the air we breathe, will not wait outside. And from art and design, we have also come to accept that there is no exemption from aesthetics. Where the Bauhäusler could stare into their lamps and chairs and see only pure function, we see an ideologically saturated aesthetic, frightening, inspiring, striking, even moving, grasping for an imaginary future. We cannot any longer see the neutrality of pure function. Unaesthetic aesthetics, unpolitical politics, and unmetaphysical philosophy all seem elusive.

But taken together the various moments of modernism (captured by the call for an *Aufbau*) do convey a vision. It was a brief and fleeting vision, one held under fire. And that vision was given ever more coherence as the left wings of the Bauhaus and the Vienna Circle faced an increasingly hostile, murderously determined world.

The American welcome, while warm and for many lifesaving, was immensely complex and transformative. Pragmatists may have been sympathetic to positivists, but that did not rule positivism to be a form of pragmatism. The politics of a cautious ecumenical liberalism altered as it received. Whether challenging the "international" in the *International Encyclopedia of Unified Science* or questioning the right interpretation of "encyclopedists," Morris and his American colleagues subtly shifted Neurath's modernist technocratic Austromarxism to a technical tool kit. And in others' hands, the new *Lebensgestaltung,* dreamt of as a modernist *Aufbau,* slowly reduced to an academic specialty within the philosophy of science. The ideal of an *Aufbau,* as understood in mid-1920s Vienna, was conjointly architectural, political, and philosophical; it did not as such move across the Atlantic.

3. Summary and Conclusion

The problem is this: for any thesis about the association of two arenas of culture — whether it is telegraphy and electrodynamic theory, pneumatics and parliamentary politics, Weimar antirationalism and quantum uncertainty, thermonuclear weapons and particle physics, or architecture and philosophy — the question arises as to the status of cultural links.[28] One possibility would be that a claim that two groups are part of the same enterprise would amount to an assertion that there exists a relation of *logical* implication. For example, to claim that a Cartesian must oppose the intensely hierarchical Aristotelian cosmos follows from the basic Cartesian cosmological assumption that the universe is homogeneous. This does not seem appropriate to the philosophy-architecture links described here. Nothing in the tenets of logical positivism *implies* subscription to the basic assumptions of Bauhaus architecture.

Another possibility would be that one movement *caused* the other to take the form it did: either the Vienna Circle's form of constructivism caused that of the Bauhaus, or the constructivism of the Bauhaus engendered that of the Vienna Circle. This too seems overly confining. Clearly there were logical positivists who had little interest in architecture one way or the other. Worse yet — for a linear causal account of the link between architecture and logical positivism — the rallying cry to create an *Aufbau* became a slogan of the fascists in the mid-1930s, which would hardly suggest a unique bond between the two movements.

A far weaker assumption abdicates any form of explanation and simply identifies the two arenas as having developed in parallel. Unlike the logical or unicausal accounts, this structuralist one seems too weak; it loses touch entirely with the *work* that went into the establishment of relations between the two groups, their mutual legitimation, their common causes, and their enemies.

One possible alternative to the extremes of logical implication, unicausal determinism, and parallelism is to develop a notion of cultural meaning. That is, in the German-speaking world during the 1920s, a vocabulary and reservoir of images evolved that were shared by both the positivists and the Bauhäusler. This repository included a notion of transparent construction in which each element had its place and function. It included the aspiration of both groups to harmonize their own discipline (and interior life more generally) with the rationalism each saw embodied in what they perceived as the regularity, intelligibility, and functionality of contemporary technology. It included an emphasis on the role of the collective action of peoples. But these meanings and associations of meanings had to be *put* together and *held* together — there was nothing inalienably "rational" about a hydroelectric dam. The dam could have been (and was) deployed to signify the power of the unified nation under fascism, the productive energy of communism under Stalin, and the regenerative strength of America during the Works Progress Administration.

The point of examining these images of technology as *meanings* is twofold. On the one hand, meanings are not entirely aleatory. Groups cannot arbitrarily choose to make the icons of a society signify what they will. On the other hand, meanings are not fixed forever. The cultural meaning of *Aufbau,* Bauhaus architecture, and logical positivism all shifted radically — often to their practitioners' horror — as the ideas moved outside the community in which they were forged.

To pursue a notion of cultural meaning then becomes at once a conceptual and historical task. It involves the articulation of the chains of associations that terms carried in a specific cultural location. *Aufbau,* to return to our example, not only maintains the age-old notion of building; it came, through the socialist city-state of Red Vienna in the post–World War I period, to include much more than mere bricks. An *Aufbau* reformed life, restructured values away from the decorative and superfluous and toward the practical; *Aufbau* in 1917–26 incorporated rationalism and in general reached all the way down to the very form of life: *Lebensgestaltung, Lebensordnung,* and *Gestaltung der Lebens.* It was, as many authors emphasized, more than a revolution.

The historical-philosophical task of establishing the link between two arenas of culture (here logical positivism and modernist architecture) then becomes one of tracking the specific modes by which the two groups came to *share* cultural meanings: in this instance, through Neurath's work on *Aufbau* with leading city planners and architects; through Josef Frank's lectures and collaboration with Neurath; through Carnap's, Reichenbach's, and Neurath's visits to the Dessau Bauhaus; through the mutual political support offered between the logical positivists and Meyer's Bauhaus. If I could borrow a metaphor from chemistry, I would say that the Vienna Circle and the Dessau Bauhaus shared meanings the way two atoms share valence electrons. The philosophical-cultural-historical task of establishing a link becomes one of articulating the conditions under which concepts like *Aufbau* fit simultaneously into the meaning valences of both philosophy and city building. This relationship is, as advertised, not one of unilateral causation, logical implication, or deracinated *Zeitgeist.*

Indeed, without the experienced context of a postwar Vienna, a collapsed empire with its now illusory signs of aristocratic power, without the explosive building boom of Red Vienna and the ideological program that accompanied it, Americans did not and could not see the notion of *Aufbau* quite the way post–World War I Vienna did. It is not, I would argue, a mere accident that Rolf George translated Carnap's magnum opus as *Logical Structure,* not *Logical Construction.* Red Vienna had vanished, and the New World had its own fascination with universal, ahistorical, and apolitical *structures.*

In a different context, the great architectural historian Nicholas Pevsner once used the phrase "architecture of reason" to capture the faith many builders had that they could embody rationality in the very stuff of an edifice (Pevsner 1975). In a sense, the architects of the Bauhaus and Red Vienna aspired to a world that was the mirror image of the one sought by the Vienna Circle. Where the architects wanted to make buildings proceed like reason itself, excluding the decorative and dysfunctional, the philosophers of the Vienna Circle espoused spare, cool, and rational modern architecture as an emblem of their systematic construction of concepts, a reason of architecture. The *Aufbau,* the purifying moment of construction, stood for reformation in the domain that the Bauhaus wanted to substitute for art and in the field that the Vienna Circle hoped would replace philosophy.

In the first instance, the particular affiliation of the Vienna Circle and the Dessau Bauhaus depended on specific cultural complexes in Austria and Germany just after World War I. But, more than that, their *continued* association was tightly bound to the cultural location of both groups and the efforts they made to maintain the sphere of meanings associated with the new unaesthetic aesthetics and unphilosophical philosophy — and it was here that terms like *Aufbau* functioned both as symbol and cement of that association. But when the external environment changed radically, through Nazism and emigration, the cultural meaning of the two movements shifted and constricted. After a brief moment together in the New Bauhaus in 1937–38, the two groups ceased to coexist in the same cultural sphere of early twentieth-century modernism. An international style of building in the early 1960s carried with it precious little of the resonance that Hannes Meyer might have wanted. And philosophy of science in the 1960s no longer self-consciously sought to participate in the reformation of the totality of culture, politics, and reason; nor did it seek to induce a reformation of daily life. Once in each other's valence of meanings, modern architecture and logical positivism had parted ways.

Notes

This chapter is an expanded version of P. Galison, "The Cultural Meaning of Aufbau," in F. Stadler, ed., *Scientific Philosophy: Origins and Developments* (Dordrecht: Kluwer, 1993). For helpful comments and suggestions, I would like to thank Caroline A. Jones. Research Assistants Richard Beyler, Cathy Carson, and Nani Clow were a great help. Jean Titilah assisted with the preparation of the manuscript. My thanks go as well to the Pittsburgh Archives for Scientific Philosophy (PASP), especially to Gerald Heverly. I am much indebted to the archivists at the University of Chicago (USMP designates the Unity of Science Movement Papers, Department of Special Collections, the University of Chicago Library; and PEP, the Peirce Edition Project, Indiana University–Purdue University at Indianapolis, which kindly provided access to many documents and letters relevant to Morris and the Vienna Circle. Material from the Carnap Collection of the Archives for Scientific Philosophy [ASP] is quoted by permission of the University of Pittsburgh.) Support for this research came from the Andrew W. Mellon Foundation.

1. One finds, for example, the less than immortal verses:

> Aufsteht der Deutsche.
> Befreit von der Fessel des Fröners,
> reckt er die sehnigen Arme.
> Angreift er sein Werk
> nach seinem Gesetz.
> Lasten bewegt er,
> die schienen zu spotten
> jedweder Kraft.
> Berge versetzt er,
> die wehrten den Weg ihm.
> Und Spöttern zum Hohn,

zum Staunen der Welt,
wirkt er gewaltiges Werk.
(*Aufbau und Erbe* 1938, 6)

2. On *Siedlungen* and the new architecture, see Lane 1968, 58, 84ff.

3. For an excellent discussion of Neurath's involvement in the Bavarian revolutionary government and the links between Neurath's philosophy and his politics, see Cat, Chang, and Cartwright 1996.

4. On Neurath's involvement in the housing movement, see Hoffmann 1982, 140–48.

5. For details of the interactions between the Vienna Circle and the Bauhaus, see Galison 1990, 709–52; see also Galison 1986, 271–88; P. Neurath 1982, 289–97; and Stadler 1982b.

6. For more on Schuster, Schacherl, and Tessenow, see Hautmann and Hautmann 1980.

7. Wingler 1978, 3. Also on the masthead of the Viennese *Der Aufbau* was Generaldirektor Dr. Ing. Hermann Neubacher (Vienna).

8. This discussion draws from Marcuse 1986, 558–85.

9. There is an increasingly large and impressive literature on Neurath and his political engagement. See, for example, the following important works: Hegselmann 1983, 67–74; Hegselmann 1979; and Nemeth 1981.

10. For more on the Vienna Circle's implicit and explicit politics, see Wartofsky 1982.

11. On the relation of Carnap to Kant, see, for example, Coffa 1991; Friedman 1987, 521–45; Proust 1986; Haack 1977; Moulines 1985; Richardson 1992b.

12. Schlick to Carnap, 4 January 1928, Carnap Collection, ASP 029–30–36.

13. Engelbrecht 1973, 287. For more on the establishment of the New Bauhaus, see also Jordy 1969; see also *50 Jahre New Bauhaus,* Bauhaus Archiv, Argon: Berlin, n.d.

14. Morris to Weaver, 28 January 1935, box 1, folder 15, USMP.

15. Morris to Neurath, 29 September 1941, box 2, folder 1, USMP.

16. Carnap to Morris, 29 April 1936, PEP.

17. Morris to Neurath, 16 November 1936, box 1, folder 15, USMP.

18. Morris to Neurath, 17 April 1937, box 1, folder 16, USMP.

19. Morris to Neurath, 13 July 1937, box 1, folder 16, USMP.

20. Morris to Neurath, 26 August 1937, box 1, folder 16, USMP.

21. Quoted in Morris to Professor Robert C. Woellner, 22 October 1938, box 1, folder 17, USMP.

22. Neurath to Morris, 18 November 1944, PEP.

23. Neurath to Feigl, 26 February 1943, PEP.

24. Neurath to Morris, 18 November 1944, PEP.

25. Neurath to Morris, 18 November 1944, PEP; bracketed ellipsis is in original.

26. Neurath to Morris, 1 October 1934, box 2, folder 6, USMP.

27. Neurath to Morris, 30 May 1942, PEP.

28. Obviously, I choose these examples nonrandomly and will reflect on the work of Smith and Wise 1989; Shapin and S. Schaffer 1985; Schorske 1981; Forman 1971; and others.

Michael Friedman

Overcoming Metaphysics: Carnap and Heidegger

It is well known that Rudolf Carnap (in Carnap 1932e) uses examples from Martin Heidegger as illustrations of metaphysical pseudo-sentences — including, most famously, the sentence "Nothingness itself nothings [*Das Nichts selbst nichtet*]" (Heidegger 1929b). It is tempting today for those on both sides — for those who sympathize with Carnap and those who sympathize with Heidegger alike — to view this episode with a more or less tolerant smile. Among those sympathetic to Carnap, Heidegger's sentence now appears as simply unintelligible, but hardly dangerous, nonsense: one is by no means surprised by such obvious absurdities coming from a fuzzy-minded "continental" thinker. Among those sympathetic to Heidegger, Carnap's criticism now appears as a case of simple blindness to Heidegger's point: one cannot expect a narrow-minded "analytic" philosopher even to begin to grasp such profundities. What both sides miss, I believe, is the depth and force this encounter had for Carnap and Heidegger themselves. We thereby miss the meaning and extent of the common context within which both contemporary philosophical traditions — both "continental" and "analytic" traditions — arise and develop.

The first point to notice is that Carnap and Heidegger had earlier met one another: at a celebrated disputation, or *Arbeitsgemeinschaft,* between Heidegger and Ernst Cassirer that took place during the International University Course held at Davos, Switzerland, from 17 March through 6 April 1929.[1] It was on this occasion, two years after the sensational appearance of *Being and Time,* that Heidegger first made public a radical phenomenological-metaphysical interpretation of the *Critique of Pure Reason* developed in explicit opposition to the Marburg school of neo-Kantianism with which Cassirer was closely associated — an interpretation Heidegger then wrote up in a few short weeks following the Davos university course and published as *Kant and the Problem of Metaphysics.*[2] Heidegger thereby presented himself — with extraordinary success, as it turned out — as the author of a fundamentally new kind of philosophy destined to replace the hegemony of the neo-Kantian tradition and to supplant the remaining "rationalist" tendencies in Husserlian phenomenology as well. In July 1929 Heidegger symbolically completed this ascension when he delivered his inaugural address as Edmund Husserl's successor to the chair of philosophy at Freiburg: Heidegger 1929b is the published record of this address. Carnap attended the Davos university course and reported on the occasion in his diary.[3] Like everyone else in attendance he appears to have been especially caught up in the intellectual excitement of the encounter between Heidegger and Cassirer. Moreover, Carnap was clearly impressed by Heidegger and had several philosophical conservations with him (ASP 025-73-03, entries from 18 March, 30 March, and 3 April 1929). When Carnap returned to Vienna he retained this sense of excitement and seems, in fact, to have studied *Being and Time* rather seriously. In particular, he actively participated in a discussion group in the summer of 1930 led by Heinrich Gomperz and Karl Bühler where Heidegger's book was intensively examined.[4] The first draft of Carnap 1932e was then written up in November 1930. Carnap presented it as lectures at Warsaw (November 1930), Zurich (January 1931), and Prague (November 1931) and then (in a revised version) at Berlin (July 1932) and Brünn (December 1932).[5] The published version appeared in *Erkenntnis,* the official journal of the Vienna Circle, in 1932. In §5 of Carnap 1932e, entitled "Metaphysical Pseudo-Sentences," Carnap introduces his consideration of examples from Heidegger 1929b by remarking that, although he "could just as well have selected passages from any other of the numerous metaphysicians of the present or the past," he has here chosen to "select a few sentences from that metaphysical doctrine which at present exerts the strongest influence in Germany."

The second point to notice is that Carnap's analysis and criticism of "Nothingness itself nothings" is more sophisticated and penetrating than one might have antecedently expected. For, on the one hand, Carnap's complaint is not that the sentence in question is unverifiable in terms of sense-data; nor is the most important problem that the sentence coins a bizarre new word and thus violates ordinary usage. The main problem is rather a violation of the logical form of the concept of nothing. Heidegger uses the concept both as a substantive and as a verb, whereas modern logic has shown that it is neither: the logical form of the concept of nothing is constituted solely by existential quantification and negation. On the other hand, however, Carnap also clearly recognizes that this kind of criticism would not affect Heidegger himself in the slightest; for the real issue between the two lies in the circumstance that Heidegger denies while Carnap affirms the philosophical centrality of logic and the exact sciences. Carnap accordingly refers to such Heideggerian passages as the following:

> [N]othingness is the source of negation, not vice versa. If the power of the understanding in the field of questions concerning nothingness and being is thus broken, then the fate of the dominion of "logic" within philosophy is also decided therewith. The idea of "logic" itself dissolves in a vortex of more original questioning.
>
> The supposed soberness and superiority of science becomes ridiculous if it does not take nothingness seriously. Only because nothingness is manifest can science make what exists itself into an object of investigation. Only if science takes its existence from metaphysics can it always reclaim anew its essential task, which does not consist in the accumulation and ordering of objects of acquaintance but in the ever to be newly accomplished disclosure of the entire expanse of truth of nature and history.
>
> Therefore no rigor of a science can attain the seriousness of metaphysics. Philosophy can never be measured by the standard of the idea of science.[6]

Carnap concludes, in his own characteristically sober fashion: "We thus find a good confirmation for our thesis; a metaphysician here arrives himself at the statement that his questions and answers are not consistent with logic and the scientific mode of thinking" (Carnap 1932e, 232 [72]).

Heidegger's "Postscript" to Heidegger 1929b — published in the fourth edition in 1943 — considers three types of criticism that have been directed at the original lecture. Heidegger reserves his most extensive and militant response for the third criticism: namely, that "the lecture decides against 'logic.'" The heart of his response is as follows:

> The suspicion directed against "logic," whose conclusive degeneration may be seen in logistic [that is, modern mathematical logic], arises from the knowledge of that thinking that finds its source in the truth of being, but not in the consideration of the objectivity [*Gegenständlichkeit*] of what exists. Exact thinking is never the most rigorous thinking, if rigor [*Strenge*] receives its essence otherwise from the mode of strenuousness [*Anstrengung*] with which knowledge always maintains the relation to what is essential in what exists. Exact thinking ties itself down solely in calculation with what exists and serves this [end] exclusively. (Heidegger 1943, 104 [356])

It is clear, then, that Heidegger and Carnap are actually in remarkable agreement. "Metaphysical" thought of the type Heidegger is trying to awaken is possible only on the basis of a prior overthrow of the authority and primacy of logic and the exact sciences. The difference is that Heidegger eagerly embraces such an overthrow, whereas Carnap is determined to resist it at all costs.

The above sheds considerable light, I believe, on the context and force of Carnap's antimetaphysical attitude. For, by rejecting "metaphysics" as a field of cognitively meaningless pseudosentences, Carnap is by no means similarly rejecting all forms of traditional philosophy. He makes this perfectly clear, in fact, in his "Remarks by the Author" appended to the English translation of Carnap 1932e in 1957:

> *To section 1, "metaphysics."* This term is used in this paper, as usually in Europe, for the field of alleged knowledge of the essence of things which transcends the realm of empirically founded, inductive science. Metaphysics in this sense includes systems like those of Fichte, Schelling, Hegel, Bergson, Heidegger. But it does not include endeavors toward a synthesis and generalization of the results of the various sciences. (Carnap 1932e, [80])

In Carnap's reply to Paul Henle in Schilpp 1963 the point is made even more explicitly:

> Note that the characterization as pseudo-statements does not refer to all systems or theses in the field of metaphysics. At the time of the Vienna Circle, the characterization was applied mainly to those metaphysical systems which had exerted the greatest influence upon continental philosophy during the last century, viz., the post-Kantian systems of German idealism and, among contemporary ones, those of Bergson and Heidegger. On the basis of later, more cautious analyses, the judgment was not applied to the main theses of those philosophers whose thinking had been in close contact with the science of their times, as in the cases of Aristotle and Kant; the latter's epistemological theses about the synthetic a priori character of certain judgments were regarded by us as false, not as meaningless.[7]

So Carnap is primarily concerned with "overcoming" a very particular kind of "metaphysics": the main target is the post-Kantian German idealism he views as dominating recent European thought, and he views Heidegger, in particular, as the contemporary embodiment of such metaphysical dominance.

When Carnap emigrated to the United States in December 1935, he was therefore especially relieved to have finally left this European metaphysical tradition behind:

> I was not only relieved to escape the stifling political and cultural atmosphere and the danger of war in Europe, but was also very gratified to see that in the United States there was a considerable interest, especially among the younger philosophers, in the scientific method of philosophy, based on modern logic, and that this interest was growing from year to year.
>
> In 1936, when I came to this country, the traditional schools of philosophy did not have nearly the same influence as on the European continent. The movement of German idealism, in particular Hegelianism, which had earlier been quite influential in the United States, had by then almost completely disappeared. Neo-Kantian philosophical conceptions were represented here and there, not in an orthodox form but rather influenced by recent developments in scientific thinking, much like the conceptions of Cassirer in Germany. Phenomenology had a number of adherents mostly in a liberalized form, not in Husserl's orthodox form, and even less in Heidegger's version. (Carnap 1963a, 34, 40)

Carnap's sense of liberation in thus escaping the "stifling" political, cultural, and philosophical atmosphere in Central Europe is palpable.

It is important, then, to understand Carnap's antimetaphysical attitude in its philosophical, cultural, and political context. Carnap's concern for this broader context is characteristic of him and, in fact, formed one of the main bonds uniting him with his more activist friend and colleague Otto Neurath. Neurath himself, as is well known, contributed an extremely engaged, neo-Marxist perspective to the Vienna Circle. Indeed, he had served as economics minister in Ernst Toller's short-lived Bavarian Soviet Republic in 1919 and received an eighteen-month sentence when it was crushed. As is also well known, an especially striking example of Carnap's own attitude toward the relationship between the philosophical work of the Vienna Circle and this wider cultural and political context is the preface to the *Aufbau,* dated May 1928. After calling for a radically new scientific, rational, and anti-individualistic conception of philosophy that is to emulate the slow process of mutual cooperation and collaboration typical of the special sciences, Carnap continues:

> We cannot hide from ourselves the fact that trends from philosophical-metaphysical and from religious spheres, which protect themselves against this kind of orientation, again exert a strong influence precisely

at the present time. Where do we derive the confidence, in spite of this, that our call for clarity, for a science that is free from metaphysics, will prevail? — From the knowledge, or, to put it more cautiously, from the belief, that these opposing powers belong to the past. We sense an inner kinship between the attitude on which our philosophical work is based and the spiritual attitude that currently manifests itself in entirely different spheres of life. We sense this attitude in trends in art, especially in architecture, and in the movements that concern themselves with a meaningful structuring [*Gestaltung*] of human life: of personal life and the life of the community, of education, of external organization at large. We sense here everywhere the same basic attitude, the same style of thinking and working. It is the orientation that is directed everywhere towards clarity yet recognizes at the same time the never entirely comprehensible interweaving of life, towards care in the individual details and equally towards the greater shape of the whole, towards the bonds between men and equally towards the free development of the individual. The belief that this orientation belongs to the future inspires our work. (Carnap 1928a, x–xi [xvii–xviii])

And, as Carnap explains in his diary, he is here expressing precisely the attitude that he and Neurath share.[8]

Carnap suggests that his and Neurath's orientation has much in common with that of modern architecture and, in particular, with that of the Dessau Bauhaus — a point that is borne out by the recollections of Herbert Feigl:

> Carnap and Neurath also had a great deal in common in that they were somewhat utopian social reformers — Neurath quite actively, Carnap more "philosophically." . . . I owe [Neurath] a special debt of gratitude for sending me (I think as the first "emissary" of the Vienna Circle) to Bauhaus Dessau, then, in 1929, a highly progressive school of art and architecture. It was there in a week's sojourn of lectures and discussions that I became acquainted with Kandinsky and Klee. Neurath and Carnap felt that the Circle's philosophy was an expression of the *neue Sachlichkeit* which was part of the ideology of the Bauhaus. (Feigl 1969, 637)

Carnap's basic philosophical-political orientation is thus best expressed by the *neue Sachlichkeit* (the new objectivity, soberness, matter-of-factness): a social, cultural, and artistic movement committed to internationalism, to some form of socialism,[9] and, above all, to a more objective, scientific, and anti-individualistic reorganization of both art and public life inspired equally by the new Russian communism and the new American technology.[10]

Carnap and Heidegger are therefore at opposite ends of the spectrum not only philosophically but also in cultural and political terms.[11] And I think there is no doubt that this cultural and political dimension of their disagreement represents at least part of the explanation for Carnap's choosing Heidegger for his examples of metaphysical pseudosentences in Carnap 1932e. Indeed, particularly when read in the context of such programmatic statements as the preface to the *Aufbau,* this is already suggested by the sentence from §5, quoted above, where Carnap explains that he has here chosen to "select a few sentences from that metaphysical doctrine which at present exerts the strongest influence in Germany." Such a wider cultural and political context is also suggested by a passage in §6, where Carnap explains that the method of logical analysis has both a negative aspect (antimetaphysical) and a positive aspect (constructive analysis of science): "This negative application of method is necessary and important in the present historical situation. But the positive application — even already in contemporary practice — is more fruitful."[12] Carnap expresses this last idea in stronger and more militant terms, however, in the second version of his paper, the one presented at the lectures in Berlin and Brünn in July and December 1932. In this version the lecture closes with a discussion of the positive task of the method of logical analysis (that is, the clarification of the sentences of science) and, in particular, with the following words:

> These indications [are presented] only so that one will *not* think that the *struggle* [*Kampf*] *against metaphysics is our primary task.* On the contrary: in the meaningful realm [there are] many tasks and difficulties, there will always be enough struggle<?>. The struggle against metaphysics is only necessary because of the historical situation, in order to reject hindrances. There will, I hope, come the time when one no longer needs to present lectures against metaphysics. (ASP 110-07-19, p. 4)

One can imagine that this statement, coming at the very end of Carnap's lecture at Berlin in July 1932, had a much more dramatic impact than the more subtle suggestions buried in the published version.[13]

Neurath, for his part, dispenses entirely with all such subtleties. He never tires, for example, of characterizing "metaphysicians" and "school philosophers" — among whom Heidegger is a prominent representative — as enemies of the proletariat:

> Science and art are today above all in the hands of the ruling classes and will also be used as instruments in the class struggle against

> the proletariat. Only a small number of scholars and artists place themselves on the side of the coming order and set themselves up as protection against this form of reactionary thought.
>
> The idealistic school philosophers of our day from Spann to Heidegger want to rule, as the theologians once ruled; but the scholastics could support themselves on the substructure of the feudal order of production, whereas our school philosophers do not notice that their substructure is being pulled out from beneath their feet.[14]

Neurath was particularly ill-disposed toward attempts by such thinkers as Heinrich Rickert, Wilhelm Dilthey, and Heidegger to underwrite a special status for the *Geisteswissenschaften* in relation to the *Naturwissenschaften* (the humanities in relation to the natural sciences) — which attempts, according to Neurath, constitute one of the principal obstacles to rational and scientific social progress.[15]

That Carnap was in fact in basic agreement with Neurath here emerges clearly in a conversation he records after his final lecture presentation of Carnap 1932e at Brünn in December 1932:

> My lecture "Die Überwindung der Metaphysik" (Il ... had added: and the world-view of modern philosophy) in the banquet hall. Pretty well attended, lively participation, 1¼ hours. Afterwards various questions. *Erkenntnis* is here completely unknown. Then to a cafe. Prof. B. . . . , chemist, gives philosophical courses at the popular university, will report on my lecture in the socialistic ... newspaper. Marxist, is pleased with my Marxist views on how metaphysics will be overcome through reformation [*Umgestaltung*] of the substructure. (ASP 025-73-03, entry for 10 December 1932)

There can be very little doubt, therefore, that Carnap's attack on Heidegger, articulated and presented at an extraordinarily critical moment during the last years of the Weimar Republic, had more than purely philosophical motivations — or, perhaps better: that Carnap, like Neurath, conceived his philosophical work (and the attack on Heidegger in particular) as a necessary piece of a much larger social, political, and cultural struggle.[16]

It is noteworthy, finally, that Heidegger was aware of Carnap's attack and, indeed, explicitly responded to it: in a part of his 1935 lecture course "Introduction to Metaphysics" that does not appear in the published version in 1953.[17] Heidegger explains how, with the collapse of German idealism in the second half of the nineteenth century, the philosophical understanding of Being degenerated into a consideration of the "is" — that is, a logical consideration of the propositional copula. He continues in a memorable paragraph that is worth quoting in full:

> Going further in this direction, which in a certain sense has been marked out since Aristotle, and which determines "Being" from the "is" of the proposition and thus finally destroys it, is a tendency of thought that has been assembled in the journal *Erkenntnis.* Here the traditional logic is to be for the first time grounded with scientific rigor through mathematics and the mathematical calculus, in order to construct a "logically correct" language in which the propositions of metaphysics — which are all pseudo-propositions — are to become impossible in the future. Thus, an article in this journal (2:1931–32, 219ff.) bears the title "Überwindung der Metaphysik durch logische Analyse der Sprache." Here the most extreme flattening out and uprooting of the traditional theory of judgment is accomplished under the semblance of mathematical science. Here the last consequences of a mode of thinking which began with Descartes are brought to a conclusion: a mode of thinking according to which truth is no longer disclosedness of what exists and thus accommodation and grounding of Dasein in the disclosing being, but truth is rather diverted into *certainty* — to the mere securing of thought, and in fact the securing of mathematical thought against all that is not thinkable by it. The conception of truth as the securing of thought led to the definitive profaning [*Entgötterung*] of the world. The supposed "philosophical" tendency of mathematical-physical positivism wishes to supply the grounding of this position. It is no accident that this kind of "philosophy" wishes to supply the foundations of modern physics, in which all relations to nature are in fact destroyed. It is also no accident that this kind of "philosophy" stands in internal and external connection with Russian communism. And it is no accident, moreover, that this kind of thinking celebrates its triumph in America. All of this is only the ultimate consequence of an apparently merely grammatical affair, according to which Being is conceived through the "is," and the "is" is interpreted in accordance with one's conception of the proposition and of thought. (Heidegger 1983, 227–28)[18]

Thus Heidegger, in terms no more subtle than Neurath's, once again expresses a rather remarkable agreement with Carnap concerning the underlying sources of their opposition — which, as is now clear, extend far beyond the purely philosophical issues between them.

•

These philosophical issues, as we have seen, are based in the end on a stark and profound disagreement about the nature and centrality of logic. Thus Carnap criticizes "Nothingness itself nothings" primarily on the grounds of logical form: modern mathematical logic shows that the concept of nothing is to be explained in terms of existential quantification and negation and can therefore by no means function either as a substantive (individual constant) or as a verb (predicate). For Heidegger, by contrast, such a purely logical analysis misses precisely his point: what he calls nothingness is prior to logic and hence prior, in particular, to the concept of negation. In tracing out the roots of this fundamental disagreement over the philosophical centrality of logic, it turns out, we need to appreciate the extent to which the thought of both philosophers arises from the neo-Kantian tradition that dominated the German-speaking world at the end of the nineteenth and beginning of the twentieth century — a tradition within which both men received their philosophical training.

There were in fact two quite distinguishable versions of neo-Kantianism that were dominant at the time: the so-called Marburg school of neo-Kantianism founded by Hermann Cohen and then continued by Paul Natorp and (at least until about 1920) Ernst Cassirer, and the so-called Southwest school of neo-Kantianism founded by Wilhelm Windelband and systematically developed by Heinrich Rickert. The former school emphasized the importance of mathematics and natural science and, in fact, saw the true achievement of the *Critique of Pure Reason* as a laying of the groundwork for Newtonian mathematical physics. The latter school, by contrast, emphasized the distinctive importance of the *Geisteswissenschaften* and, accordingly, devoted considerable philosophical efforts to articulating a sharp methodological distinction between the latter and the *Naturwissenschaften.* Heidegger studied with Rickert at Freiburg (before Rickert succeeded Windelband at Heidelberg) — completing his habilitation under Rickert in 1915. Carnap, for his part, studied Kant at Jena with Bruno Bauch — another student of Rickert's from Freiburg — and, in fact, completed his doctoral dissertation under Bauch in 1921.[19] It is clear, moreover, that Carnap carefully studied both versions of neo-Kantianism and, in particular, the writings of Natorp, Cassirer, and Rickert.[20]

Common to both versions of neo-Kantianism is a certain conception of epistemology and the object of knowledge inherited from Kant.[21] Our knowledge or true judgments should not be construed, according to this conception, as representing or picturing objects or entities that exist independently of our judgments — whether these independent entities are the "transcendent" objects of the metaphysical realist existing somehow "behind" our sense-experience or the naked, unconceptualized sense-experience itself beloved of the empiricist. In the first case ("transcendent" objects), knowledge or true judgment would be impossible for us, since, by hypothesis, we have absolutely no independent access to such entities by which we could verify whether the desired relation of representation or picturing holds. In the second case (naive empiricism), knowledge or true judgment would be equally impossible, however, for the stream of unconceptualized sense-experience is in fact utterly chaotic and intrinsically undifferentiated: comparing the articulated structures of our judgments to this chaos of sensations simply makes no sense. How, then, is knowledge or true judgment possible? What does it mean for our judgments to relate to an object? The answer is given by Kant's "Copernican Revolution": the object of knowledge does not exist independently of our judgments at all; on the contrary, this object is first created or "constituted" when the unconceptualized data of sense are organized or framed within the a priori logical structures of judgment itself. In this way, the initially unconceptualized data of sense are brought under a priori "categories" and thus first become capable of empirical objectivity.

Yet there is a crucially important difference between this neo-Kantian account of the object of knowledge and judgment and Kant's original account. For Kant, we cannot explain, on the basis of the a priori logical structures of judgment alone, how the object of knowledge becomes possible. We need additional a priori structures that mediate between the pure forms of judgment comprising what Kant calls general logic and the unconceptualized manifold of impressions supplied by the senses: these mediating structures are the pure forms of sensible intuition — space and time. Thus the pure logical forms of judgment only become categories when they are "schematized," that is, when they are given a determinate spatio-temporal content in relation to the pure forms of sensible intuition. The pure logical form of a categorical judgment, for example, becomes the category of *substance* when it is schematized in terms of the temporal representation of permanence; the pure logical form of a hypothetical judgment becomes the category of *causality* when it is schematized in terms of the temporal representation of succession; and so on. For Kant, then, pure formal logic (general logic) must, if it is to play an epistemological role, be supplemented by what he calls transcendental logic: with the theory of how logical forms become schematized in terms of pure spatio-temporal representations belonging to the independent faculty of pure intuition. And it is precisely this theory, in fact, that forms the heart of the transcendental analytic of the

Critique of Pure Reason: the so-called metaphysical and transcendental deductions of the categories.

Now both versions of neo-Kantianism entirely reject the idea of an independent faculty of pure intuition. The neo-Kantians here follow the tradition of post-Kantian idealism in vigorously opposing the dualistic conception of mind characteristic of Kant's own position: the dualism, that is, between a logical, conceptual, or discursive faculty of pure understanding and an intuitive, nonconceptual, or receptive faculty of pure sensibility. For the neo-Kantians, the a priori formal structures in virtue of which the object of knowledge becomes possible must therefore derive from the logical faculty of the understanding and from this faculty alone. And, in this way, epistemology or "transcendental logic" becomes the study of purely logical, purely conceptual, *and thus essentially non–spatio-temporal* a priori structures. Space and time, conceived as Kantian pure forms of sensible intuition, can no longer play a role in our explanation of how the object of knowledge and judgment becomes possible.

It is this last feature of their conception of epistemology, moreover, that associates the neo-Kantians (again in both versions) with Husserlian phenomenology and, in particular, with the polemic against psychologism of Husserl's *Logical Investigations.* For the neo-Kantians had also arrived — albeit by a different route — at a conception of pure thought or pure logic whose subject matter is an essentially nontemporal, and therefore certainly not psychological, realm: an "ideal" realm of timeless, formal-logical structures. Indeed, Husserl's conception of "pure logic" — which was generally recognized to have its sources in the earlier work of Bernhard Bolzano, Johann Herbart, Rudolf Lotze, and Alexius Meinong — can be fairly characterized as the dominant idea of the period. Accordingly, it plays a central role not only in the thought of the neo-Kantians of the Marburg and Southwest traditions but also in the early thought of both Carnap and Heidegger.[22]

We shall have to return to the relationship between Husserl and Heidegger shortly. But it is first necessary to appreciate the fundamental problems facing the epistemology of the neo-Kantians (in both versions) — problems that flow directly from their enthusiastic embrace of the idea of "pure logic." For, as we have just seen, the logical forms of judgment, in Kant's original conception, become categories — and thus make the object of knowledge possible — precisely through their prior application to the pure forms of sensible intuition. It is on this basis, and on this basis alone, that we can explain how the purely analytic forms of thought apply to the spatio-temporal world of sense so as to make synthetic knowledge of empirical objects (that is, of appearances) possible. Yet the neo-Kantians entirely reject the idea of such an intermediate faculty of pure intuition and hence the central Kantian conception of the schematism of the pure concepts of the understanding as well. How, then, are the pure forms of thought — now conceived as belonging wholly to an "ideal," essentially non–spatio-temporal realm — supposed to apply to the spatio-temporal world of sense? How do the categories make the object of (empirical) knowledge possible?

It is in attempting to answer these questions that the two schools of neo-Kantian epistemology strikingly diverge from one another. The Marburg school, as indicated above, continues to follow Kant in taking mathematical physics as the paradigm of objective knowledge. That school's most basic move, accordingly, is to "mathematize" the pure forms of thought. Beginning with Cohen's attempt to assimilate the fundamental moment of judgment to the mathematical concept of the differential, this line of thought reaches its culmination in Cassirer's *Substance and Function* — where logic is identified with the pure theory of relations developed especially (building on the work of David Hilbert, Georg Cantor, and Richard Dedekind) in Bertrand Russell's *The Principles of Mathematics.* The timeless realm of "pure thought" is thus identified with the totality of what we now call relational structures — the "objects" thereof being simply abstract "places" within such a relational structure. In empirical knowledge, on the other hand, we develop an essentially nonterminating — but in some sense converging — *sequence* of relational structures, each element of which represents the state of mathematical physics at some particular point in the methodological history of science. (That this sequence does not terminate thus constitutes the essential point of difference between pure and empirical knowledge.) The object of empirical knowledge — the sensible world — is then conceived simply as the ideal limit or infinitely distant *X* toward which the methodological sequence of science is converging.[23] The Marburg school thereby solves the problem of the categories by a kind of "logicization" of the object of empirical knowledge, and it is with good reason, then, that the school's epistemological conception becomes known as "logical idealism."

Within the Southwest school, on the other hand, logic and the realm of "pure thought" are sharply and explicitly separated from mathematics. And this fundamental divergence between the two schools emerges with particular clarity in a dispute between Natorp and Rickert in the years 1910–11. Natorp argues that the concept of number belongs to "pure thought" and thus neither to pure intuition nor to psychology (Natorp 1919). Rickert directly challenges Natorp's conception — arguing that the concept of one as a number (as the first element of the number series) cannot be derived from the logical concepts of identity and difference and is therefore "alogical" (Rickert 1911). For Rickert, the numerical concept of one (unlike the logical concepts of identity and difference) does not apply to all objects of thought as such but presupposes that we are antecedently given objects arranged in a homogeneous serial order. The numerical concept of quantity can therefore not belong to logic — where logic is here being clearly identified with traditional syllogistic logic.[24] In this approach, since we neither follow Kant in invoking a mediating faculty of pure intuition nor follow the Marburg school in "logicizing" the object of empirical knowledge after the example of mathematical physics, we are therefore left only with the forms of judgment of traditional logic on the one side and the spatiotemporal manifold of given empirical objects on the other. So, as one would expect, particularly vexing problems in attempting to explain the application of the former to the latter arise within the Southwest school.

The underlying tensions expressed in the epistemology of the Southwest school become painfully evident in the work of Emil Lask, a brilliant student of Rickert's who then held an associate professorship at Heidelberg and was killed in the Great War in 1915. The basic argument of Lask 1912 is that whereas the Kantian philosophy has indeed closed the gap between knowledge and its object, we are nonetheless left with a new gap between what Lask calls "transcendental," "epistemological," or "material" logic, on the one side, and "formal" logic, on the other. Formal logic is the subject matter of the theory of judgment — the realm of necessarily valid and timeless "senses," "objective thoughts," or "propositions in themselves" familiar within the tradition of "pure logic."[25] Transcendental or material logic, on the other hand, is the theory of the categories in Kant's sense: the theory of how the concrete object of knowledge and experience is made possible by the activity of thought. But, and here is the central idea of Lask's argument, transcendental or material logic is *not* based on formal logic, and, accordingly, we explicitly reject Kant's metaphysical deduction — the entire point of which, as indicated above, is precisely to derive the categories from the logical forms of judgment.[26] For Lask, what is fundamental is the concrete, already categorized real object of experience: the subject matter of formal logic (comprising the structures of the traditional logical theory of judgment) only arises subsequently in an artificial process of abstraction, by which the originally unitary categorized object is broken down into form and matter, subject and predicate, and so on. Moreover, since this comes about due to a fundamental weakness or peculiarity of our human understanding — our inability to grasp the unitary categorized object *as* a unity — all the structures of "pure logic," despite their undoubtedly timeless and necessary status, are in the end artifacts of subjectivity. Since the pure forms of judgment of traditional logic are now seen as entirely bereft of the capacity even to begin the "constitution" of any real empirical object, the entire realm of "pure logic" appears as nothing but an artificially constructed intermediary possessing no explanatory power whatsoever.[27]

•

Heidegger's earliest philosophical works, as noted above, fall squarely within the antipsychologistic tradition of "pure logic" and receive their primary orientation (not surprisingly) from his teacher Rickert. Accordingly, these early investigations revolve around the central distinctions between psychological act and logical content, between real thought process and ideal atemporal "sense," between being (*Sein*) and validity (*Geltung*). For, as Lotze in particular has shown, the realm of the logical has a completely different mode of existence (validity or *Geltung*) than that of the realm of actual spatio-temporal entities (being or *Sein*).[28] Moreover, as Rickert has shown, the realm of the logical (the realm of validity) is also distinct from that of the mathematical; for, although the latter is equally atemporal and hence equally ideal, it presupposes a particular object — the existence of "quantity" — and therefore lacks the complete generality characteristic of the logical. It follows that we must sharply distinguish the realm of the logical both from the given heterogeneous qualitative continuum of empirical reality and from the homogeneous quantitative continuum of mathematics.[29] In emphasizing these fundamental distinctions and, above all, in maintaining "*the absolute primacy of valid sense* [*den absoluten Primat des geltenden Sinnes*],"[30] Heidegger shows himself to be a faithful follower of Rickert indeed.

Yet, as we have just seen, Rickert's fundamental distinctions lead naturally to fundamental problems — problems that stand out especially vividly in the work of Lask. In particular, once we have delimited the realm of the logical so sharply from all "neighboring" realms, it then becomes radically unclear how the realm of the logical is at all connected with the real world of temporal being [*Sein*]: with either the realm of empirical nature where the objects of our (empirical) cognition reside or the realm of psychological happenings where our acts of judgment reside. The realm of "valid sense," which was intended as an intermediary between these last two realms wherein our cognition of objects is "constituted" and thus made possible, thereby becomes deprived of all explanatory power. Now Heidegger, for his part, was of course most sensitive indeed to the difficult position in which Rickert had become entangled; and, as a consequence, he became increasingly attracted to the more radical position of Lask,[31] which Heidegger also saw as having essential ideas in common with the emerging new phenomenology being developed by Edmund Husserl. After Rickert left Freiburg to take Windelband's chair at Heidelberg and Husserl left Göttingen to take Rickert's chair at Freiburg in 1916, Heidegger became an enthusiastic proponent of the new phenomenology and, in particular, distanced himself further and further from Rickert. This distance from Rickert became quite extreme by 1925–26, when Heidegger was completing the work on *Being and Time,* and it is graphically evident in lectures Heidegger presented at Marburg in the summer semester of 1925 and the winter semester of 1926. Here Heidegger speaks of Rickert with almost undisguised contempt, whereas Husserl appears as the leader of a new "breakthrough" in philosophy — a "breakthrough" that has decisively overcome neo-Kantianism.[32]

The first element of this "breakthrough" is a "direct realist" conception of truth as "identification" — a conception that can be seen as definitively rejecting the idea that formal logic is foundational for truth in general and thus as also overturning the "Copernican Revolution." For, according to the theory of truth articulated in volume 2 of the *Logical Investigations,* truth in general is not even propositional: it consists simply in the circumstance that an intention or meaning (whether propositional or not) is directly "identified" — in immediate intuition — with the very thing that is intended or meant. Thus, truth in general need involve none of the structures (subject and predicate, ground and consequent, and so on) studied in traditional formal logic. On the contrary, such peculiarly logical structures only emerge subsequently in the very special circumstances of "categorial intuition," where specifically propositional intentions or meanings are intuitively grasped in their most abstract — and, as it were, secondary and derivative — formal features. In this sense, then, Husserl's "direct realist" conception of the relationship between logical form and truth in general parallels Lask's view of the artificiality and subjectivity of logical form: in neither case can formal logic be in any way foundational or explanatory for truth as "relation to an object."[33]

The second element of the Husserlian "breakthrough" is just the idea of phenomenology as such — an idea that also emerges in volume 2 of the *Logical Investigations* as that of an "epistemology of the logical." For it is this idea, and this idea alone, that first opens up the possibility of bridging the gulf between the logical and the psychological created by the polemic against psychologism of volume 1. The problem, however, is to explain how such an "epistemology of the logical" —

which is to investigate the *relationship* between psychological act and logical or "essential" content — can itself avoid collapsing into psychologism. The answer is disarmingly simple: phenomenology is not empirical psychology because it aims to elucidate the underlying a priori structures or "essences" of psychological phenomena (the "essences" of perception, recollection, imaginative representation, and so on). Our investigation proceeds by means of "essential analysis [*Wesensanalyse*]" and "essential intuition [*Wesenserschauung*]" — and therefore is, in particular, entirely independent of the actual instances of the psychological structures in question that may or may not exist in the real world.[34] We are interested, that is, only in the purely ideal or "essential" structures of psychological phenomena: in "pure consciousness."[35]

Yet, for this very reason, Husserl's conception of pure phenomenology and "pure consciousness" could not be fully satisfactory from Heidegger's point of view. For Heidegger's problem — arising so painfully and vividly within the neo-Kantian epistemology of the Southwest school — was precisely that of the application of abstract and ideal "valid senses" to concrete and real objects of cognition: the problem of the application of the categories in Kant's sense to actual spatio-temporal objects. And *this* problem, it is clear, cannot be solved within the framework of Husserlian "pure consciousness"; for the latter, as we have just seen, itself belongs to the purely ideal realm of "essences" and is thus entirely independent of the existence of any and all concrete instances (whether of actual states of consciousness or of its empirical objects). It is by no means surprising, therefore, that we already find rumblings in a new and quite un-Husserlian direction in the concluding chapter added to the published version of Heidegger's habilitation in 1916. Heidegger there suggests that a genuine unification of time and eternity — of change and absolute validity — can be effected only through the concept of "living spirit [*der lebendige Geist*]" construed as a concrete and essentially historical subject. The "subjective logic" sought for by Rickert and Husserl requires a more fundamental point of view according to which the subject is no mere "punctiform" cognitive subject but an actual concrete subject comprehending the entire fullness of its temporal-historical involvements. And such an investigation of the concrete historical subject must, according to Heidegger, be a "translogical" or "metaphysical" investigation.[36] Thus, Heidegger is here already beginning to come to terms with the historically oriented *Lebensphilosophie* of Wilhelm Dilthey — an influence that will prove decisive in *Being and Time*.[37]

It is of course in *Being and Time*, completed ten years later, that Heidegger finally works out the desired "subjective logic" with a *con-*

crete subject — the so-called existential analytic of Dasein. Heidegger's concrete subject — Dasein, the concrete living human being — is distinguished from the "pure consciousness" of Husserlian phenomenology in three fundamental respects. First, Dasein necessarily exists in a world: a world of concrete spatio-temporal objects existing independently of it, which it does not create and over which it has only very limited control, and into which — without its consent, as it were — it is "thrown." Indeed, for Heidegger, Dasein essentially *is* such "being-in-the-world." Second, Dasein's relationship to this world is first and foremost practical and pragmatic rather than epistemic and contemplatively intuitive. The items in Dasein's world therefore appear to it originally as practically "ready-to-hand [*Zuhanden*]" — as environmental items to be used in the service of particular concrete projects — rather than as merely "present-to-hand [*Vorhanden*]" for theoretical inspection and consideration. Indeed, for Heidegger, theoretical cognition of the merely "present-to-hand" is a *derivative* mode of Dasein: a particular "modification" of the more basic, essentially practical and pragmatic, mode of involvement with the "ready-to-hand."[38] Finally, Dasein is essentially a historical being — in an important sense it is *the* historical being. For the essence or "being" of Dasein is "care [*Sorge*]" — roughly, the above-described orientation toward its world from the point of view of the totality of its practical involvements and projects — and the "ontological meaning of care" is *temporality,* where temporality in this sense is essentially historical and thus to be sharply and explicitly distinguished from the uniform and featureless "time" of natural science.[39]

There is no doubt, then, that Heidegger's Dasein is much more concrete than Husserl's "pure consciousness" — in the sense that the former has more of the features of a real human being than does the latter. From the point of view of Husserlian phenomenology, however, an obvious dilemma for Heidegger arises at this point. For what can Heidegger's "existential analytic of Dasein" possibly mean from Husserl's perspective? Either Heidegger is trying to describe the concrete reality of empirical human beings in their concrete and empirical character, in which case his enterprise is simply a branch of empirical anthropology having no specifically philosophical interest whatsoever; or Heidegger is trying to elucidate the "essence" or nature of the concrete human being by means of an "essential analysis" of that nature, in which case Heidegger, too, must perform the "eidetic" reduction and abstract from all questions involving the real existence of the entities under consideration. Thus, either Heidegger falls prey to the charge of naturalism and psychologism or his "existential analytic of Dasein" is in the end no closer to actual concrete reality than is Husserl's phenomenology.

It is in response to this dilemma, I believe, that Heidegger's true philosophical radicalism emerges; for it is precisely here — in attempting to construct an a priori analysis of the real concrete subject — that Heidegger fundamentally changes the terms within which the entire tradition of "pure logic" was articulated and, in fact, introduces a fundamentally new existential dimension into this tradition. For Heidegger, the "existential analytic of Dasein" can by no means be assimilated to a description of the "essence" of Dasein in the traditional meaning of this term (where "essence" or "whatness" is contrasted with "existence" or "thatness") precisely because the distinguishing feature of Dasein — as opposed to all other entities encountered within the world — is that Dasein has no "essence" of this kind at all: *What* Dasein is can be determined only by Dasein's own free choice in the face of its fundamental possibility of "being-toward-death" — a choice that can be either "authentic" and "resolute" or "inauthentic" and fallen into the "They" of everyday public existence.[40] In either case, however, since Dasein's "essence" or "whatness" depends in the end on its own free choice — and is thus in no sense simply "given" as in the case of entities encountered within the world — Dasein's "essence" cannot be meaningfully separated from its "existence" at all. Indeed, from this point of view, Dasein's "essence" *is* "existence." The dilemma just raised from the point of view of Husserlian phenomenology therefore has no force whatever from Heidegger's own point of view: by replacing phenomenological "essential analysis" with what he calls "existential-ontological analysis," Heidegger has transcended the traditional distinction between "essence" and "existence" and opened up the paradoxical-sounding possibility of an a priori analysis of concrete existence itself.[41]

At the same time, Heidegger has thereby definitively transcended the problematic of the neo-Kantian tradition as well. This comes out most clearly in §44 of *Being and Time,* entitled "Dasein, Disclosedness, and Truth," where Heidegger explicitly rejects the "Copernican Revolution" and the associated idea that truth is to be understood in terms of "valid judgment" in favor of an apparently "direct realist" account in which truth is conceived as a kind of immediate "disclosedness [*Erschlossenheit*]" or "uncoveredness [*Entdeckt-sein*]" of a being within the world — an account that explicitly invokes Husserl's notion of "identification."[42] Heidegger's "direct realism" is very special, however, for Dasein's most fundamental relation to the world is not a cognitive relation at all. Indeed, Dasein's most fundamental relation to the world is one of either "authentic" or "inauthentic" existence — in which Dasein's own peculiar mode of being (that is, "being-in-the-world") is itself either disclosed or covered over.[43] Moreover, in the moment of an "authentic" decision, Dasein must choose among possibilities already given in the world—possibilities that must themselves be somehow already "present" and available in Dasein's historically given situation. Dasein must thus appropriate its "fate" and, in precisely this way, is essentially historical.[44]

What is therefore central, for Heidegger, is the circumstance that we do not start with a merely cognitive subject together with its "contents of consciousness," but rather with a living practical subject — a subject that is essentially temporally finite and hence necessarily engaged with its historically given environmental situation. Assertion or judgment then appears as a "derivative mode of interpretation" in which the "hermeneutical 'as'" of practical understanding of the "ready-to-hand" (where an item "ready-to-hand" is understood "as" suited for a given end or purpose) is transformed into the "apophantical 'as'" of theoretical understanding of the "present-to-hand" (where an item "present-to-hand" is understood "as" determined by a given predicate). If we forget this derivative character, however, all the misunderstandings prevalent in "the presently dominant theory of 'judgment' oriented around the phenomenon of 'validity [*Geltung*]'" then arise. We end up, in particular, with Lotze's distinction between "being" and "validity." Since we begin, within this latter tradition, with a "Cartesian" subject entirely enclosed within the psychical realm of its own representations, we cannot explain truth as a relation between these representations and an object existing independently of them. Truth can therefore only mean what is constant and unchangeable in the flux of representations and is thus understood as "form" or "essence" in a quasi-Platonic sense standing over and against the realm of flux and change. We thereby arrive at the distinction between "being" and "validity," "real" and "ideal." And this distinction, by the "Copernican" conception of the object of knowledge, is now equated with the distinction between subjective and objective. Finally, since "objectivity" is thus equated with "validity" in the sense of atemporal or eternal "ideal being," "objectivity" is also equated with necessary *intersubjectivity:* with "bindingness" for all subjects.[45]

For Heidegger himself, by contrast, truth is in no way to be equated with "objectivity" in the sense of necessary and universal intersubjectivity. On the contrary, this most basic idea of the Kantian "Copernican Revolution" is definitively rejected in favor of a "direct realist" conception of truth as direct "disclosedness" to Dasein in a particular and irreducibly historical environmental situation: all truth is ultimately both particular and historical.[46] Indeed, to think otherwise is to refuse to acknowledge the essential particularity of an "authentic" decision in the face of "being-toward-death" and to seek refuge instead in the public everydayness of the "They."[47] In the end, therefore, it is Heidegger's radical transformation of the neo-Kantian tradition within which he was trained that underwrites his equally radical rejection of the priority and centrality of logic: his claim that the traditional theory of judgment based on the "is" of predication is itself necessarily derivative from a properly philosophical point of view.

•

We observed above that Carnap also received his philosophical training within the neo-Kantian tradition and, in fact, that he completed his doctoral dissertation in 1921 under Rickert's student Bruno Bauch — a dissertation in which Kantian themes predominate.[48] It was in the years immediately after finishing his dissertation, in 1922–25, that Carnap undertook most of the work on the project that was eventually to issue in *Der logische Aufbau der Welt* (Carnap 1963a, 16–19). And it was on this basis, moreover, that Carnap caught the attention of the "Philosophical Circle" that had gathered around Moritz Schlick at the University of Vienna. Carnap had become acquainted with Schlick in the summer of 1924 and was invited to give lectures to Schlick's circle in the winter of 1925. These lectures on the *Aufbau* project — which was at the time entitled "Entwurf einer Konstitutionstheorie der Erkenntnisgegenstände" (Outline of a constitutional theory of the objects of cognition) — were extremely well received: Carnap returned to Vienna as assistant professor, with his "Entwurf" (then being eagerly read within Schlick's circle) serving as his habilitation.[49] A revised version was finally published in 1928 under the now familiar title.

The aim of the *Aufbau,* as recent scholarship has made increasingly clear, is by no means exclusively to represent the point of view of phenomenalistic or extreme empiricist "positivism." Indeed, Carnap himself explains the relationship between "positivism" and neo-Kantianism as follows:

> Cassirer ([*Substanzbegr.*] 292ff.) has shown that a science having the goal of determining the individual through lawful interconnections [*Gesetzseszusammenhänge*] without its individuality being lost must apply, not class ("species") concepts, but rather *relational concepts;* for the latter can lead to the formation of series and thereby to the establishing of order-systems. It hereby also results that relations are necessary as first posits, since one can in fact easily make the transition from relations to classes, whereas the contrary procedure is only possible in a very limited measure.

> The merit of having discovered the necessary basis of the constitutional system thereby belongs to two entirely different, and often mutually hostile, philosophical tendencies. *Positivism* has stressed that the sole *material* for cognition lies in the undigested [*unverarbeitet*] experiential *given;* here is to be sought the *basic elements* of the constitutional system. *Transcendental idealism,* however, especially the neo-Kantian tendency (Rickert, Cassirer, Bauch), has rightly emphasized that these elements do not suffice; *order-posits* [*Ordnungssetzungen*] must be added, our "basic relations."[50]

Carnap does not intend simply to supplant neo-Kantianism by "positivism" in the *Aufbau:* he hopes, on the contrary, to retain the insights of *both* views.[51]

As Carnap suggests, the influence of Cassirer's *Substance and Function* is especially important to the *Aufbau.* This is not surprising, for the agreement between the two conceptions actually extends far beyond the emphasis on the significance of the modern logical theory of relations stressed here. According to *Substance and Function,* as indicated above, the theory of knowledge consists of two parts. On the one hand, we have the theory of the *concept* (part 1), which, for Cassirer, is given by the totality of pure relational structures provided by the new logic. On the other hand, however, we have the theory of *reality* (part 2), in which pure relational structures are successively applied in the methodological progress of mathematical natural science in such a way that a never completed — but convergent — sequence results. Thus, whereas pure mathematics is given by the collection of all pure or abstract relational structures, applied mathematics (mathematical physics) is given by an infinite methodological series of such structures. And it is this methodological series of abstract structures that, for Cassirer (and for the Marburg school more generally), represents the empirical side of knowledge given by "sensation." The concrete empirical world of sense-perception is not a separate reality existing somehow outside of this methodological series: it is simply the fully determinate and complete "limit theory" toward which this series is converging.

Now Carnap, in the *Aufbau,* also represents empirical knowledge by a serial or stepwise methodological sequence. This sequence is intended to represent, not so much the historical series of mathematical-physical successor theories, but rather the epistemological progress of a single individual or cognitive subject — in which its knowledge extends from the initial subjective sensory data belonging to the *autopsychological* realm, through the world of public external objects constituting the *physical* realm, and finally to the intersubjective and cultural realities belonging to the *heteropsychological* realm. Carnap's methodological series is thus a "rational reconstruction" intended formally to represent the "actual process of cognition."[52] For Carnap, as for Cassirer, we thereby represent the empirical side of knowledge by a methodological sequence of formal structures. For Carnap, however, this is not a sequence of historically given mathematical-physical successor theories but *a sequence of levels or ranks in the hierarchy of logical types* of Russell's and Whitehead's *Principia Mathematica* — a sequence of levels *ordered by type-theoretic definitions.* Objects on any level (other than the first) are thus defined as classes of objects (or relations between objects) from the preceding level.[53]

The construction or "constitution of reality" begins with "elementary experiences" — holistic momentary cross-sections of the stream of experience — ordered by a (holistically conceived) "basic relation" of remembrance-of-part-similarity-in-some-arbitrary-respect. The main formal problem within the autopsychological realm is then to differentiate — on this initially entirely holistic basis — the particular sense qualities and sense modalities from one another. After grouping elementary experiences into classes (and classes of classes . . .) thereof via the one given basic relation and a complex procedure of "quasi-analysis," Carnap is in a position to define the *visual field* as the unique sense modality possessing exactly five dimensions (two of spatial location and three of color quality).[54] On this basis we can then define the "visual things" in the physical realm: after embedding the visual fields of our subject in a numerical space-time manifold ($\mathbf{R}^4$), we project colored points of these visual fields along "lines of sight" onto colored surfaces in such a way that principles of constancy and continuity are satisfied. And, in an analogous fashion, we can then define the "physical things" or objects of mathematical physics: we coordinate purely numerical "physical state magnitudes" with sensible qualities in accordance with the laws and methodological principles of the relevant science (e.g., the electro-dynamic theory of light and color).[55] Finally, we can constitute the heteropsychological realm by, first, constructing other subjects of experience analogous to the initial subject (that is, systems of elementary experiences coordinated to "other" human bodies) and, second, constructing an "intersubjective world" common to all such subjects through an abstraction (via an equivalence relation) from the resulting diversity in "points of view."[56]

In this way Carnap's "constitution of reality" achieves a "logicization" of experience or the sensible aspects of reality parallel to that of the Marburg school. For the entire point of Carnap's method of "purely structural definite descriptions" (like that of the visual field sketched above) is to individuate the objects in question in purely formal-logical terms, making no reference whatever to their intrinsic or ostensive phenomenal qualities. The constitutional system thereby demonstrates that objective — that is, intersubjectively communicable — knowledge is possible *despite* its necessary origin in purely subjective experience.[57] Carnap characterizes the resulting kinship between his constitutional system and the "logical idealism" of the Marburg school as follows:

> Constitutional theory and *transcendental idealism* agree in representing the following position: all objects of cognition are constituted (in idealistic language, are "generated in thought"); and, moreover, the constituted objects are only objects of cognition *qua* logical forms constructed in a determinate way. This holds ultimately also for the basic elements of the constitutional system. They are, to be sure, taken as basis as unanalyzed unities, but they are then furnished with various properties and analyzed into (quasi-)constituents (§116); first hereby, and thus also first as constituted objects, do they become objects of cognition properly speaking — and, indeed, objects of psychology.[58]

Indeed, there is one important respect in which Carnap's conception is even more radical than that of the Marburg school. Cassirer's *Substance and Function,* for example, retains an element of dualism between pure thought and empirical reality: the contrast between the pure relational structures of logic and mathematics, on the one hand, and the historical sequence of successor theories representing the methodological progress of empirical natural science, on the other. For Carnap, by contrast, empirical reality simply *is* a particular logical structure: a type-theoretic structure (representing the epistemological progress of an initial cognitive subject) erected on the basis of a single, primitive, nonlogical relation.[59]

The sense in which Carnap has here gone even further than the "logical idealism" of the Marburg school stands out especially clearly in §179 — entitled "The Task of Science." According to the Marburg school, as we have seen, the real individual object of empirical cognition is as a matter of fact never actually present in the methodological progress of science at all: this real empirical object remains always a never completed *X* toward which the methodological progress of science is converging. But Carnap, in §179, explicitly rejects this "genetic" view of knowledge:

> According to the conception of the Marburg school (cf. Natorp [*Grundlagen*] 18ff.), the object is the eternal X; its determination is an incompleteable task. In opposition to this it is to be noted that

> finitely many determinations suffice for the constitution of the object — and thus for its unambiguous description among the objects in general. Once such a description is set up the object is no longer an X, but rather something unambiguously determined — whose complete description then certainly still remains an incompleteable task.[60]

Carnap thus rejects the idea that the object of empirical knowledge, in contradistinction to the purely formal objects of mathematical knowledge, is to be conceived as a never-ending, necessarily incompleteable progression.[61] For Carnap, all objects whatsoever — whether formal or empirical, ideal or real — are rather to be defined or "constituted" at *definite finite ranks* within the hierarchy of logical types: there are no objects in the constitutional system that remain necessarily incomplete.[62]

In this sense Carnap completes the "logicization" of experience that the Marburg school had begun and, at the same time, arrives at an even more radical transformation of the Marburg tradition. For in the constitutional system of the *Aufbau,* epistemology is transformed into a logical-mathematical constructive project: the purely formal project of actually writing down the required structural definite descriptions within the logic of *Principia Mathematica.* This purely formal exercise is to serve, in particular, as a *replacement* for traditional epistemology in which we represent the "neutral basis" common to all traditional epistemological schools. All such schools are in agreement, according to Carnap, that

> cognition traces back finally to my experiences, which are set in relation, connected, and worked up; thus cognition can attain in a logical progress to the various structures of my consciousness, then to the physical objects, further with their help to the structures of consciousness of other subjects and thus to the heteropsychological, and through the mediation of the heteropsychological to the cultural objects. (Carnap 1928a, §178)

Since the constitutional system precisely represents this common ground of agreement within the neutral and uncontroversial domain of formal logic itself, all "metaphysical" disputes among the competing schools — disputes, for example, among "positivism," "realism," and "idealism" concerning which constituted structures are ultimately "real" — are thereby dissolved.[63] The fruitless disputes of the epistemological tradition are replaced by the seriousness and sobriety of the new mathematical logic, and philosophy (once again) becomes a science: for Carnap, a purely technical subject.[64]

By precisely representing some of the central ideas comprising the "logical idealism" of the Marburg school within the new mathematical logic of *Principia Mathematica,* Carnap has thereby injected the *neue Sachlichkeit* into philosophy itself. Philosophy becomes, in particular, an "objective" discipline capable (like the exact sciences) of cooperative progress and, in principle, of universal agreement as well; indeed, it has now become a branch of mathematical logic — the most "objective" and universal discipline of all. We have thus arrived at a conception of philosophy that, in Carnap's eyes, best serves the socialist, internationalist, and anti-individualistic aims of that cultural and political movement with which he most closely identifies.[65] And this "objectivist" and universalist conception of philosophy (based on the new mathematical logic) of course stands in the most extreme contrast with the particularist, existential-historical conception of philosophy we have seen Heidegger develop (based on an explicit rejection of the centrality of logic) — a conception that, in Heidegger's eyes, best serves the neoconservative and avowedly German nationalist cultural and political stance favored by the latter philosopher.[66] But what is of most interest, from our present point of view, is the extent to which both philosophers develop their radically new conceptions of philosophy by rigorously thinking through the ideas of late nineteenth- and early twentieth-century neo-Kantianism. By pushing these neo-Kantian ideas to their limits in two opposite directions, as it were, Carnap and Heidegger thereby contribute decisively toward defining and shaping the contemporary opposition between "analytic" and "continental" philosophical traditions with which we began.

Notes

I am indebted for helpful discussions and advice to Sandra Bartky, Susan Cunningham, Graciela De Pierris, Lynn Joy, Theodore Kisiel, Alison Laywine, Alan Richardson, Werner Sauer, Thomas Uebel, and Kathleen Wright. All translations from the German are my own.

1. For eyewitness accounts see the report of L. Englert in Schneeberger 1962, 1–6; Pos 1949; and T. Cassirer 1981 — relevant parts of which are reprinted in Schneeberger 1962, 7–9.

2. Heidegger 1929a. This work appears in Heidegger 1991 together with appendices containing Heidegger's notes for his Davos lectures and a protocol of the Cassirer-Heidegger debate prepared by O. Bollnow and J. Ritter. These materials are also found in the English translation of Heidegger 1929a.

3. I am indebted to Thomas Uebel for first calling my attention to the fact that Carnap attended the Cassirer-Heidegger lectures and debate at Davos. Carnap reports on the occasion in ASP 025-73-03, entries from 18 March through 5 April 1929.

4. Heinrich Gomperz, the son of the famous historian of Greek philosophy Theodore Gomperz, was a professor of philosophy at the University of Vienna and the author of *Weltanschauungslehre* (1905). Karl Bühler was an important psychologist and psycholinguist; he founded the Psychological Institute at the University of Vienna in 1922. For Carnap's participation see ASP 025-73-03, entries for 24 May and 14 June 1930.

5. The two versions of these lectures, including reports on the discussions, are ASP 110-07-21 and ASP 110-07-19, respectively. I am indebted to Brigitte Uhlemann of the University of Konstanz for providing me with transcriptions from Carnap's shorthand.

6. Heidegger 1929b, 14, 18 [107, 111–12]. Carnap quotes selections from these passages in Carnap 1932, 231–32 [71–72]. (Pagination of the English translations appears in brackets.)

7. Carnap 1963c, 874–75. On the following page Carnap continues: "I think, however, that our [antimetaphysical] principle excludes not only a great number of assertions in systems like those of Hegel and Heidegger, especially since the latter says explicitly that logic is not applicable to statements in metaphysics, but also in contemporary discussions, for example, those concerning the reality of space or of time." Compare the remarks in Carnap 1963a, 42–43: "It is encouraging to remember that philosophical thinking has made great progress in the course of two thousand years through the work of men like Aristotle, Leibniz, Hume, Kant, Dewey, Russell, and many others, who were basically thinking in a scientific way."

8. See ASP 025-73-03, entry for 26 May 1928: "In the evening with Waismann at Neurath's. I read the Preface to the 'Logischen Aufbau' aloud; Neurath is astonished and overjoyed at my open confession. He believes that it will affect young people very sympathetically. I say that I still want to ask Schlick whether it is too radical." (Schlick did indeed think it was too radical: see entry for 31 May.)

9. Carnap became attracted to the antimilitarist internationalism of the "socialist worker's movement" already during the Great War (Carnap 1963a, 9–10).

10. For a general cultural and political history of this orientation see Willett 1978. For a specific discussion of the relationship between the Vienna Circle and the Dessau Bauhaus see Galison 1990. Not all members of the Vienna Circle shared in this orientation, however. Schlick, in particular, was attracted neither to Marxism nor to anti-individualism more generally. Thus, for example, Feigl poignantly describes Schlick's reaction when he was presented with the manifesto (Hahn, Neurath, and Carnap 1929) — which calls for a new internationalist and collaborative form of philosophy and, in keeping with this spirit, is not even signed by its authors — on his return from Stanford in 1929: "Schlick was moved by our amicable intentions; but as I could tell from his facial expression, and from what he told me later, he was actually appalled and dismayed by the thought that we were propagating our views as a 'system' or 'movement.' He was deeply committed to an individualistic conception of philosophizing, and while he considered group discussion and mutual criticism to be greatly helpful and intellectually profitable, he believed that everyone should think creatively for himself. A 'movement,' like large scale meetings or conferences, was something he loathed" (Feigl 1969, 646).

11. The literature on Heidegger's own political involvement is now enormous. See, in particular, Ott 1988; Farias 1987; and Schneeberger 1962. Wolin 1991 is a very useful selection — including a translation of Heidegger's notorious *Die Selbstbehauptung der deutschen Universität,* delivered in celebration of the new Nazi regime when he assumed the rectorate at Freiburg in May 1933. A particularly interesting contribution, locating Heidegger's involvement in the context of that of the other German philosophers of the time, is Sluga 1993.

12. Carnap 1932e, 238 [77]. Compare also the preface to *Logical Syntax of Language,* dated May 1934: "In our 'Vienna Circle' and in many similarly oriented groups (in Poland, France, England, USA and *in isolated cases even in Germany*) the view has currently grown stronger and stronger that traditional metaphysical philosophy can make no claim to scientific status" (Carnap 1934b, iii [xiii]; emphasis added). For Carnap's view (from Prague) of the situation in Germany and Central Europe in 1934 see Carnap 1963a, 34: "With the beginning of the Hitler regime in Germany in 1933, the political atmosphere, even in Austria and Czechoslovakia, became more and more intolerable.... [T]he Nazi ideology spread more and more among the German-speaking population of the Sudeten region and therewith among the students of our university and even among some of the professors."

13. In ASP 025-73-03, entry for 5 July 1932, Carnap triumphantly records the fact that there were 250 people in the audience at Berlin.

14. Neurath 1932a (reprinted in Neurath 1981, 572–73). Orthmar Spann was an especially virulent Austrian-Catholic right-wing ideologue of the time.

15. See, e.g., Neurath's remarks (made in 1933) in Neurath 1981, 597n: "Here [in Austria] there is not an exclusive dominance by metaphysics as it is practiced by Rickert, Heidegger, and others — through which those of a new generation become well-known through *geisteswissenschaflicher Psychologie, geisteswissenschaftlicher Soziologie* and similar things." Compare Carnap's remarks on Neurath's commitment to physicalism, unified science, and Marxism in Carnap 1963a, 22–24.

16. Compare also the following retrospective remarks of Neurath's (made in 1936): "The strong metaphysical trends in Central Europe are probably the reason that within the Vienna Circle the antimetaphysical attitude became of central significance and was purposefully practiced — much more, for example, than would have been the case with the adherents of similar tendencies in the United States, among whom a particular, more neutral common-sense empiricism is very widespread and where metaphysics could not exert the influence that it did in Germany, say.... It is entirely understandable that a Frenchman is at first surprised when he hears how the adherents of the Vienna Circle distance themselves in sharp terms from 'philosophers' — he thinks perhaps of Descartes and Comte in this connection, the others however of Fichte and Heidegger" (Neurath 1981, 743).

17. As is well known, *Introduction to Metaphysics* depicts Germany as Europe's last hope for salvation from Russian communism, on the one side, and American technological democracy, on the other, and contains Heidegger's notorious remark about the "inner truth and greatness" of the National Socialist movement (Heidegger 1953, 152 [166]).

18. I am indebted to Kathleen Wright for first calling my attention to this passage. As Wright also first pointed out to me, the noted Heidegger scholar Otto Pöggeler comments on *Introduction to Metaphysics* as follows: "Heidegger had sufficient taste not to deliver a previous version of his lecture in which Carnap's emigration to America was put forth as confirmation of the convergence between Russian communism and the 'type of thinking in America'" (quoted from Wolin 1991, 218–19). Given that Carnap did not emigrate until December 1935, however, whereas Heidegger's lectures were held in the summer of that year, Heidegger cannot be here referring to Carnap's emigration. It is more likely, for example, that he is referring to Schlick's trip to Stanford in 1929, which is prominently mentioned in the foreword to Hahn, Neurath, and Carnap 1929. The remark about Russian communism, on the other hand, almost certainly refers to Neurath's activities.

19. The dissertation appeared as Carnap 1922. Carnap defends a modified version of the Kantian synthetic a priori according to which the topological — but not the metrical — properties of space are due to the form of our spatial intuition. Bauch was influenced not only by his teacher Rickert but also by the more scientifically oriented neo-Kantianism

of the Marburg school — as well as by his colleague at Jena, the logician Gottlob Frege (who also greatly influenced Carnap, of course). Some discussion of Bauch in relation to both Frege and Carnap can be found in Sluga 1980. Sluga (1993) explains the depth and centrality of Bauch's involvement with Nazism — in comparison with which Heidegger's own engagement somewhat pales. Curiously, Carnap himself never mentions Bauch's political involvement.

20. Carnap explains his neo-Kantian philosophical training in Carnap 1963a, 4, 11–12. Writings of Cassirer, Natorp, and Rickert (as well as Bauch) play an important role in the *Aufbau:* see Carnap 1928a, §§5, 12, 64, 65, 75, 162, 163, 179.

21. Some of the most important epistemological works of the two traditions are Cohen 1902; Natorp 1910; E. Cassirer 1910; and Rickert 1892. Rickert 1909 and Natorp 1912 are very useful summary presentations of the two traditions.

22. See Husserl 1900–1901. For Husserl and the neo-Kantians see, for example, Natorp 1912, 198; and Rickert 1909, 227. Husserl's notion of *Wesenserschauung* plays a central role in Carnap's conception of "intuitive space" in Carnap 1922. All of Heidegger's earliest works fall squarely within the antipsychologistic tradition of "pure logic" and, accordingly, are dominated by the thought of Rickert and Husserl: these include "Neure Forschungen über Logik" (1912), his doctoral dissertation *Die Lehre vom Urteil im Psychologismus* (1913–14), and his habilitation *Die Kategorien- und Bedeutungslehre des Duns Scotus* (1915–16) — all of which are reprinted in Heidegger 1978. There is, of course, a close relationship between this "pure logic" tradition and the work of Frege: indeed, as is well known, it was Frege's review of Husserl's earlier *Philosophie der Arithmetik* that inspired the antipsychologistic polemic of the *Logical Investigations.* It is interesting to note also that Heidegger comments very favorably on Frege's work in his "Neure Forschungen über Logik" (Heidegger 1978, 20).

23. This "genetic" view of the object of empirical knowledge is common to Cohen, Natorp, and Cassirer. It is articulated with particular force by Natorp (in the works cited in note 21 above). Cassirer's achievement, in *Substance and Function,* is to make the view precise by finally articulating a coherent conception of logic (as the theory of arbitrary relational structures) — something that had eluded both Cohen and Natorp. Thus, for example, whereas Cohen and Natorp self-consciously attempt to align logic more closely with mathematics, they still continue to make essential use of the traditional classification of the forms of judgment.

24. Natorp (1912) then replies to Rickert's criticism. E. Cassirer (1929, 406 [348]) brings out the issue between Rickert and the Marburg school here with particular clarity: "Rickert's proof-procedure, in so far as it is simply supposed to verify this proposition [that number is not derivable from identity and difference], could have been essentially simplified and sharpened if he had availed himself of the tools of the modern logical calculus, especially the calculus of relations. For *identity* and *difference* are, expressed in the language of this calculus, *symmetrical* relations; whereas for the construction of the number series, as for the concept of an ordered sequence in general, an *asymmetrical* relation is indispensable."

25. Here Lask cites, among others, the theories of Herbart, Bolzano, Husserl, Rickert, Meinong, and (Heinrich) Gomperz (see Lask 1912, 23–24). Note that what Lask calls "formal logic" coincides with what Rickert calls "transcendental logic."

26. Lask (1912, 55) writes: "The 'form' of judgment, concept, inference, etc. is a completely different thing from form in the sense of the category. One best distinguishes these two kinds of form as structural form and contentful form." Kant's metaphysical deduction, by contrast, rests entirely on the idea that "the same understanding, through precisely

the same action whereby it brought about the logical form of a judgment by means of analytic unity, also brings about, by means of synthetic unity, a transcendental content in its representations in virtue of which they are called pure concepts of the understanding" (A79/B195).

27. For a discussion of Lask's argument from the point of view of the Marburg school see E. Cassirer 1913, 6–14. I am indebted to Werner Sauer for first calling my attention to this essay and for emphasizing to me, in this connection especially, the crucially important differences between Cassirer and the Southwest school.

28. The reference is to Lotze 1874, §§316–20. See, for example, Heidegger 1978, 170.

29. See Heidegger 1978, 214–89. As Heidegger notes, his discussion here is based on Rickert 1911.

30. Heidegger 1978, 273. The realm of valid sense enjoys this primacy because *all* realms of existence as such (the natural, the metaphysical-theological, the mathematical, and the logical itself) become objects of our cognition only through the mediation of the logical (Heidegger 1978, 287).

31. Heidegger's judgment of the superiority of Lask over Rickert emerges already in his dissertation of 1913–14 (Heidegger 1978, 176–77n).

32. Heidegger spent the years 1923–28 as associate professor at Marburg, after serving as Husserl's assistant at Freiburg from 1916 through 1922. It was at Marburg, through his lectures, that Heidegger established a reputation as one of the most brilliant and exciting young philosophers in Germany even before the appearance of *Being and Time.* The Marburg lectures in question appear as Heidegger 1979 and Heidegger 1976, respectively. To be sure, Husserl's own overcoming of neo-Kantianism is by no means complete from Heidegger's point of view — a point to which we shall return below.

33. Husserl's discussion of truth as "identification" occurs in Husserl 1900–1901, vol. 2, §§36–39. The discussion of "categorial intuition" then follows in §§40–58. For Heidegger's assessment of the relationship between these ideas and the work of Lask — which Heidegger sees as together destroying once and for all the Kantian "mythology" of a synthesis of understanding and sensibility, form and matter — see Heidegger 1979, 63–90.

34. When Husserl speaks of the realm of the logical and asks after an "epistemology of the logical" he has in mind the entire realm of a priori "essences" accessible to *Wesensanalyse* and *Wesenserschauung* (which include, for example, the a priori "essences" of spatial phenomena studied by geometry, of color phenomena studied by the a priori "eidetic science" of color, and so on). The very special structures studied by formal logic properly so-called (subject and predicate, and so on) represent, as we have just seen, only a tiny fraction — the most abstract part — of this "essential" realm.

35. This idea of phenomenology as a pure or "transcendental" psychology becomes fully explicit only in Husserl 1911; it is developed in elaborate detail in Husserl 1913. In the first edition of volume 2 of the *Logical Investigations* Husserl had misleadingly characterized phenomenology as "descriptive psychology" — which, as he himself immediately recognized, concealed precisely the "transcendental" relation in which he intended phenomenology to stand to (empirical) psychology. See Husserl 1911, 318n [115–16n].

36. See Heidegger 1978, 341–411. Heidegger there links his conception of "subjective logic" with the problem of the application of the categories on p. 407: "If anywhere, then precisely in connection with the problem of the *application* of the categories — insofar as one admits this in general as a *possible* problem — the *merely* objective-logical treatment of the problem of the categories must be recognized as one-sided." The attached footnote then emphasizes the importance of Lask 1912. For Husserl himself, on the other hand, since he developed the idea of phenomenology entirely independent of the Kantian and

neo-Kantian traditions, this problem of the application of the categories was never a problem. Husserl's own problem was always rather the relationship between the logical and the psychological — a problem that need not involve the relationship in general between the abstract and the concrete.

37. For Heidegger's assessment of Dilthey's conception of the subject as "living person with an understanding of active history" in contrast to Husserl's more formal conception of the subject, see Heidegger 1979, 161–71. The influence of Dilthey is further exhibited in 1916 in Heidegger's preface to his habilitation, with its call for philosophy to become *weltanschaulich* — that is, engaged in the concrete historical events of the time (Heidegger 1978, 191; and cf. 205 n. 10). This call contrasts sharply with Husserl's own arguments (in Husserl 1911, 323–41 [122–47]) — contra Dilthey — that philosophy *as a science* must be eternally valid and thus essentially unhistorical. It seems clear, moreover, from the remarks on Emil Lask's "distant soldier's grave," that Heidegger's call for a *weltanschaulich* philosophy here is directly connected with his attitude toward the Great War.

38. The idea of "being-in-the-world," together with the idea that the theoretical orientation toward the "present-at-hand" is founded on the more basic practical orientation toward the "ready-to-hand," is presented in Heidegger 1927, §§12–13 and is then developed in detail in the remainder of division 1.

39. For "Care as the Being of Dasein," see Heidegger 1927, §§39–44; for "Temporality as the Ontological Meaning of Care," see §§61–66; for "Temporality and Historicality" see §§72–77. In developing this conception of the essential "historicality [*Geschichtlichkeit*] of Dasein," Heidegger is, as emphasized in note 36 above, self-consciously following the work of Dilthey — work that he explicitly opposes to the "superficial" and "merely methodological" attempt to distinguish the *Geisteswissenschaften* from the *Naturwissenschaften* (based on the distinction between "generalizing" and "individuating" modes of concept formation) developed within the school of Windelband and Rickert. See the comments on Rickert in Heidegger 1927, §72, and compare the polemic called "Die Trivialisierung der Diltheyschen Fragestellung durch Windelband und Rickert" presented in Heidegger 1979, 20–21.

40. The analysis of "being-toward-death" and the ensuing possibility of "authentic" existence are presented in Heidegger 1927, §§46–60 — an analysis that is intended to present the "being of Dasein" (which was presented only fragmentarily, as it were, in the preceding sections) for the first time as a unitary and unified whole.

41. For the priority of "existence" over "essence" in the analytic of Dasein, see Heidegger 1927, §9. For Heidegger's diagnosis of the failure of Husserlian phenomenology as resting on a neglect of the question of the *existence* of "pure consciousness," see Heidegger 1979, 148–57, in particular 152: "Above all, however, this conception of ideation [that is, *Wesenserschauung*] as abstraction from real individuation rests on the belief that the What of any being is to be determined in abstraction from its existence. If, however, there were beings *whose What is precisely to exist and nothing but to exist,* then this ideational mode of consideration with respect to such a being would be the most fundamental misunderstanding."

42. The footnote to Heidegger 1927, §44 (p. 218), refers us to the sections on truth and "categorial intuition" in volume 2 of the *Logical Investigations* discussed above, along with the work of Lask. Heidegger warns us against relying exclusively on the first volume of the *Logical Investigations,* which appears merely to represent the traditional theory of the proposition (in itself) derived from Bolzano.

43. See Heidegger 1927, §44 (p. 221): "Dasein *can* understand *itself* as understanding from the side of the 'world' and the other or from the side of its ownmost

possibility-for-being [*aus seinem eigensten Seinkönnen*]. The last-mentioned possibility means: Dasein discloses itself to itself in and as its ownmost possibility-for-being. This *authentic* disclosedness shows the phenomenon of the most original truth in the mode of its authenticity. The most original and authentic disclosedness in which Dasein as possibility-for-being can be, is the *truth of existence*."

44. The temporality of "authentic" existence is articulated in Heidegger 1927, §§61–66, the temporality of everyday "inauthentic" existence in §§67–71, the temporality of "historicality" in §§72–77. How the temporality of Dasein is actually the prior ground of the "ordinary conception of time" (namely, the all-embracing public time within which events are dated and ordered) is explained in §§78–81.

45. See Heidegger 1927, §33, "Die Aussage als abkünftiger Modus der Auslegung," and compare the discussion of Lotze's theory of "validity" in Heidegger's lecture course on logic from 1925–26 (Heidegger 1976, 62–68). This discussion clarifies the relationship Heidegger perceives between the "Cartesian" predicament of the world-less subject enclosed within its own contents of consciousness and the "Husserlian" predicament of the ideal subject isolated from all questions of real existence. In Husserlian terminology, it clarifies the relationship between the "phenomenological" reduction that withdraws our attention from the external world and focuses on the contents of consciousness themselves and the more radical "eidetic" reduction that then focuses only on the "essence" or formal structure of these conscious phenomena — arriving, in the end, at "pure" or "absolute" consciousness. Heidegger's idea is that if we once start with the "Cartesian" predicament, but nonetheless demand a kind of objectivity, then all we have left, as it were, is the contrast between change and constancy, the real and the ideal. We thus arrive at a conception of truth or objectivity on which truth is fundamentally necessary, "essential," or eternal truth; and, in this way, the denial of "naive realism" leads to "essentialism." And it is this last form of "essentialism" that is Heidegger's ultimate target.

46. Thus Heidegger 1927, §44 contains such provocative assertions as, "Before Newton's laws were uncovered they were not 'true' "; and "[these] laws became true through Newton, with him a being became accessible in itself for Dasein" (226–27). Given Heidegger's fundamentally historical conception of truth, together with his "existential conception of science" (§69, pp. 362–64), the meaning of these assertions is relatively straightforward: Newton arrived at the laws of motion by means of an "authentic projection" of a particular scientific framework in a given historical situation — the context of the scientific revolution of the sixteenth and seventeenth centuries (cf. §3, pp. 9–10, on "scientific revolutions"). Outside of this historical context, on the other hand, Newton's "discovery" and accompanying "assertion" of the laws of motion simply make no sense. For Heidegger, there is then no "valid sense" or "proposition in itself" beyond Newton's (and our) actual historical "assertions" capable of serving as a "vehicle" of "eternal truth." Nevertheless, *what* Newton discovered of course existed before Newton: "With the uncoveredness the being showed itself precisely as that being that was already there before. So to uncover is the mode of being of 'truth' " (p. 227).

47. See the remarks on the "objectivity [*Objektivität*]" of authentically historical truth in Heidegger 1927, §76, p. 395: "In no science are the 'universal validity' of standards and the pretensions to 'universality' that the They and its common sense require *less* possible criteria of 'truth' than in authentic history."

48. See n. 19 above.

49. Carnap 1963a, 20–22. An outline of Carnap's lecture to the Circle on 21 January 1925, bearing the title "Gedanken zum Kategorien Problem: Prolegomena zu einer Konstitutionstheorie," appears as ASP 081-05-03. The "Entwurf" manuscript has not yet been

found. A table of contents, bearing the dates 17 December 1924 and 28 January 1925 (a revision after the lecture in Vienna), appears as ASP 081-05-02.

50. Carnap 1928a, §75. The passage from Cassirer's *Substance and Function* to which Carnap is here referring (E. Cassirer 1910, chap. 4, §9) is a criticism of Rickert's well-known argument that concepts in the *Naturwissenschaften* cannot individuate (cf. note 39 above). Cassirer diagnoses Rickert's error here as stemming from a neglect of the essentially relational mode of concept formation of modern mathematics and logic. Carnap (1928a, §12) points out, again referring to this discussion of Cassirer's (and also to Rickert, Windelband, and Dilthey), that the "logic of individuality" desired in the *Geisteswissenschaften* can be attained precisely in the modern theory of relations.

51. For recent work on Kantian and neo-Kantian aspects of the *Aufbau,* see, for example Haack 1977; Moulines 1985; and Sauer 1985, 1989 (which particularly stress the importance of Cassirer and the passage from Carnap 1928a, §75); Friedman 1987, 1992b, and Richardson 1992b (which also emphasizes the importance of Cassirer and the Marburg school); and Webb 1992. Two recent extended treatments of the development of logical positivism, Coffa 1991 and Proust 1986, are also worth consulting in this connection.

52. See Carnap 1928a, §§100, 143. Cf. the remarks in Carnap 1963a, 18: "The system [of the *Aufbau*] was intended to give, though not a description, still a rational reconstruction of the actual process of the formation of concepts."

53. See especially the discussion of "ascension forms [*Stufenformen*]" in Carnap 1928a, pt. 3.B. There are exactly two such "ascension forms": namely, class and relation extensions (§40). As Carnap (1963a, 11) explains, he first studied *Principia Mathematica* — whose type-theoretic conception of logic pervades the *Aufbau* — in 1919.

54. See Carnap 1928a, pt. 4.A, and cf. §§67–94. The procedure of "quasi-analysis" is a generalization (to nontransitive relations) of the "principle of abstraction" employed by Frege and Russell in the definition of a cardinal number (see §73).

55. See Carnap 1928a, pt. 4.B, for the constitution of the physical realm — including the qualitative realm of ordinary sense-perception (§§125–35) and then the quantitative realm of mathematical physics (§136). As Carnap makes clear in §136, the constitution of the latter realm is based on his earlier methodological studies (Carnap 1923, 1924).

56. See Carnap 1928a, pt. 4.C. According to Carnap, *only* the purely abstract world of physics — and not the qualitative world of commonsense perceptual experience — "provides the possibility of a unique, consistent intersubjectivization" (§136; cf. §133).

57. For the independence of the definition of the visual field, in particular, from all phenomenal qualities, see Carnap 1928a, §86. For the importance of purely structural definite descriptions, see pt. 2.A, especially §16: *"[E]very scientific statement can in principle be so transformed that it is only a structural statement.* But this transformation is not only possible, but required. For science wants to speak about the objective; however, everything that does not belong to structure but to the material, everything that is ostended concretely, is in the end subjective." "From the point of view of constitutional theory this state of affairs is to be expressed in the following way. The series of experiences is different for each subject. If we aim, in spite of this, at agreement in the names given for the objects [*Gebilde*] constituted on the basis of the experiences, then this cannot occur through reference to the completely diverging material but only through the formal indicators of the object-structures [*Gebildestrukturen*]." For a fuller discussion, as well as detailed arguments against an empiricist-phenomenalist interpretation of Carnap's motivations, see my articles cited in note 51 above.

58. Carnap 1928a, §177. Section 116 presents the actual constitution of *sensations* —

defined via a purely structural definite description containing only the basic relation as a nonlogical primitive.

59. Carnap's type-theoretic sequential construction therefore takes the place of the "general serial form" Cassirer sees as expressing the essence of empirical knowledge. Carnap agrees with Cassirer, however, that this kind of methodological sequence is the ultimate "datum" for epistemology and, in particular, that the contrast between "being" and "validity" — which, as we have seen, generates fundamental problems for the Southwest school — therefore has only a *relativized* meaning in the context of such a sequence (see Carnap 1928a, §42; and cf. E. Cassirer 1910, 412–13 [311]). In Friedman 1992b, §3, I mistakenly read §42 as a *criticism* of the Marburg school and, in general, failed to draw the crucial distinction between the Marburg view of this question and that of the Southwest school. I am indebted to Alan Richardson and Werner Sauer for rightly protesting against this assimilation (cf. nn. 27 and 51 above).

60. The reference is to Natorp 1910, chap. 1, §§4–6; cf. E. Cassirer 1910, chap 7, especially 418–19 [315]. I am indebted to Alison Laywine for emphasizing to me the importance of this aspect of Natorp's view in the present connection.

61. Cf. E. Cassirer 1910, 337 [254]: "In contrast to the mathematical concept, however, [in empirical science] the characteristic difference emerges that the construction [*Aufbau*], which within mathematics arrives at a fixed end, remains in principle *incompleteable* within experience."

62. It is worth noting, in this connection, that the well-known technical problems afflicting the constitution of the physical or external world in the *Aufbau* appear in fact to undermine Carnap's attempt to distinguish himself from the Marburg school here. It appears, in particular, that Carnap's rules for assigning colors to points of space-time ($\mathbf{R}^4$) never close off at a definite set (that is, a definite relation between space-time points and colors) located at a definite rank in the hierarchy of logical types: for this assignment is to be continually revised as we progress to higher and higher ranks (Carnap 1928a, §§135, 136, 144). And this means, from the point of view of Carnap's own constitutional system, that the Marburg doctrine of the never completed *X* appears after all to be fully correct — at least so far as physical (and hence all higher-level) objects are concerned.

63. See again Carnap 1928a, §178: *"[T]he so-called epistemological tendencies of realism, idealism, and phenomenalism agree within the domain of epistemology. Constitutional theory presents the neutral basis* [neutrale Fundament] *common to all. They first diverge in the domain of metaphysics and thus (if they are to be epistemological tendencies) only as the result of a transgression of their boundaries*." All other properly philosophical disputes are similarly dissolved. Thus, for example, both sides in the debate over the relationship between the *Geisteswissenschaften* and the *Naturwissenschaft* are correct: cultural objects are constructed out of heteropsychological objects and the latter, in turn, out of physical objects; in this sense the theses of physicalism and the unity of science are correct. On the other hand, however, cultural objects nonetheless belong to a distinct "object sphere" in the type-theoretic hierarchy; in this sense the thesis of the autonomy and independence of the cultural realm is equally correct. See §56 and also §§25, 29, 41, 151. Cf. nn. 15, 39, and 50 above.

64. See the (first edition) preface to Carnap 1928a: "The new type of philosophy has arisen in close contact with work in the special sciences, especially in mathematics and physics. This has the consequence that we strive to make the rigorous and responsible basic attitude of scientific researchers also the basic attitude of workers in philosophy, whereas the attitude of the old type of philosophers is more similar to a poetic [attitude]. . . . [T]he individual no longer undertakes to arrive at an entire structure of philosophy by a [single] bold stroke. Instead, each works in his specific place within

the *single* total science." Cf. also the discussion in Carnap 1963a, 13, on the impact of reading Bertrand Russell's *Our Knowledge of the External World* in 1921: Carnap is most impressed by Russell's description of "the logical-analytic method of philosophy"— together with its accompanying call for a new "scientific" philosophical practice.

65. In this respect, Carnap's identification with the *neue Sachlichkeit* is even more radical than Neurath's. For Neurath, unlike Carnap, makes no attempt to turn philosophy itself into an "objective" — purely technical — discipline. See the remarks on Neurath in Carnap 1963a, 22–24, 51–52, and cf. Uebel 1996 for an illuminating discussion of the relationship between Neurath's philosophy and his politics. This significant difference between Carnap and Neurath seems to be missed in the otherwise quite useful discussion of the relationship between the Vienna Circle and the *neue Sachlichkeit* in Galison 1990, which generally ignores the important areas of disagreement between the two philosophers.

66. Heidegger himself is perfectly explicit about the connection between his political engagement and his philosophical conception of the necessary "historicality of Dasein" in a well-known conversation (in 1936) reported by Karl Löwith (Wolin 1991, 142). Curiously, this crucial connection seems to be missed in Bourdieu 1988, an otherwise very interesting study of the relationship between Heidegger's philosophy and German neoconservatism.

Nancy Cartwright and Jordi Cat

Neurath against Method

1. The Protocol-Sentence Debate

There are two central aspects of Carnap's *Aufbau.* One is the logical construction of science upwards; the second is the assumption that the foundations of the construction can be epistemologically secure, for science can in principle begin with sense-data reports to which the individual knower has privileged access. Neurath entered the protocol-sentence debate with views firmly opposed to both features of Carnap's construction. He left the debate with a view that is much stronger — a view that loosens the connection between the superstructure as a whole and its foundation. In the end, not only is logical construction ruled out, but so too are the falsificationism, the inductivism, the confirmation theories, and the hypothetico-deductive method with which Carnap's successors would try to replace it.

At the beginning of the protocol-sentence debate Neurath was both a holist and a pluralist: he believed that only whole complexes of hypotheses — hypotheses cum auxiliary hypotheses cum theories of how and by whom the measurement was carried out, and so on — can confront the data. He also believed that no matter how much data, there will always be more than one whole complex of hypotheses that can account for them. By the end of the debate Neurath would undermine the very notion of accounting for the data. In Neurath's pre–protocol-debate work the choice among competing complexes of hypotheses could not be made scientifically but was fixed by external factors of history, tradition, and will; later, even the question of what theory complexes are in the competition becomes a matter of choice.

The central aim of this essay is to show how in making this shift, Neurath brought together his earlier epistemological views with his long-standing Marxist views — views that for Neurath had been not just theoretical but indeed critical for his engagement in political practice.[1]

Before the two stood for the most part side by side; later they would interact to create a radically new and important philosophical stance. The key to the interaction, we will argue, is Neurath's concept of *Ballungen.*

The protocol-sentence debate is about what kind of data science is supposed to be responsible to: Can we follow Schlick and the early Carnap and take as data individual sense-experiences that can be known immediately and infallibly? Or must we, as Neurath insisted, admit only data that are intersubjectively available? These of course were not the terms in which the debate was fought. Because of the well-known "linguistic turn" of the Vienna Circle, talk about the real world outside our description and experience of it was forbidden; it was to be replaced by talk about our representations of the world. So the debate focused not on the form of the data but rather on the form of the sentences describing the data. Carnap was willing to admit sense-data reports, sentences like "Red, here, now." Neurath insisted on sentences about publicly accessible events occurring in space and time. The closest he could come to a sense-data report was a sentence about a sense-data report: "Otto says, 'Red, here, now.' " *Protokoll* in German means "record" or "report"; for Neurath, then, the data-sentences read like the minutes of a faculty meeting or a statement made to the police.

Neurath was always a materialist: he believed in one world, the material world, where events and interactions take place in space and time. The objective of his materialism was to rule out idealism, religion, superstition, and traditional moral canons — all doctrines that served, he believed, to mislead and disempower the working classes. With the added sophistications of the linguistic turn, Neurath did not make assumptions about a single physical world but rather insisted on a single physicalist language. Science studies what happens in space and time. During this period, Neurath conflated, perhaps as much for polemical reasons as for any other, empiricism and materialism. In the protocol-sentence debate both doctrines emerge. For a materialist a starting sentence — a data-recording sentence — could be something like: "There was an alpha-particle track in the upper-left hand corner of the chamber." But for an empiricist what is to count as data must be the subject of human experience. For Neurath protocols about experience serve simultaneously the demands of materialism and of empiricism. "Otto says, 'Otto sees an alpha-particle track' " is a report of a space-time occurrence that nevertheless incorporates Otto's experience. Moreover, it incorporates the experience in a way that makes clear our options for integrating it into our total hypothesis system, what Neurath calls our "encyclopedia": we can assume that Otto is truthful, that he observes accurately, that the instrument is working correctly, and that an alpha-particle has indeed passed by, in which case tracks of that form had better be consistent with the rest of our theory; or we can assume that Otto observed correctly but that the machine misrecorded because the fluid in the chamber was impure, in which case it would be a good idea to produce an additional protocol, like "Otto says, 'Otto's found the leak' "; or we can assume Otto is hallucinating, in which case we should like to find someone who reported "I saw Otto taking some suspicious-looking pills"; and so forth.

2. Of Boats and *Ballungen*

Although Neurath's motives for his physicalist protocols were materialist, this could hardly serve as a lever to shift Carnap, for Carnap repeatedly tried to convince him that materialism was as much a metaphysical doctrine as any other claim about the "real" world (presumably this means, "as much as any other claim about the Kantian *Ding an Sich*"). Indeed, Neurath's arguments against Carnap's phenomenalist language were different, invoking, first, his claim that scientific concepts and practices are always subject to revision; and, second, the position that no true language can be private. This is a story nicely told by Thomas Uebel (1992b), so we will not pursue these claims here. What is important for our project is what follows once Neurath has introduced real speakers in real situations into the basic data-sentences of science. The key is Neurath's new idea of *Ballungen.* A *Ballung* is a congestion or a concentration. A *Ballungsgebeit,* for instance, is an overcrowded area — the way a modern city looks on a map, dense and fuzzy-edged. The verb is *ballen* — basically, to press together to a *Ball,* a swollen or blown-up body.[2] Protocol sentences in the form that Neurath wanted always involve the concepts of everyday life; and these inevitably have the character of *Ballungen* — congested, heterogeneous, and fuzzy-edged, resisting attempts to make them totally precise:

> In the interest of scientific work, more and more formulations in the unified language of unified science are becoming increasingly precise. No term of unified science, however, is free from imprecision, since all terms are based on terms that are essential for protocol statements whose imprecision must be immediately obvious to everyone.
>
> The fiction of an ideal language composed of neat atomic statements is as metaphysical as the fiction of Laplace's spirit. . . . The statement "Otto observes an angry man" is less precise than the statement "Otto observes a thermometer registering 24 degrees" inasmuch

as "angry man" must be less precisely defined than a "thermometer reading 24 degrees"; but "Otto" itself is in many respects an imprecise term; the statement "Otto observes" can be replaced by the statement, "The man whose carefully taken photo is No. 16 in the file observes"; but the term "photo No. 16 in the file" has not been replaced yet by a system of mathematical formulas that is unambiguously coordinated to another system of mathematical formulas that takes the place of "Otto," of "angry Otto," of "kind Otto," etc. (Neurath 1932b, 91)

Neurath then recalls that *Ballungen* are the starting point of our scientific knowledge: "What is first given [to] us is our historical ordinary language with a multitude of imprecise, unanalysed terms [*Ballungen*]" (ibid.). The argument concludes that not only are *Ballungen* not entirely removable but any comprehensive collection of the totality of our scientific knowledge is bound to contain them. Inevitably in unified science both ordinary and more precise advanced scientific language overlap. This opens a break in Neurath's earlier ideal that the language of science can be exact. The exact language of science is now replaced by the more practical "universal jargon": "In a scientific treatise that touches upon the whole range of unified science . . . only a 'jargon' that contains terms of both languages will do" (ibid., 92).

Neurath closed this part of his discussion of *Ballungen* with the incorporation of a new idea into the ship metaphor. (Recall that Neurath had three important uses of the boat metaphor: 1913a, 1921, and 1932b.)[3] This is the Third Boat:

There is no tabula rasa. We are like sailors who have to rebuild their ship on the open sea, without ever being able to dismantle it in dry-dock and reconstruct it from the best components. Only metaphysics can disappear without trace. Imprecise "verbal clusters" [*Ballungen*] are somehow always part of the ship. If imprecision is diminished at one place, it may well reappear at another place to a stronger degree." (Neurath 1932b, 92)

Notice how the question of the nature of the boards is raised and compare the Second Boat. Again Neurath opens with the claim about the tabula rasa: "We cannot start from a tabula rasa as Descartes thought we could. We have to make do with words and concepts that we find when our reflections begin" (Neurath 1921, 198). He then gets to the critical feature of the Second Boat:

Duhem has shown with special emphasis that every statement about any happening is saturated with hypotheses of all sorts and that these

> in the end are derived from our whole worldview. We are like sailors who on the open sea must reconstruct their ship but are never able to start afresh from the bottom. Where a beam is taken away a new one must at once be put there, and for this the rest of the ship is used as support. In this way, by using the old beams and driftwood, the ship can be shaped entirely anew, but only by gradual reconstruction. (ibid., 198)

There are two major elements missing from the (First and) Second Boat that are crucial to the Third Boat. (1) Both boats contain the coherence principle for the unified ship of science. But in regard to the Second Boat there is no talk of comparing only sentences with sentences. In this period Neurath unself-consciously talks of "the facts" and "the phenomena" and requires — as Duhem does — that a good theory account for them. This change is clearly a reflection of the linguistic turn of the Vienna Circle. (2) No problem is raised about what the connection between fact and theory is supposed to be. Throughout Neurath's earlier work he was prepared to talk about different total theories "fitting" the facts or not. This changed with the Third Boat. On account of the *Ballungen,* the facts themselves — or, more accurately given the linguistic turn, our concepts of the facts — become fuzzy-edged, vague, and congested. They are neither strictly compatible nor strictly incompatible with a given hypothesis-complex.

The sailors on the Second Boat did not question whether a given board would fit a given gap or not. It either did, or it did not. The point was that if it did not, it could still be built into the ship by casting out some of the surrounding boards and thus reshaping the gap. But gaps on the Third boat are not so well defined. The wood is soft, and the edges are irregular and ill defined. Chunks of rather different shapes can be forced into the same gap, or, conversely, the edges of the gap can be bent about a bit to make it accommodate a given piece.

Ballungen may seem harmless enough, but in fact they ground the attack on scientific method: the very ideal of scientific testing becomes *pseudorational.* The terms of advanced science are (relatively) precise; those of everyday life are congestions; and between the two no determinate relations can obtain. The hypothetico-deductive method will not work because statements involving the fuzzy-edged *Ballungen* cannot be deduced from the exact claims of science. Induction can fare no better, nor can confirmation. All three standard accounts of scientific method — falsification, induction, confirmation — must go since the connection between a hypothesis-system and the data is never so strict as they require. With the new stress on *Ballungen* it becomes clear for Neurath that determining whether a fact and a hypothesis-system are mutually compatible is not simple. In the earlier works it appeared as a straightforward matter: either they were or they were not compatible; if they were not, we had to decide (though, of course, not arbitrarily) which to keep. But now choice enters at an even more fundamental level. Their content and syntax have not already determined whether they are or are not compatible: that too requires a decision. Thus Neurath proposes giving up the language of induction, falsification, and confirmation and replacing it with the term "shaking": no datum can falsify a system of hypotheses; it can only shake one's confidence in it. This provides an independent source of the pluralism that Neurath praised in science. Starting from the very same original boat, systematically different choices about which pieces fit which gaps can lead to very different structures in the end.

3. *Ballungen* and How Marxists Think of History

Where do *Ballungen* come from for Neurath? Duhem's picture incorporates the idea that the concepts of science are precise and hence cannot fit in any exact way onto the imprecise and more qualitative concepts of everyday life. But what about the congested aspect of *Ballungen* — their density and swollen-upness? It is no surprise that science is exact given the German Enlightenment. But how does "reality" get congested? And in what context does Neurath come to see how crucial the congestion and inexactness are?

Neurath's first public use of the concept *Ballungen,* so far as we know, was on 4 March 1931 in a discussion titled "Besprechung über Physikalismus," which he gave to the Vienna Circle — not one of their regular Thursday meetings, but the day before. The idea is developed as one of many connected themes in two essays that we know from Neurath himself were written in 1931 in the context of the protocol-sentence debate — "Physicalism" and "Sociology in the Framework of Physicalism" (Neurath 1931a, 1932a). Throughout the year Neurath was deeply engaged with the problems of protocol sentences. But he was also following a debate in *Der Kampf* to which he contributed at the end of the year, involving another left-wing member of the Vienna Circle, Edgar Zilsel. The starting point of this debate was an article by Zilsel — "Materialismus und marxistische Geschichtsauffassung" (Materialism and the Marxist conception of history) — reviewing Max Adler's book *Lehrbuch der materialischen Geschichtsauffassung* (Textbook of the materialist conception of history) (M. Adler 1930–32 [1964]; Zilsel 1931a).

Max Adler was the intellectual leader of the social-democratic movement and one of the visible heads of the neo-Kantian, Austro-Marxist movement. He was also a cofounder of *Der Kampf,* along with Neurath's friend Otto Bauer. In his book Adler pictured the classical philosophical mind-body problem as parallel to the Marxist distinction between ideology (or superstructure) and economic basis (or substructure) and used it to argue from a Kantian position against a materialist philosophy that reduced spirit to matter. On these grounds Adler defended the compatibility between Marxism and religion. Zilsel on his part argued that Marxism was to be scientific and materialist and that the mind-body problem was at any rate only a fussy philosophical problem that had nothing to do with real questions of how entire social and economic structures come about.

The debate went on between Adler and Zilsel (Adler's "Wozu schreibt man Bücher?" [Why does one write books?] [1931]; Zilsel's "Partei, Marxismus, Materialismus, Neo-Kantismus" [Party, Marxism, materialism, neo-Kantianism] [1931b]), with an intervention in defense of the compatibility of Marxism and religion by Wilhelm Frank ("Ist die marxistische Geschichtsaufassung materialische?" [Is the Marxist conception of history materialistic?] [1931]), and a final article by Neurath ("Weltanschauung und Marxismus" [Worldview and Marxism] [1981 (1931d)]). Neurath's ideas in this article were, not surprisingly, the same as the ones formulated in the two articles already mentioned ("Physicalism" and "Sociology in the Framework of Physicalism") where *Ballungen* appear. Neurath's principal point in this article is that Marxism itself is no *Weltanschauung;* it is — as he had long and deeply held — a science. Nor is Marxism compatible — or, for that matter, incompatible — with any *Weltanschauung,* for there is no such thing as a *Weltanschauung.* There is only science; all the rest is metaphysics — and likely to be politically insidious. He also in this article points to his own solution to the problem of historical materialism, which he mentions in "Sociology in the Framework of Physicalism" as well. Briefly put, it is that the distinction between the economic basis, on the one hand, and ideological factors, on the other, and the interaction between them must all be cast in the framework of strict physicalism. But "physicalism is perfectly monistic" (Neurath 1931c, 56); this applies to Marxist sociology as well.

How does this connect with Neurath's simultaneously developing thoughts about the concept of *Ballungen?* To trace this influence, we have first to consider the question of historical materialism in the context of a much broader and earlier discussion. If the Little Debate of 1931 was the locus, the Big Debate begun in the 1890s was, for Neurath, the real source. Its main protagonists were Friedrich Engels; the German revisionist Eduard Bernstein; Antonio Labriola, the Italian heterodox; and Georgy Plekhanov, the Russian orthodox. For his direct connection with Neurath we will take, again, Max Adler as starting reference and intellectual link between the two debates.

In his book of 1930, Adler, as we said before, argued from the standpoint of Kantian idealism against the reductionist materialist philosophy commonly ascribed to Marxism. It is impossible, he said, to derive consciousness from physical motion. And causal explanation of social phenomena is not in conflict with the existence of a human will. He explicitly called into question the traditional distinction between "material" and "spiritual" (ideological) factors in this historical process and stressed that economy and ideology "are parts of a single cultural system of human life."

Adler referred to Engels's doctrine of "mediate causes" to give textual support to his interpretation. Specifically he quotes from two famous letters by Engels, one to J. Bloch (21–22 September 1890), the other to Heinz Starkenburg (25 January 1894). In the extract from the first one, Engels remarks that "there are innumerable intersecting forces, an infinite group of parallelograms of forces which give rise to a resultant, the historical event. This may again be viewed as the product of a force which works as a whole, unconsciously and without volition, for what each individual wills is obstructed by everyone else, and what emerges is something that no one willed" (Marx and Engels 1935, 476). In the second letter Engels points out that "in the last resort, political, juridical, philosophical, religious, literary, artistic, etc. development is based on economic development. But all these react upon one another and also on the economic base" (ibid., 517). As a matter of fact, we should say that these letters were the true triggers of the discussions and various interpretations given within the Marxist community of the thesis of historical materialism. Adler then refers to the pamphlet by Bernstein of March 1889, *Die Voraussetzungen des Sozialismus and die Aufgaben der Sozialdemokratie* (translated as *Evolutionary Socialism*) (Bernstein 1897 [1961]), and warns against Bernstein's interpretation of Engels's letters as grounds for dualism.

Adler incidentally mentions the case of Plekhanov, whom he places in an intermediate position between Bernstein and himself. He states that Plekhanov, precisely because he is "not really materialist, has always determined to work out the human nature of economic relationships" (M. Adler 1930–32 [1964], 103n.). The immediate question is: What did Plekhanov have to say about Engels's letters? Plekhanov, we think, is the key path to the origin of *Ballungen.*

In 1908 Plekhanov published his book *Fundamental Problems of Marxism* (Plekhanov 1908 [1969]). There he presents the same criticism that Adler offers to Bernstein's interpretation. The editions of this book usually include as an appendix an earlier essay by Plekhanov from 1897, "The Materialist Conception of History," where he reviewed Labriola's newly published essay under the same title. Adler reviewed the German translation in *Der Kampf* in 1910 (M. Adler 1910); later that year he wrote again about Plekhanov and his views of the dialectic interaction between thought and object. In between is a brief letter of Kautsky's discussing Plekhanov and answering a question about one of Neurath's favorite philosophers, Ernst Mach: Are Mach's teachings compatible with Marxism? Kautsky's answer is exactly the thesis of Neurath's 1931 essay. Of "Marxism," Kautsky says, "I understand it not as a philosophy, but as an experiential science, a particular conception of history. The conception is certainly incompatible with idealistic philosophy, but not incompatible with Machian epistemology" (Kautsky 1909, 452). His later remark about Marx in the letter is also one close to Neurath's own thought: "Marx is no philosopher; rather he has announced the end of all philosophy" (ibid.). Neurath did not contribute to this series of articles. But he surely read them and had them to turn back to in 1931. We bring this up because Neurath's contribution to *Der Kampf* in 1931 would be — in both content and language — far more appropriately placed in the 1910 skirmish in the Big Debate than in the Little Debate of 1931. Let us just cite the most striking feature. Neurath's 1931 piece, we have seen, is about Marxism as a *Weltanschauung.* But this concept played essentially no role in the 1931 exchanges between Adler and Zilsel. It is central in the Big Debate. Adler's discussion, titled "Marxismus und Materialismus" (Marxism and materialism) (M. Adler 1910), focused on Plekhanov's opening statement, "Marxism is an entire *Weltanschauung.* It is, briefly expressed, the modern materialism which represents the most highly developed *Weltanschauung* to date" (M. Adler 1910, 564). The term *Weltanschauung* does play a role in Wilhelm Frank's 1931 contribution, unlike Adler's and Zilsel's. But his very title, like Zilsel's, directs the attention of the 1931 reader back to the topic of the Big Debate — both refer to the materialist conception of history.

Plekhanov's own work of 1897 was a review of a newly published essay by Antonio Labriola, whom he viewed as the first true Italian Marxist (Plekhanov's and Labriola's essays were both titled "The Materialist Conception of History"). In the essay Labriola attacked the "theory of factors." His view was largely a critique of what he considered vulgar interpretations of Marxism as the theory that the "economic factor" dominates in history. Like Engels (as he interpreted him), Labriola rejected philosophical materialism. He believed rather in the interrelations of human activities. The historical process, he argued, develops organically and pluralistically as a "complex": multiple "factors" intervene that cannot be disregarded or considered autonomously. In an earlier essay, "Problems of the Philosophy of History" (1887), he had complained that Hegel's monistic philosophy of history had become a Procrustean bed for the historical sciences, which needed to concern themselves with various forms of social life such as law, language, and art: "The original centres of civilisation are many in number and cannot be reduced by any sleight of hand. . . . [T]he consideration of so many separate series of events, so many factors that resist simplification, so many unintended coincidences, . . . makes it seem highly improbable . . . to suppose that there is at the root of everything a real unity" (quoted in Kolakowski 1978, 2:287).

In "The Materialist Conception of History" Labriola warned that these factors are not to be taken as real. The study of history begins from the *complexus:*

> The narrator finds himself, in a word, confronted with a complexus of accomplished facts and of facts on the point of being produced, which in their *totality* present a certain aspect. . . . Yet into this complexus he must introduce a certain degree of analysis, resolving it into groups and into aspects of facts, or into concurrent elements [factors], which afterwards appear at a certain moment as independent categories." (Labriola 1908 [1897], 141–42; emphasis added)

In his review Plekhanov stressed this aspect of the "doctrine of factors" clearly and explicitly:

> A historical factor is an *abstraction,* and the idea of it originates as a result of a process of abstraction. Thanks to the process of abstraction, various *sides* of the social *complex* assume the form of separate *categories,* and the various manifestations and expressions of the activity of social man — morals, law, economic forms, etc. are converted in our minds into separate forces which appear to give rise to and determine this activity to be its ultimate causes, . . . a multiform reflection . . . of the single and indivisible history. (Plekhanov 1908 [1897], 108)

This is precisely the density aspect of *Ballungen* that is not included in Duhem's views on scientific abstractions. And this is exactly the view on sociohistorical phenomena that Neurath develops in 1931 in his text

Empirical Sociology: "The materialist conception of history begins with the total process of life. If one continues to use the traditional delimited terms such as 'religion,' 'art,' 'science,' 'law,' and so on, these formations appear as 'interwoven' into the total social process" (Neurath 1973 [1931a], 352). He also states that "peoples, states, age groups, religious communities, all are complexes built up of single individuals. . . . The separate features of these complexes are not independent of each other but are related" (ibid., 386–87). This the same view that Neurath — in the context of the debate with Adler — incorporated in the articles of 1931, "Physicalism" and "Sociology in the Framework of Physicalism." On the linguistic level, in "Physicalism" Neurath talks about a "complex of statements of science" (Neurath 1931c, 53). In "Sociology in the Framework of Physicalism" he publishes on *Ballungen* for the first time; and he argues that

> for the prediction of the behaviour of a group in some respect it is often necessary to know the whole life of the group. The individual ways of behaviour that can be lifted out of the totality of events, the construction of machines, the building of temples, the rules of marriage are, in their changes, not "autonomously" computable; they have to be regarded as parts of the complex that is investigated at the time. (Neurath 1932c, 75–76)

The conceptual symmetry between complexes and *Ballungen* is no coincidence. These are just a few samples of the similarities of Neurath's thought on sociology, economy, and philosophy in 1931 with his own favored version of materialism. We hope it is enough to make clear the process of cross-fertilization between the debate on historical materialism and the protocol-sentence debate.[4]

Notes

1. For Neurath's political engagements, see Cat, Chang, and Cartwright 1996.
2. On this cf. D. Drosdowski et al. 1976.
3. On Neurath's uses of the boat metaphor see Uebel's contribution in this volume and part 2 of Cartwright et al., forthcoming.
4. We shall argue for this thesis at greater length in part 3 of Cartwright et al., forthcoming.

Thomas E. Uebel

The Enlightenment Ambition of Epistemic Utopianism: Otto Neurath's Theory of Science in Historical Perspective

The systematic analysis of Otto Neurath's theory of science reveals a thinker who has remained surprisingly topical. After briefly indicating one contemporary relevance of Neurath's "anti-philosophical" program, I will explore the question of how it was possible for Neurath to develop his postpositivist conception of science in the Vienna Circle, the reputed high church of twentieth-century positivism. Neurath's originality cannot be accounted for solely by stressing that Vienna Circle philosophy must not be mistaken for the caricature current among many postpositivists. We must also attend to the differentiating characteristics of Neurath's intellectual development within the newly recovered "real existing" Vienna Circle: Neurath's social-science-oriented overcoming of logical positivism realized a distinctively activist understanding of the Enlightenment program he shared with his colleagues on the left wing of the Vienna Circle.

1. Naturalistic Theory of Science à la Neurath

In the Vienna Circle, Neurath advocated an explicitly "anti-philosophical" position (Neurath 1931b [1983], 48). This position may best be understood as an epistemological naturalism that ultimately, however, goes beyond W. V. O. Quine's later understanding of a program of that type.[1]

> "School philosophy," with its definite conception of the fundamental basis of being or thinking, presumes to sit in judgment on science as if it were a court of last resort, and this presumption the representatives of the scientific world-conception summarily reject. (Neurath 1931b [1983], 49; translation altered)

> [Any] "theory of knowledge" . . . could only consist of defense actions against metaphysics, i.e., unmasking meaningless terms. Some problems of the theory of knowledge will perhaps be so transformable

> into empirical questions so that they can find a place within unified science.... The possibility of science becomes apparent in science itself.... [I]t is impossible to separate the "clarification of concepts" from the "pursuit of science" to which it belongs. (Neurath 1932c [1983], 67, 61, 59; translation altered)

Naturalism generally requires that the practice of science be the focus of epistemological investigations. The sciences as we know them are paradigmatic instances of defensible knowledge claims. Neurath's naturalism is characterized by a particularly radical anti-apriority. All knowledge claims are viewed as requiring justification — including those of the theory of science itself. Besides advocating the use of principles stemming from the formal disciplines of logic and mathematics, Neurath rejects transempirical principles of explanation or reconstruction. In particular, philosophical epistemology is to be replaced by an interdisciplinary scientific effort. This naturalism does not brook a sharp separation of the contexts of discovery and justification such that theories of justification could abstract from what are deemed natural conditions of human cognition. The claim is that the rationality of science be demonstrated "from within": all the epistemological concepts thus employed can be explicable by scientific concepts in turn. Terms like "justification" or "epistemic priority" must thus be explicated by concepts applicable to the explanation of processes we antecedently deem natural—processes occurring in the spatio-temporal world.[2]

The question arises how reductive such a naturalism must be. It would seem that Neurath's naturalism meant the explication of "philosophical" by causal or correlational concepts, but it should be noted that naturalistic epistemology may also employ all the explanatory kinds used in the special sciences (Neurath 1932c [1983], 75). The envisaged defensibility of knowledge claims need *not* be understood in a reductive sense. Still more momentous was Neurath's conclusion that a naturalistic theory of science need concern itself not with the justification of true beliefs but rather with the *acceptance* of knowledge claims.[3] Granting epistemological primacy to the public domain shifted the naturalistic focus from a protocomputational model of cognition to a discourse model. This was the "linguistic turn" with a difference: Neurath explored the sociological dimension of nonreductionist reasoning about scientific reason and was less concerned about the individual psychological dimension. Among other things, this demanded that the contexts be taken account of within which the practice of claim acceptance takes place. Together with the recognition of the historicity of the concept of knowledge and of the general social determinants of the practice of science, this leads to including history and sociology in the naturalistic theory of science. Theory acceptance is understood as determined not only by internal methodological conventions and biological, psychological, and linguistic facts but also by factors such as the organization of science, the potential use of its research results, the funding of research programs, as well as general sociohistorical conditions like the state of the development of the productive forces. This sociopolitical dimension of Neurath's conception of the "science of science" is clearly lacking in Quine's. It was doubly important therefore — not only in keeping with the Circle's campaign against "metaphysics" — that this theory of science also provided the means to criticize inadequate knowledge claims and did not preclude taking a normative stand. Given the rejection of apodictic reasoning, the conditional norms of instrumental rationality were appealed to in order to "ground" epistemic normativity.[4]

Neurath's program for a nonreductive and socialized naturalism reflects his "image of science," which, in turn, is most graphically rendered by "Neurath's Boat" or what Quine called the "parable of the mariners": "We are like sailors who have to rebuild their ship on the open sea, without ever being able to dismantle it in dry-dock and reconstruct it from the best components" (Neurath 1932b [1983], 92).[5] What is more, the program provides a framework for explicating Neurath's ill-understood proposal concerning the nature (form and content) and the status of scientific data, that is, of protocol statements: "A complete protocol statement might for example be worded like this: 'Otto's protocol at 3:17 o'clock: [Otto's speech-thinking at 3:16 o'clock was (at 3:15 o'clock there was a table perceived by Otto)]' " (Neurath 1932b [1983], 93). Fittingly, Neurath's discourse-theoretical response to the problem of the empirical basis of knowledge is deeply antifoundationalist yet still claims to isolate the specific warrant that observational evidence possesses.

According to Neurath, a protocol statement may be understood as a repeated, indeed a threefold, embedding of a statement containing only observable predicates and spatio-temporal coordinates.[6] The (nontransparent) embeddings represent different sets of partial conditions and their independence and interdependence, conditions that concern institutional behavior, individual intentionality, and, ultimately, neurophysiological input from the environment. The fulfillment of these conditions governs the acceptability of the embedded factual statement over and above the acceptability of the protocol as such. For instance, we can distinguish between the acceptance of a statement "as a protocol statement" and the acceptance of a protocol statement as "valid." When we accept a protocol statement "as a protocol statement," we accept that the institutional-behavioral condition is fulfilled: somebody offered (or certifiably could do or have done so) the embedded statement as a scientific protocol statement. When we accept a protocol statement as "valid," all four conditions are fulfilled. Only if all four conditions are jointly satisfied — the institutional-behavioral, the belief-desire-psychological, the neurophysiological, and the "factual" condition — can the innermost embedded observation statement be accepted as stating a scientifically certified datum. (Typical reasons for rejection — if already accepted "as a protocol" — would be that the designated person never had the thought in question or was not stimulated in the appropriate fashion or that contrary descriptions of the observational state of affairs have already been accepted as "facts.")

So much for the "simple" theory of protocol statements. The "extended" theory not only concerns observational data but also serves to elucidate the process of testing scientific hypotheses and theories. The extension consists in the additional condition that certain intra- and extratheoretical conditions of coherence have to be fulfilled. To accept a protocol statement as "binding" means both accepting it as valid and being prepared to revise or reject not only previously accepted observational knowledge claims but also hypotheses and whole theories, if they are contradicted by the newly accepted valid protocol statement. A decision to cancel the presumption of counting accepted protocols as binding — as pointing to a scientific "anomaly," to speak with Kuhn — depends on pragmatic considerations: it can be taken, but it will not be taken lightly. Neurath's antifoundationalism runs deep indeed: the class of protocol statements gains its distinctive role not by reflecting "reality" in a particularly pure form without any reliance upon antecedent theory but because the acceptance of those statements recommends itself once certain conditions are fulfilled. These acceptance conditions define the role of observation statements in the language game of science (and so help to characterize the latter).

Within the framework outlined, the theory of protocol statements constitutes part of a theory of science as an in part discursive practice. It may be noted therefore that in Neurath we not only find anticipated ideas from Quine's later naturalistic epistemology but also find elements that transgress the limitations of the latter. As one instance of this, note that in Neurath's proposal we meet a basic element of the theories of science developed by Ludwik Fleck and Bruno Latour. With Latour, Neurath's proposal assumes that the recognized epistemic status of a scientific statement finds expression in the prefixes it carries (for example, "It has been claimed that . . . " — prefixes signaling qualifications of the knowledge claim that Latour calls "modalities"). Similarly, Neurath's proposal is consistent with Fleck's view that the most desirable epistemic status consists in carrying no such qualifications altogether (Neurath's embeddings are devices for the explication of the conditions of scientific practice, not prescriptions for linguistic reform).[7] Compared to the still common descriptions of logical positivism, particularly of its relation to observation as "unproblematical," this is surely a surprising result. We are prompted to ask how such "postpositivist anticipations" were possible. I will argue here that their attribution does not only meet the internal interpretative conditions of textual coherence but that Neurath's "anticipations" can also be explained from a historical perspective.

2. Neurath and the First Vienna Circle

Neurath's theory of protocol statements was the product of his engagement in the Vienna Circle, or what he would have called the "net result" of his debate with Carnap and Schlick. The conceptual basis for his underlying conception of science had already been established for a long time, however: Neurath's antifoundational holism and pragmatic naturalism found characteristic expression in his image of knowledge. Above I quoted "Neurath's Boat" of 1932, but this simile was first employed in Neurath's writings in 1913.

What then do we know about the early Neurath? What were the pressing philosophical problems that prompted him to adopt antifoundational holism and pragmatic naturalism as early as 1913? The first half of the development I ascribe to him begins with Mach and leads via the French conventionalists. That is a relatively traditional path for a member of the Vienna Circle, of course, yet in the Circle he remained relatively isolated as a naturalist. Where, when, and why did he begin to differ from what was to become the Viennese orthodoxy? Philipp Frank's remarks about Mach's legacy characterize succinctly the intellectual situation of the discussion group around Hans Hahn — the "first" Vienna Circle that Frank and Neurath belonged to (see Haller 1985) — and illumine those intellectual aspects that Neurath shared with them.

Let us consider first how they understood the Enlightenment, a tradition in which Frank located Mach and the first Vienna Circle. One of its "essential characteristics" was its "protest against the misuse of merely auxiliary concepts in general philosophical proofs" (P. Frank 1917 [1949], 73). Frank even discerned a "tragic feature" in the Enlightenment philosophy: "It destroys the old system of concepts, but while it is constructing a new system, it is already laying the foundation for new misuse. For there is no theory without auxiliary concepts, and every such concept is necessarily misused in the course of time" (ibid., 78). "Hence," Frank concluded, "in every period a new Enlightenment is required in order to abolish this misuse." Mach himself had seen his role to be not to fight "against the Enlightenment of the 18th century but rather to continue its work" (ibid., 73, 75). In this sense, it became the task of the first Vienna Circle in turn to study and criticize Mach and what his theories had become.

What then were Mach's "auxiliary concepts" that had to be reassessed? For the first Vienna Circle, they were primarily Mach's theory of elements, his apparent phenomenalism. What remained of his theory of science was his antimetaphysical thrust (which turned him into the patron saint of the second Vienna Circle). As Frank noted elsewhere, Mach's main mistake lay in his neglect of the logical problems that the mathematicization of physics had entailed: the reductionism of his theory of elements was not to be upheld (P. Frank 1941, 7). Yet that only revealed the following problem: "We admitted that the gap between the descriptions of facts and the general principles of science was not fully bridged by Mach, but we could not agree with Kant, who built this bridge by forms or patterns of experience that could not change with the advance of science" (P. Frank 1949b, 8). The first Circle was united in this conclusion. The problem that it faced was how to account for the creativity of scientific work without jeopardizing its empirical character. In the later Vienna Circle, there were, roughly speaking, two answers to this problem. The first was the conception of science as a set of sentences with a bipartite conception of single theories and a hierarchical structure for unified science (subsequently overthrown by the antipositivists). The second was a less clearly developed conception that focused on the scientific practice in the individual disciplines and in unified science. Neurath's Boat may serve as its guiding image.

Consider how the Circle proceeded, once Mach's mistakes had been diagnosed. For Frank, the confrontation between Mach and the French conventionalists ultimately led to what we now know as the "orthodox" conception of scientific theories. According to it, the formulation of scientific theories requires two languages, a so-called observation language in which observable spatio-temporal processes are recorded and a so-called theoretical language, which was not considered fully interpreted because only select terms of it were tied to the observational language by partial operational definitions (correspondence rules). This orthodox conception could, if one were so inclined, be used to support a moderate instrumentalism, an instrumentalism toward the theory proper, that is, toward the unification of experiential regularities that was formulated in the theoretical language. Observation statements would thus remain unaffected. The position expressed by Neurath's Boat entails a much more radical instrumentalism (or what appears to be such): it rules out what the orthodox position does not. In 1913 — as later in 1932 — Neurath denied that we could start from a "tabula rasa" upon which a "clean" structure of scientific concepts and hypotheses could be erected. There were no secure foundations to be had, neither descriptive nor normative, neither first- nor second-order ones. Thus there existed no observation statements that could not be doubted. Clearly, Neurath contradicted the orthodox Viennese solution to the problem of the creativity of scientific knowledge already in 1913. But what prompted him to do so at such an early date?

It has to be stressed that both approaches — the orthodox one as well as Neurath's — represent attempts to renew the Enlightenment ideas in Mach's epistemology, most notably the concern to render all scientific knowledge claims scrutable and rationally criticizable. The Viennese "orthodoxy" and Neurath were united in their Enlightenment ambition. What divided them was their estimation of the possibilities for a naturalistic theory of knowledge. At least in part, this difference in turn was founded in the different scientific interests of the theorists in question and reflects their different areas of expertise. To see this, we need only compare some champions of the orthodox conception with Neurath. For someone like Carnap, who came from the formal sciences (particularly the Fregean camp), any ultimately empirical strategy of explaining knowledge of his speciality was obviously suspect. But even for someone like Frank, who came from the natural sciences (especially the Machian camp), there was no need to experiment with radical naturalism: the all-important empirical scrutability of scientific theories was supported by their orthodox rational reconstruction. Neurath, by contrast, came from the social sciences, and this background rationalizes his distinctive epistemological stance. In the social sciences, theory construction alone can have causal consequences in its domain irrespective of whether what the theory said was true as antecedently observed (such that false or indeterminate knowledge claims could become the cause of their truth at a later date). Already in 1909 Neurath drew attention to the fact that monetary calculation viewed as a simple measure of "productivity" has the "strange" characteristic of influencing the production process itself (Neurath 1910, 602), and in 1911 he formulated the idea of self-fulfilling prophesies (Neurath 1911, 113; 1913b, 443). Obviously it became important for him to achieve an understanding of the actual processes of the formation of scientific concepts and theories. The sharp distinction between the contexts of discovery and justification and the dismissal of the former as irrelevant for the theory of science—which did not seem to affect metatheories like Carnap's or Frank's adversely—could not serve Neurath's interests as a social scientist.

Still another disadvantage of the orthodox view was that the distortions of the picture of scientific knowledge fostered by the sharp distinction of contexts are not conducive to fulfilling the expectations of the social-scientific Enlightenment. Neurath was attracted to the naturalistic conception of science not just by its promise to render transparent the pursuit of science generally—or at least to render it more transparent than it was — but specifically by its pragmatic potential to render more transparent the pursuit of social science and so to allow it to better realize its transformative potential. The point bears stressing, I believe, for it is here that Neurath's theory of science connects in a distinctive way with his personal biography: the systematic perspective of the former converges with his long-standing practical engagement for social transformation. As the son of a controversial protagonist of the turn-of-the-century Viennese Enlightenment (the economic reform theorist Wilhelm Neurath was well respected by Mach and the inventor and utopian economist Josef Popper-Lynkeus),[8] Otto Neurath identified with this tradition from early on.[9] Later, as the director of various museums and institutes of education, he considered it his Enlightenment task to make vital statistical information about economic and social matters available to the general public in a visually encoded, easily comprehensible form. Similarly, as a theorist of economic organization, he considered it to be his Enlightenment task to offer blueprints of alternative economic orders to the public, specifically the working class and its representatives, for discussion (and, if desired, implementation). Yet it is important to note that these activities did not exhaust Neurath's Enlightenment ambitions: he also regarded it as his task to develop a theory of science that made it possible to align scientific activity with the project of promoting social emancipation — without either calling the objectivity of science into question or foreshortening the Enlightenment project.

3. The First Boat: Naturalizing Conventionalism

The thesis then is this: Neurath's nonreductive naturalism represents his considered response to the epistemological concerns of the first Vienna Circle, given his distinctive academic focus on social science and personal roots in the Viennese Enlightenment. Neurath's "postpositivism" accordingly resulted from his pursuit of a radically antifoundationalist theory of scientific knowledge that seeks to make publicly available not only the conceptual tools for predicting and manipulating nature but also those required for the political and economic transformation of society.

Put in radical terms, the point of Neurath's naturalistic theory of science was to allow conventions to be viewed as natural. Despite this quasi-paradoxical formulation, the idea was not to view that which was determined by convention as determined by nature. The point was rather to consider recourse to conventional determinations to be natural for human cognition and thus to open up this dimension for scientific inquiry. Neurath was concerned to develop an alternative to the different neo-Kantianisms that sought to impose order upon the seeming caprice of scientific cognition by means of synthetic a prioris.

It would be no exaggeration to say that for the generation of scholars that preceded Neurath natural explanations and reference to conventions were viewed as sharp contrasts. For them, naturalism often meant the reduction of principles of ethical and social life to natural-scientific states of affairs, typically biological ones. For conventions to become natural in the sense intended by Neurath, a change in the concept of nature (among other things) was required. The obvious problem lay in how the intellectual productions of human beings were viewed: as something immaterial or as something that was part of the natural world. The key lay in how language was understood. Neurath's naturalism begins with the fact that for the symbolizing species "human being," the linguistic mediation of its cognitive activity is wholly natural, for it is altogether indispensable. In the absence of unconditional truths, what could be more "natural" for Neurath than to develop — in place of a philosophical epistemology — a theory that tries to understand what we call "scientific knowledge" as, in part, the product of historically developed and highly specific forms of human discourse? The need for conventions to order not only this type of discourse was a "natural fact" of the human world.

Speaking as generally of convention as I just did risks the ambiguity that, for instance, the French conventionalists may be viewed to be guilty of. Frank and Neurath viewed their conventions — especially those fixing "the general principles of science" — as conceptual determinations that, once made, were fundamental for a given conceptual scheme or system but that in and of themselves could have turned out differently: "[T]hey are free creations of the human mind" (P. Frank 1949b, 11–12). Yet whether these determinations were effected consciously or unconsciously (they just happened) was of no concern to the conventionalists. Neurath, however, attempted to overcome precisely this ambiguity. He was concerned to distinguish between the more or less unconsciously accepted conventions of social life and natural language and the conventions adopted for the production of scientific knowledge. Neurath strove for a conventionalism that would manage to secure the rationality of science on the basis of explicit, conscious decisions on the part of those who pursued science and who took an interest in it. In other words, Neurath tried to assign constitutive responsibility for the conduct of science to those who do science and for whom science is done. The pursuit of science was to be ordered by conventions, which were to be judged on the basis of pragmatic, functional criteria.

Consider now Neurath's first employment of his boat simile. Since its setting makes admirably clear not only the problematic to which it was addressed but also the continuity in Neurath's outlook from then onward, I will quote *in extenso* the relevant passage from an early, far-ranging essay on the formation of concepts, theories, and disciplines in social science (especially political science):

> Of course, someone could espouse the idea to suspend all political economists from their business, as was demanded by Kant for the field of philosophy in his *Prolegomena.* But this is not the way of science, much less, by the way, that of philosophy. It is always a question of the totality of problems with which we struggle throughout the centuries. We are never in the position to place certain indisputable sentences at the very top and then clearly and accurately display the whole chain of ideas, be it in logic or in physics, in biology or in philosophy. What is unsatisfactory seeps through the whole of the realm of ideas, it is detectable in the first premises as much as in the later ones. It is of no use to exercise the care of supposedly renouncing knowledge already gained in order to proceed from a tabula rasa and improve things henceforth, as Descartes had the audacity to try. Such attempts only end with rough masquerades of insight which tend to be worse than all that preceded them. We cannot but declare truthfully that the current state of knowledge has been presupposed and that we shall try to improve matters by making changes here and there. Our thinking is of necessity full of tradition, we are children of our time, even if we fight against it as we may; there are only ages which recognize this more clearly than others. What good did it do for Kant to try to tear himself away? Despite his eminent genius we epigones are often able to show how some of his trains of thought can only be explained by reference to the thinking of his contemporaries and elders, but impossibly so by reference to an unprejudiced view of the world. We are like sailors who are forced to totally reconstruct their boat on the open sea with beams they carry along, by replac-

> ing beam for beam and thus changing the form of the whole. Since they cannot land they are never able to pull apart the ship entirely in order to build it anew. The new ship emerges from the old through a process of continuous transformation. (Neurath 1913b, 456–57)

Neurath's antifoundationalism represents a response to the following problems of the theory of empirical science. Are there descriptive foundations of the first order? Are there unconditional values (or prescriptive foundations)? Are there any foundations of the second, metatheoretical order? Neurath's answer is a threefold no.

The normative antifoundationalism is perhaps most easily understood and accepted — if not by us, then at least by Neurath himself, who as a historian and economist had learned about the unscientific nature of absolute value judgments from Ferdinand Tönnies and Georg Simmel. (This does not affect the scientific standing of investigations into the psychosociological facts of belief in such absolute values, of course.) With this view, Neurath was again in clear agreement with the rest of the Vienna Circle. It was Neurath's descriptive antifoundationalism that marked his separate path of development. This antifoundationalism forbade first-order certainties of any sort and in all sciences. (This already meant a twofold extension of the "antifoundationalism" Neurath was familiar with from French conventionalists like Duhem.)

Yet Neurath's antifoundationalism also forbade all metatheoretical certainties. That in particular opened up possibilities for a "true" conventionalism (one that even the French conventionalists did not envisage). What remains is a thoroughgoing pragmatism. That is more than one might think, however: Neurath does not surrender to tradition, even though he would not dare deny its influence on himself. Quite the contrary, Neurath envisaged a conceptual voluntarism to redeem the promise of the Enlightenment work of Mach. Given the failure of Mach's reductionism that sought to found scientific knowledge on an ontologically neutral but quasi-individualist basis, Neurath in fact proposed to hand over the conceptual management of science to its practitioners. (Later it was this voluntarism that united Neurath and Carnap — until the contrast between the former's then fully fledged naturalism and the latter's purely logically oriented rational reconstructionism began to appear unbridgeable.) To bring out this trend in Neurath's thought more clearly, I will now consider Neurath's metatheoretical antifoundationalism and the suggestions for his conceptual voluntarism — his radical (or true) conventionalism — in greater detail.[10]

Neurath's conceptual voluntarism — the positive moral of Neurath's parable — represents an alternative solution to the problems raised by the overcoming of Mach's naturalism. Neurath's solution put conscious decisions in place of the biologizing explanation with the help of which the general, nonreducible principles of science tended to be "legitimated" at the time.[11] To show how the problem posed itself for Neurath, I must return to Mach before commenting about a problem of conventionalist metatheory.

According to Frank, Mach's lasting achievement was the rejection of the dichotomy of either renouncing the unification of science or introducing metaphysical concepts into the sciences. Mach "proclaimed, rather, the unification of science by means of the elimination of metaphysics" (P. Frank 1938 [1949], 89). The sciences could be unified only if they stuck with strictly positive knowledge. For Mach, not only the first-order sciences but also scientific metatheory had to offer such positive knowledge: the principle of economy of thought not only represented a normative desideratum of scientific theory construction but also constituted a basic fact of human concept formation. Language, for instance, was an economic tool that made possible the representation of the similarities that abstracted from the uniqueness of single experiences. Science was a late development of a type of intelligence that was functional for survival and like less theoretical thought had to be comprehended pragmatically.

Once Mach's reductionism was rejected, the question of the status of scientific metatheory arose. Consider the thesis of the unity of science and the principle of economy and their justification. For Mach, the unity of science was as good as a proven fact: his projected reduction of the content of scientific statements to the facts of perception supported, and was supported by, the very monistic viewpoint that found expression in his theory of (apersonal) elements. Yet the conventionalists Poincaré and Duhem put an end to that reductionism: How then were the thesis of the unity of science and the principle of economy to be justified? Merely by convenience? These questions did not seem to find satisfactory answers from the conventionalists.[12] Moreover, how "free" were such "creations"? Poincaré once remarked that the guide for the adoption of constitutive postulates in the sciences was not individual but rather "ancestral experience." It was "by natural selection [that] our mind has adapted itself to the conditions of the external world, that it has adopted the geometry most advantageous to the species; or in other words, the most convenient" (Poincaré 1902 [1946], 91). This, as well as talk of an "instinct of simplicity" (ibid. 139), suggests evolutionarily acquired dispositions as the ground of scientific conventions. For Duhem, moreover, the economy of thought meant different things for different kinds of thinkers. The desire for a systematic and abstract unification of physics (to which he and Poincaré had reduced the unity thesis) "results from an innate feeling of ours and cannot be justified by purely logical considerations" (Duhem 1906 [1962], 102). The relevant "type of mind" — no universalism here! — was determined by hereditary qualities. For Poincaré and Duhem, then, "biologically determined convention" was not a contradiction in terms, just as an "instinct of classification" was not for Ludwig Boltzmann (Boltzmann 1905) nor a "biologically determined synthetic a priori" for Simmel. For Simmel, the very dualism of noumenon and phenomenon was "resolved in that the forms of thought which create the world as concept are determined by the practical effects and countereffects fashioned by our mental constitution — nothing other than the body — according to its evolutionary needs" (Simmel 1895 [1982], 71).

Until about 1910, Neurath did not seem to have rejected such biologizing out of hand (Uebel 1993b). Yet by 1911, he refuted the argument that a monetary economy was superior to a moneyless economy — an argument others had based on the notion that "a people without money misses the standard by which its national wealth could be measured" and cannot feel the "stimulation . . . to increase [their] useable property." Neurath objected that "the monetary form of economical organization can only then be judged for its power of increasing wealth if one employs a method of calculating wealth which is independent of money" (Neurath 1911, 89–90). In particular, he argued that references to possibly biologically determined dispositions — which "natural" (as opposed to "cultured") peoples were said not to possess — could not decide how wealth was to be conceptualized in economics. Still more generally, he remarked in 1913:

> Some are of the opinion that to start with, one could reflect, and then when reflection fails, turn to instinct: this view misuses instinct by consciously introducing it as a mere stop-gap. . . . Precisely if one values the significance of instinctive action so highly one should not misuse it like that. One should clearly realise that instinct must fail with respect to the complex rational relationships created by the consciously shaped institutions of the social order and modern technology. (Neurath 1913a [1983], 5)

Conscious thought, which according to Mach emerged out of instinctual life, cannot hand back the reins to that from which it once wrested them. Much less, of course, could scientific or social conventions be defended by reference to instinct as rationally binding.

Against reliance on instinct Neurath bet on consciously determined convention. The conventionally determined domain of science included not only the so-called auxiliary concepts and generally agreed upon principles of explanation but, significantly, also what Neurath called "auxiliary motives." Two points must be made about the conception of auxiliary motives. First, its conative associations reflect a general feature of Neurath's holism, the close connection of theory and practice. As for Mach, so for Neurath: theoretical reason does not differ in principle from practical reason. (Neurath's fifth condition on adopting or rejecting a protocol as a control instance of a given theory — that the protocol be held to be binding — indicates that theory choice is a pragmatic decision.) The second point turns on what auxiliary motives are. We may understand them to be subsidiary epistemic goals or norms beside the truth and/or the predictive reliability of the theories concerned, namely, metatheoretical values like simplicity, far-reaching unification, and so on. This pragmatic conventionalism helps in deciding between different theories with more or less equal probability; for instance: an auxiliary motive is "an aid to the vacillating" (ibid., 4). We can choose between the different theories not only on the basis of their efficiency for predicting certain processes, namely, those that concern our overall cognitive aim, but also on the basis of their conformity to the standards set by the auxiliary motives.

Already in 1913 Neurath attacked as "pseudorationalist" the refusal to accept the view that it is only by including reflections about a candidate theory's contexts of use (intra- and extrascientifically) that rational theory choice can become possible. Accordingly, the economy of thought and the unity of science were not to be viewed as basic truths of nature, as it were, but as conventions, as consciously arrived at auxiliary motives. (Here it becomes plain why Neurath did not need to worry about the pluralism of his later encyclopedism as detrimental to the cognitive dignity of science.) How then were general metascientific principles to be justified? By reference to the function they were intended to fulfill and for which they were adopted by the deliberations of the "scientific collective." Neurath's conception took seriously the recognition of the fact of conceptual contingency and explored the possibilities of conceptual plasticity. Yet how was the reason of that "scientific collective" determined? Did the scrutability of science not find an inglorious end here? Once again the renewal of Mach's pragmatism played an important role. Neurath did not view science as an instrument for survival only in a biological and evolutionary, but especially in a historical and social, perspective. Principles like the unity of science were to be understood as adopted by a joint decision of the scientific collective. Such principles were not eternal laws scientific reason had to obey but perhaps only provisional rules scientific reason set for itself under specific historical conditions. Their justification lay in their instrumental power to yield results that answered to the cognitive interests of the collective of scientists. Set thus in a matrix of means-ends relations, the scrutability of metascientific principles is upheld, and convention is "naturalized."

4. The Socialization of Conceptual Voluntarism: Metatheory as Utopian Practice

From Neurath's first employment of his maritime simile and early pragmatic conventionalism there was still a long way to go to his discourse-theoretical proposals in the later Vienna Circle.[13] Of the contributing factors one may be briefly highlighted here, given not only its relevance to the political dimensions of Neurath's theory of (social) science but also its discourse-theoretical turn.[14] This particular impetus came from Otto Bauer after World War I at about the time of Neurath's return to Vienna. (As foreign minister of Austria's first Republic, Bauer was responsible for the early remission of Neurath's sentence for participation in the Munich *Räterepublik* and his deportation to Austria.) A leading figure in the Austrian socialist movement and (like Neurath) a proponent of socialization plans (albeit less radical ones), Bauer had important things to say about the problematic relation between the theories of Mach and Marx. This was an issue that Neurath could no longer neglect, given his recent turn to socialism. In a manuscript written while he was a prisoner of war and not published until 1924, but, to be sure, discussed with friends like Neurath after his return to Vienna, Bauer noted:

> The dissolution of the mechanistic conception of nature is completed by the epistemologies of the moderns, especially Mach's. But inasmuch as this dissolution has led us to place the systems based upon a mechanistical conception of nature in their social and historical context, it has led us beyond the moderns, even beyond Mach.... Marxism requires a theory of knowledge that Mach and Avenarius, Poincaré and James cannot provide. Such a theory would demonstrate in detail the procedures, the mental process, by means of which human beings create their worldview according to the example of their own labor, as a reflection of the social order in which they live, or according to the social order for which they fight, and according to

the requirements of their economic and social, political and national struggles. (Bauer 1924, 462f., 464)[15]

According to Bauer, any theory of knowledge that does not consider the sociopolitical dimension of knowledge production simply will not do. Neurath agreed: after his experiences as a "social engineer" in Munich, he became once more concerned to integrate into his theory of science the reflexivity of an investigator who realizes that he is "part of the social scene" investigated (Neurath 1931a [1973], 406).

Neurath's sketch of a theory of science fits the program laid out by Bauer. It sought to "demonstrate in detail" some of "the procedures . . . by means of which human beings create their worldview according to the example of their own labor, as a reflection of the social order in which they live, or according to the social order for which they fight." The gradual articulation of the discourse-theoretical model may be viewed as marking Neurath's gradual realization of concrete ways of "rationalizing" the processes of knowledge production such that it retained its scrutability while allowing for the possibility of interventions that restructured its constitutive mechanisms. Neurath's discourse-theoretic turn was not merely descriptive but also designed to provide a model for an alternative practice of science.

It is highly significant then that Neurath, who was so concerned to integrate sociology into his theory of science, did not use the expression "scientific community" but preferred "republic of scholars."[16] More than the Enlightenment associations of "the republic of letters" are at issue here (though they were surely not unwelcome).[17] As a social scientist whose mentor was Ferdinand Tönnies — author of the seminal *Gemeinschaft und Gesellschaft* (Tönnies 1887)[18] — Neurath was long familiar with the distinction of community and society, and this becomes clear in his choice of words. Neurath's project was to comprehend scientific rationality as a social practice and form of discourse that was distinctive in that it was constituted by certain rules concerning intersubjective control that could be examined by all. Scientific rationality was *not* to be viewed as given or intuitive, like the rules of a form of discourse traditionally legitimated by the forms of communitarian life. Only the public constitution of scientific rationality promised to create the intellectual space for intersubjectively intelligible interventions in the knowledge production process, and only a theory of science that accounted for such a public constitution could serve as the "superstructure of the great revolutionary transformations of the social and economic order" (Neurath 1932a [1981], 576) that Neurath and Bauer sought to promote.

Marx's "Theses on Feuerbach" may be recalled in this context as well. In the years of the Kautskyan orthodoxy of the German socialist movement, these were virtually the last remains of the young Marx and so of prime importance for voluntarist opponents (like Neurath)[19] to the mechanist interpretation of historical materialism. After exhorting readers to comprehend "reality" as a "sensuous, human activity" and "practice"—and to comprehend human activity as "object-directed activity" — Marx considers the issue of the "education of the educators": "The coincidence of the changing of circumstances and of human activity or the changing of self can only be grasped and rationally comprehended as revolutionizing practice" (Marx 1845, 3d thesis). The standpoint of this "new materialism" is no longer the individualism of bourgeois society, Marx then remarks, but "human society or social humanity" (ibid., 10th thesis). The new materialism required also—here we need now only recall Bauer's prescription—the overcoming of that individualism in epistemology. Neurath responded with his discourse-theoretical model and his theory of protocols: scientific observation was analyzed as a social practice of a specific form. Intuitive communitarian "understanding" in science was tolerated only as long as explicit and intersubjectively intelligible legitimations for the specific forms of the practice in question were given (or could be provided in principle at least).

At this point it may be asked whether Neurath's theory of science can still qualify as naturalistic. To answer this it is first helpful to note what it means to understand scientific rationality as a form of social discourse that is constituted publicly by means of agreed upon rules. It not only means holding that the forms of a social discourse like science are subject to public scrutiny and legitimation in principle. It also means viewing scientific rationality as constructible in practice — indeed, it means assigning to the scientists themselves the task of constructing a "controllable rationality." But is that the historically given practice of science? It seems that with Neurath's socialized conceptual voluntarism we have been led to epistemic utopianism. That Neurath's projected theory of science presents a somewhat utopian conception of science does not disqualify it as non-naturalistic, however. To see this we must consider Neurath's understanding of utopianism.

We may again begin with Marx (though, as indicated, Neurath's deepest intellectual roots lie elsewhere). Human practice can become revolutionary—so we can conclude from the "Theses on Feuerbach"—when it is a reflexive social enterprise. The reflexivity of revolutionary practice requires new concepts and conceptions, of course, and it is just here that utopianism enters the picture. Neurath's utopias are possibilities conceived in thought; he describes as utopias "all orders of life which exist only in thought and image but not in reality" (Neurath 1919a [1973], 151). Not any old alternative ideas are of interest to him, however: "Through the unscientific and unhistorical utopianism we reached the antiutopian historicism, which is now being replaced by a scientific and historically grounded utopianism" (Neurath 1922, 55). Utopias thus are twice presupposed by the project of socioeconomic transformation: once as pictures of possible organizations of society and once as pictures of forms of science that are relied on in promoting such processes of transformation. Neurath's theory of science, it follows, is utopian insofar as it is revolutionary and aims to contribute to the "changing of self." Neurath's epistemic utopianism may be viewed in analogy to his economic utopianism.[20] His conception of science as a social practice whose form is determined by its practitioners is similar to his schemes for the total socialization of a nation's economy in this respect: both represent "drafts" for the purpose of discussion, "sketches" for the conscious reorganization of conceptual fields and schemes, namely, those around the notion of economic value or scientific knowledge.[21] Neurath's epistemic voluntarism aimed not only to legitimate conceptual and methodological innovation but also to master the tradition as much as it ever could be: staying with traditional ways required justification as much as embarking on new ones. Instead of contradicting naturalism's concern with practice, Neurath's utopian theory of science draws out the critical potential that a naturalistic and pragmatic understanding of the scientific enterprise possesses.

Yet here another worry arises. We may applaud Neurath's Enlightenment ambition, but is his epistemic utopianism not a feature of his thought that, unlike his "postpositivist" theory of protocols, marks him as a characteristically modernist progressivist? Neurath would be the last to deny that people do science mostly as Marx once remarked they make history, namely, "under circumstances not of their own choosing" (Marx 1852). What prevents the discourse of science from being "free of domination" (to use an anachronistic term) was an issue for the sociology of science to investigate. In careful moments, moreover, Neurath denied the intrinsic teleology that his economic models seemed to share with less enlightened versions of historical materialism.[22] The same reasoning applies to his epistemic utopias. To begin with, viewed as drafts for discussion, his pictures of possibilities were highly unlikely to be implemented as first put forth; moreover, their goals had to be agreed upon by the majority concerned. Appropriately enough, Neurath entitled his later picture-statistical universal history *Modern Man in the Making*

(Neurath 1939) — given Neurath's voluntarist outlook, the outcome of the struggle was not assured.

Neurath's view can be further illustrated by comparison with two fellow "utopians" from the Bavarian revolution. Karl Ballod, once a teacher of Neurath's in Berlin and later his collaborator in the Munich experiment, expressed the idea of scientific utopianism in terms of the creation of a "synthesis . . . from already known and individually used elements that during the reign of the capitalist system could not have been combined in one structure for it would have run counter to the interests of ruling circles" (Ballod 1919, 6). For Gustav Landauer, the visionary anarchist writer murdered when the *Räterepublik* was suppressed, by contrast, a utopia is a state of affairs that is permanently "not yet" realized. It is the ever frustrated idea of a better future and is to be contrasted with — and viewed as a radical alternative to — what Landauer called "topias" (*Topien*), more or less stable social realities. "The utopia thus does not become an external reality at all, and the revolution is simply the age of transformation from one topia to another" (Landauer 1907 [1974], 17). Neurath's utopianism may be viewed as combining elements of both conceptions. There is, on the one hand, his dour resolve to put the recent experience with large-scale war economies to practical use: the seemingly organic market structures of capitalism had been pulverized and broken up into their elements whose operation was now to be centrally reorganized. On the other hand, Landauer's notion of utopia as something never fully realized connects well with Neurath's understanding of the forever unfinished Enlightenment task.[23]

Neurath's resolve to be a utopian theorist of science and his economic plans show a similar rootedness in experience *and* openness toward future development. His "scientific world-conception" was an attitude that rejected closure: "It will be interesting when we new critics of language will be criticized with the tools we have devised" (Neurath 1941 [1983], 215). So even if the scheme of science being determined by its practitioners, say, was never to be realized, it need not remain without consequences altogether: the very discussion of the possibility of alternatives can reshape an entire intellectual landscape. Just as economic utopias were intended to "serve reality" (Neurath 1919b, 138), so the scientific world-conception was to "serve life" (Carnap, Hahn, and Neurath 1929 [1973], 318). In either domain, Neurath noted, utopias "can preserve us from many false steps and make the mind flexible and free it from accidental notions." Significantly, he added: "One fully recognizes the real only when one surveys the possible as well" (Neurath 1919a [1973], 154–55).

5. Conclusion

In "The Unity of Science as a Task," his aptly titled keynote address to the Prague preparatory conference for the First Congress for the Unity of Science in Paris 1935, Neurath claimed: "We advocates of 'scientistism'... try, through planned debates, to overcome old obscurities in our 'logical empiricism' [and] to detect new ones, in order to start new clarification" (Neurath 1935 [1983], 115). As I hope I have shown, it is not only Neurath's self-reflexive attitude as a mature metatheorist, or his post–World War I understanding of the development of utopianism, but also the Vienna Circle's pre–World War I conception of the Enlightenment task that could be called "dialectical": all scientific conceptions, especially auxiliary concepts and motives that once were liberating, must be questioned anew by every generation of thinkers — only then could enlightenment be hoped for. (I can only note in passing that the lasting message of Horkheimer and Adorno's World War II *Dialectic of Enlightenment* shows a remarkable convergence with this view.)[24] Neurath's discourse-theoretical approach to the problem of the empirical basis of science accordingly represents his contribution to the renewal of Enlightenment reason, his *Aufhebung* of Mach's naturalism in a nonreductive context.[25] Thinking of science as "revolutionizing practice" meant developing a metatheory that not only reflected existing scientific practice but also constituted an intervention toward reorganizing it. Before World War I, Neurath set the "culturalization" of cognitive principles against apparently biologically grounded conventions. After World War I, ultimately against increasingly abstract formalist reconstructions of scientific thinking, he propagated the concrete "socialization" of metatheoretical conventionalism. Throughout, Neurath's engagements throw into distinctively activist relief an Enlightenment concern shared by several of his Vienna Circle colleagues. This was the concern to ensure the "controllable rationality" of science — a notion that itself partly reflects the practice and partly the utopia of scientific thought.

Notes

1. For a discussion of the interpretative tradition of Neurath as a naturalist — initiated by Haller 1979, 1982 — see Uebel 1991.

2. Neurath's private language argument gives fitting evidence of and back-up for his naturalistic attitude; see Uebel 1992a, 1992b (chaps. 6–7, 10), and 1992c.

3. What in Neurath may be mistaken for a coherence theory of truth is, if anything, a coherence theory of justification — and of claim acceptance, not merely of a belief (see Uebel 1991, 1992b, chaps. 6 and 12).

4. See Uebel 1991, 1992b, chap. 12. Relevant paths have been explored in Dewey 1939b and Feigl 1950.

5. For a conception of an "image of science/knowledge" making explicit still more dimensions than my application plays on, see Elkana 1981.

6. In a later paper, Neurath splits into two the innermost sentence in the example quoted (the one in parentheses); see Uebel 1992b, chap. 11, and 1993a for detailed discussions of Neurath's theory of protocols.

7. Fleck 1935, Latour 1987. Latour, of course, tends to concentrate on the investigation of the interactions and reconfigurations of the social structures within and without the scientific community that bring about the change in status of a scientific statement. (For some remarks on the absence of a direct connection despite the parallelism between some of Fleck's and Neurath's views, see Uebel 1991.)

8. On the relationship between Wilhelm Neurath and Otto Neurath, see Uebel 1993b and 1995.

9. On Neurath's Enlightenment orientation in general, see Haller 1979 and Hegselmann 1979; on the Viennese Enlightenment ambience (which Neurath draws on), see Stadler 1982a.

10. On Neurath's arguments for descriptive and normative antifoundationalism, see part 2 in Cartwright et al. 1996.

11. It may be noted that Schlick also once briefly dealt with the biologistic strategy (Schlick 1915 [1979], 169–70).

12. I must sidestep here the related question of whether the first Circle and Neurath in particular noted what recently has been called Poincaré's "structural realism" (Worrall 1989). As with the case of Wittgenstein and the second Circle, the question is not one of straightforward exegesis but of how a doctrine was understood by its recipients. To judge from some oblique remarks (e.g., Neurath 1915 [1973], 102–3, 112; 1916 [1983], 28–29), it would seem that if it was noticed at all, Poincaré's structuralism was understood by Neurath in an antirealist fashion. In any case, antirealism is supported by Duhem's admission that the system of natural classification that theory seemingly anticipates lies beyond the bounds of what science is able to establish (Duhem 1906 [1962], 27).

13. For the reasoning that went into Neurath's "second Boat" of 1919/21, see part 2 of Cartwright et al. 1996.

14. For the discussion of another instance of the interrelation of Marxist debates and epistemological issues in Neurath's thought, see the contribution by Cartwright and Cat in this volume and part 3 of Cartwright et al. 1996.

15. Neurath refers to Bauer's essay — alongside his own *Lebensgestaltung und Klassenkampf* of 1928 and Zilsel's "Philosophische Bemerkungen" in *Der Kampf* of 1929 — in the annotated bibliography appended to Carnap, Hahn, and Neurath 1929 under the heading "Sociological Basis of the Development of the [Scientific] World-Conception" (reprinted only in 1981, 315ff.).

16. While the phrase "republic of the sciences" made a relatively late appearance in Neurath's publications (e.g., 1944, 13), *Gelehrtenrepublik* had long been used in his correspondence, for instance, with Carnap throughout the 1930s (see the Carnap Papers, Hillman Library, University of Pittsburgh).

17. On these terms' Enlightenment history, see Daston 1991.

18. Having first met Tönnies at a summer academy in Salzburg in 1903, Neurath corresponded intensely with him until the final years of World War I; for more on Tönnies's influence on Neurath, see part 2 of Cartwright et al. 1996.

19. See, for example, the opening passage of Neurath 1920a.

20. On this point, see also Nemeth (1982a, 1982b, 1994), who has long stressed Neurath's "orientation towards a future practice" (Nemeth 1981, 199).

21. Neurath characterizes his socialization plans thus in 1920b, 227. For the relevance of the notion of planning to Neurath's later encyclopedia project, see Reisch 1994.

22. Compare the sober final passages of Neurath 1920a (in Weber's *Archiv*) with the idealistic tone of the end of 1920c (a popular brochure).

23. This illustration is not at all meant to suggest full agreement with either of these utopians: Neurath is likely to have been rather skeptical of Landauer's underlying mysticism, for instance.

24. See Wellmer 1985 on Horkheimer and Adorno 1955, but note also M. Frank 1983 for late nineteenth-century anticipations of a related perspective in Nietzsche — a notable convergence with P. Frank's interpretation (1917) — and Schnädelbach 1979 for anticipations of this reflexive understanding in the critical contemporary reception of the German Enlightenment philosophy.

25. Koppelberg 1987 introduced dialectical terminology into Vienna Circle scholarship. For him Quine represents the *Aufhebung* or sublation of the antithesis of Carnapian formalism and Neurathian naturalism. The dialectic I project here is different: Neurath represents the *Aufhebung* of the antithesis of naturalistic reductionism and antireductionist conventionalism.

PART II

SCIENCE, PHILOSOPHY, AND SCIENTIFIC PHILOSOPHY

Don Howard

Relativity, Eindeutigkeit, *and Monomorphism: Rudolf Carnap and the Development of the Categoricity Concept in Formal Semantics*

The following essay treats several topics. In part, it is concerned with a crucial moment in the development of Rudolf Carnap's constructionist program, which culminated in the publication of *Der logische Aufbau der Welt* (Carnap 1928a). This moment is Carnap's elucidation of the notion of "monomorphic" axiom systems and concepts in his 1927 essay, "Eigentliche und uneigentliche Begriffe" (Carnap 1927), a monomorphic axiom system (and, by extension, a monomorphic concept) being one that determines the class of its possible models up to the point of isomorphism; this is the property of an axiom system that we now standardly designate as "categoricity." In part, therefore, this essay is also concerned with the history of the concept of categoricity in formal semantics. And, finally, it is concerned, in part, with tracing one of the many ways in which Albert Einstein's views on methodological problems arising in conjunction with the development of relativity theory, especially general relativity, influenced the later development of logic, the foundations of mathematics, and the philosophy of science in the 1910s, 1920s, and 1930s.

These three topics are interrelated. Take monomorphic concepts and general relativity. We can understand Carnap's elucidation of the notion of monomorphic concepts in 1927 only if we first understand the related problem of *Eindeutigkeit* (univocalness) as it arose in the context of the so-called *Lochbetrachtung* (hole argument) in general relativity around 1915. It was Einstein's obsession with the hole argument that moved the problem of *Eindeutigkeit* to the forefront of discussions in the foundations of mathematics and general methodology in the late 1910s and early to mid-1920s, and it was his peculiar resolution of the hole argument that pointed the way toward the essential clarification of the concept of *Eindeutigkeit* that we find embodied in Carnap's definition of monomorphic concepts.

Or take categoricity and general relativity. Recognizing the role of *Eindeutigkeit,* the hole argument, and relativity in the background to

Carnap's elucidation of the concept of monomorphism suggests a very different way of telling the history of the categoricity concept than what is standard in those histories that trace it back primarily to David Hilbert's *Grundlagen der Geometrie* (Hilbert 1899), Oswald Veblen's meditations thereon (Veblen 1904), and the issues in the foundations of set theory that were brought to a head in the mid-1920s by the Lowenheim-Skolem theorem. This alternative history will also associate the development of the categoricity concept with the Cauchy problem and with the problem of correctly defining the abstract topological manifold, both of which underwent essential clarifications in the course of various thinkers' working through the problem of the hole argument in general relativity.

Or consider, finally, monomorphism, categoricity, and the history of the philosophy of science. Carnap was not alone among philosophers of science in the late 1920s in feeling the need to get clear about the conditions under which a concept, a system of concepts, or a whole theory can be unambiguously associated with a unique model or an isomorphic class of models. Various influences led a whole generation of thinkers to worry about whether or not concepts and theories had within themselves the capacity to pick out for themselves a unique model or picture of the reality they pretend to describe. Hermann Weyl and Ernst Cassirer will provide us with examples of two other thinkers from very different philosophical traditions who shared this concern with Carnap. Simplistic histories of the Vienna Circle and logical empiricism, which systematically misinterpret the antimetaphysical manifestos of the movement's founders, have blinded us to the genuinely semantic and ontological concerns that were driving Carnap in his thinking about the constructionist project during the 1920s. Understanding those concerns and their place in a broader intellectual context will contribute to our appreciating the aim of the *Aufbau* project and its relations with other, contemporary movements, the most important of them being Marburg neo-Kantianism.

1. Relativity and *Eindeutigkeit:* The *Lochbetrachtung* and the Point-Coincidence Argument

The *Lochbetrachtung,* or "hole argument," first made its appearance in late 1913 or early 1914 in the context of Einstein's struggle to formulate a theory of gravitation that would also be a generalization of the special theory of relativity. At the time, Einstein had wrongly convinced himself that the gravitational field equations would have to have somewhat restricted covariance properties. There were technical reasons for

Einstein's confusion, having to do, among other things, with whether or not the field equations would yield the correct Newtonian limit. But Einstein also rationalized the restricted covariance of what has come to be called the *Entwurf* theory (after the title of the paper in which it was first propounded, Einstein and Grossmann 1913) by inventing the *Lochbetrachtung* or "hole argument," which, he claimed, proved that no fully generally covariant theory of gravitation would be acceptable because it would necessarily violate a general methodological principle, what was then known as "Das Gesetz der Eindeutigkeit" (the law of univocalness).

According to the hole argument, any generally covariant set of field equations (plus boundary conditions) supposedly will fail to have a unique solution for "holes" in space-time, regions devoid of matter and energy. By means of an ingenious construction, Einstein showed how, given fully generally covariant field equations and a solution, $g_{uv}(x)$, of those equations inside the hole, one could transform that solution into another solution, $g'_{uv}(x)$, that was different from $g_{uv}(x)$ in the sense that, to one and the same point x of the manifold, $g_{uv}(x)$ and $g'_{uv}(x)$ assign different values of the metric tensor.[1] Here is how Einstein represented the conclusion of the hole argument in one of his last published presentations of it:

> *Events* [das Geschehen] *in the gravitational field cannot be determined univocally* [eindeutig festgelegt] *by means of generally covariant differential equations for the gravitational field.*
>
> If we demand, therefore, that the course of events [*der Ablauf des Geschehens*] in the gravitational field be completely determined [*vollständig bestimmt*] by means of the laws that are to be established, then we are obliged to restrict the choice of the coordinate system. (Einstein 1914, 1067)

If correct, Einstein's hole argument would indeed have been a fatal objection to the ideal of fully generally covariant gravitational field equations. And so, for two years, Einstein continued to refine the less than fully generally covariant *Entwurf* theory, carefully defining exactly what kinds of covariance properties it did possess and attempting to show that such restricted covariance was, after all, covariance enough. But late in 1915, Einstein finally discovered his mistake regarding the Newtonian limit. He quickly reasserted the fully generally covariant field equations that he had written down but rejected two years earlier. What to do, however, about the hole argument, which pretended to demonstrate the unacceptability of such a theory?

After some careful thought, Einstein discovered (to his great relief) that the hole argument was itself flawed.[2] The mistake turned upon the way in which one individuates the points of the space-time manifold. At the time of the hole argument, Einstein wrongly assumed that the assignment of a coordinate map was sufficient for individuating the points of the manifold, which is to say, sufficient for endowing the points of the manifold with identities invariant under the class of general relativistic transformations. Now he realized that since a coordinatization was not, itself, an invariant, neither could point-identity via coordinatization be invariant. An invariant sense of point-identity would be forthcoming only from a way of defining and individuating the points of the manifold that assumes an invariant construction. In what has come to be known as the "point-coincidence" argument, Einstein chose the intersection or "coincidence" of world-lines as his invariant method for defining and individuating the points of the manifold.[3] This is how Einstein interpreted his new point of view about individuation in a letter to Paul Ehrenfest on 26 December 1915, one month after his publication of the final, fully generally covariant version of the field equations:

> In §12 of my work of last year, everything is correct (in the first three paragraphs) up to that which is printed with emphasis at the end of the third paragraph. From the fact that the two systems $G(x)$ and $G'(x)$, referred to the same reference system [the same x], satisfy the conditions of the grav. field, no contradiction follows with the univocalness of events [*Eindeutigkeit des Geschehens*]. That which was apparently compelling in these reflections founders immediately, if one considers that (1) the reference system signifies nothing real, (2) that the (simultaneous) realization of two different g-systems (or better, two different fields) in the same region of the continuum is impossible according to the nature of the theory.
>
> In place of §12, the following reflections must appear. The physically real in the world of events [*Weltgeschehen*] (in contrast to that which is dependent upon the choice of a reference system) consists *in spatio-temporal coincidences.** Real are, e.g., the intersections of two different world-lines, or the statement that they *do not* intersect. Those statements that refer to the physically real therefore do not founder on any univocal [*eindeutige*] coordinate transformation. If two systems of the g_{uv} (or in general the variables employed in the description of the world) are so created that one can obtain the second from the first through mere space-time transformation, then they are completely equivalent [*gleichbedeutend*]. For they have all spatio-temporal point-coincidences in common, i.e., everything that is observable.

These reflections show at the same time how natural the demand for general covariance is.

*and in nothing else! (EA [Einstein Archive] 9-363).

A few days later, on 3 January 1916, Einstein said much the same in a letter to Michele Besso:

> In the *Lochbetrachtung,* everything was correct up to the final conclusion. There is no physical content in the existence of two different solutions $G(x)$ and $G'(x)$ with reference to the *same* coordinate system K. Attributing two different solutions to the same manifold is senseless, and the system K has, indeed, no physical reality. The following consideration takes the place of the *Lochbetrachtung.* Physically *real* is nothing but the totality of spatio-temporal point-coincidences. If, e.g., physical events [*das physikalische Geschehen*] were to be built up solely out of the movements of material points, then the meetings of the points, i.e., the points of intersection of their world-lines, would be the only reality, i.e., observable in principle. These points of intersection are naturally preserved under all transformations (and no new ones are added), if only certain uniqueness conditions [*Eindeutigkeitsbedingungen*] are maintained. Thus, it is most natural to demand of the laws that they do not determine *more* than the totality of the spatio-temporal coincidences. According to what has been said, this is already achieved by generally covariant equations. (EA 7-272; Speziali 1972, 63–64)

One should note that in both the hole argument and the point-coincidence argument, Einstein implicitly invokes one and the same methodological principle, which is that a necessary condition to be fulfilled by any acceptable theory is that it uniquely or univocally determine the reality it aims to describe. In the case of a theory of gravitation, that reality is the metric tensor defined over a manifold of space-time points. In the hole argument, the generally covariant form of the theory was faulted (wrongly) for its failure thus to determine physical events [*das physikalische Geschehen*] univocally. In the point-coincidence argument, the now generally covariant form of the theory is praised (rightly) precisely for its success in so determining univocally the course of physical events as determined by the field equations' assignment of a value of the metric tensor to every point of the space-time manifold. In both cases, as it turned out, the root problem concerned whether or not the space-time points themselves were individuated in a univocal way. The lesson of the point-coincidence argument is that this is possible only if we individuate the points of the manifold in an invariant fashion as points of intersection of world-lines.

What is the principle that Einstein is invoking here? From a mathematical point of view, he is at least concerned with the Cauchy problem, for the question is, in part, whether or not a set of partial differential equations — the gravitational field equations — plus boundary conditions uniquely determine a solution in a given domain. And the resolution of the hole argument forced mathematicians, led by Hilbert, to an essential clarification of what it means to have a well-posed Cauchy problem. For what Hilbert pointed out in discussing this problem in the second of his two papers on general relativity is that we can no longer demand simple uniqueness of the solution we are seeking, but at most uniqueness up to the point of isomorphism, Hilbert's post-November 1915 gloss on the two "different" solutions of the field equations, $g_{uv}(x)$ and $g'_{uv}(x)$, being that they are isomorphic images of one another under the transformation that generates the latter from the former (Hilbert 1916, 59–63).

Einstein's approach to the point-coincidence argument takes us, however, beyond the question of the uniqueness of the solutions of the field equations to the deeper question of the univocal individuation of the points of the manifold on which those solutions are defined. Indeed, Einstein is demanding that reality itself, inasmuch as it is represented to us via the field equations of general relativity, be univocally determined by the theory. Hence his insistence, in the above-quoted letter to Ehrenfest, that in a correct, generally covariant theory of gravitation,

> the physically real in the world of events [*Weltgeschehen*] (in contrast to that which is dependent upon the choice of a reference system) consists in spatio-temporal coincidences.* Real are, e.g., the intersections of two different world-lines, or the statement that they do not intersect. Those statements which refer to the physically real therefore do not founder on any univocal [*eindeutige*] coordinate transformation.
>
> *and in nothing else! (EA 9-363).

The reality described by general relativity is to consist of space-time points defined as spatio-temporal coincidences (and the invariant geometrical structures defined on those points — the metric tensor and various other structures constructed out of it) because that is what is determined univocally by the theory.

In his letters to both Ehrenfest and Besso, Einstein linked the univocal determination of space-time points defined as coincidences with questions of observability. He did the same thing in a famous and oft-quoted passage from his well-known 1916 review article, "Die Grundlage der allgemeinen Relativitätstheorie":

> All our space-time verifications [*Konstatierungen*] invariably amount to a determination of space-time coincidences. If, for example, events consisted merely in the motion of material points, then ultimately nothing would be observable but the meetings of two or more of these points. Moreover, the results of our measurings are nothing but verifications of such meetings of the material points of our measuring instruments with other material points, coincidences between the hands of a clock and points on the clock dial, and observed point-events happening at the same place at the same time. (Einstein 1916a, 776)

Such remarks have given rise to considerable confusion among later scholars who, unaware of the background in the hole and point-coincidence arguments and misled by the persuasive misreadings of early logical empiricists eager to find heroic precursors of their own views, read into these passages anticipations of a verificationist criterion of meaningfulness or at least a demand for observability.[4]

Similar confusions have arisen from the fact that Einstein often chose to characterize the lesson of the point-coincidence argument as being that "time and space thereby lose the last vestige of physical reality" (Einstein to Schlick, 14 December 1915, EA 21-611; see also Einstein 1915, 831; Einstein 1916a, 776; and numerous other similar remarks). It is easy to read such remarks as expressions of an antimetaphysical positivism. Of course what Einstein really means is not an endorsement of some kind of antirealism, but rather a more focused point about what it is that we can and cannot regard as real. To say that space and time are not physically real is not to say that nothing is. On the contrary, Einstein says explicitly that something is real, namely, space-time events construed as the coincidences of world-lines or geodesics. Why are these the elements of our fundamental ontology rather than space and time as traditionally conceived (or even Minkowskian space-time, if this is characterized via coordinate charts)? Because it is only points individuated as coincidences of world-lines, and not points individuated via space-time coordinatizations, that are determined univocally by the field equations.

Thus, the issue is not primarily an epistemological question about observability, nor is it a question of some kind of blanket antirealism. In fact, as Einstein himself makes quite clear, the issue is frankly ontological: Does the theory determine univocally a model of the reality it pretends to describe?

This sounds awfully much like what we now call categoricity, a concept that had first been given its name in 1904 by Veblen (following a suggestion from John Dewey), who was himself thinking about Hilbert's discussion in the *Grundlagen der Geometrie* of the question whether or not a given set of axioms univocally determines the point-structure of a line (see Veblen 1904; see also Huntington 1902). But Hilbert and Veblen were not Einstein's source for the principle of univocalness. For not only is there no evidence of Einstein's having read the relevant literature in the foundations of geometry by the early 1910s; there is also the fact that the particular vocabulary that Einstein employs is to be found in an entirely different literature, one almost entirely forgotten today.

2. The Methodological Background: Petzoldt and Schlick

The methodological principle that Einstein was invoking in both the hole argument and the point-coincidence argument was well known by the early 1910s under the name that had first been given to it by the minor positivist Joseph Petzoldt in an 1895 essay bearing the name as its title: "Das Gesetz der Eindeutigkeit" (the law of univocalness) (Petzoldt 1895). Petzoldt's original problem had been to make sense of the neo-Humean conception of causality first introduced by Ernst Mach in his essay on the conservation of energy (Mach 1872) and elaborated in the first two editions of the *Mechanik* (Mach 1883, 1888) and to defend this view against criticisms advanced by Wilhelm Wundt (Wundt 1894). Petzoldt proposed that all causal relations could be reduced to the quasi-mathematical notion of a many-one functional dependence, or "eindeutige Funktion" (Petzoldt 1895, 161–62, 171). Petzoldt went on in this essay to identify a number of other contexts in which a principle of *Eindeutigkeit* (or univocalness) is to be encountered, including least action or variational principles (185–88), the law of inertia (188–94), and the logical principles of identity and contradiction (199–200).

Petzoldt refined and developed the law of univocalness in many of his later writings. Thus, for example, the whole of chapter 3 of volume 1 of his *Einführung in der Philosophie der reinen Erfahrung* (Petzoldt 1900, 1904) — an "introduction" to Richard Avenarius's notoriously difficult *Kritik der reinen Erfahrung* (Avenarius 1888, 1890) — is devoted to an exposition of "Das Gesetz der Eindeutigkeit," adding to the earlier discussion a helpful new characterization:

> For every process means of determination may be found that determine it univocally [*eindeutig*], in such a manner that for every variation in this process that one can conceive as being determined by the same means, one can find at least one other variation that would then be determined in the same way as, and thereby be equivalent to, the first, and thus would have, as it were, the same *right* to be realized as the first. We shall designate this principle as the law of univocalness [*Eindeutigkeit*]. (Petzoldt 1900, p. 39)

The point here is to stress the univocal character of events and processes in nature, the lack of any ambiguity in the determination of real events, which is the feature of nature that Petzoldt had in mind when, earlier, he appropriated Mach's motto, "Die Natur ist nur *einmal* da" (Nature is given only *once*) (Mach 1883, 455; Petzoldt 1890, 57; 1895, 171). The example that Petzoldt now gives involves a ball situated on a smooth, flat surface that is struck by another ball. Which way does the first ball move? In the same direction as the ball that struck it, because if we were to consider any alternate trajectory, say ten degrees to the right, there would be another, equivalent trajectory — in this case ten degrees to the left — with no "means of determination" favoring the one over the other. Only for the first trajectory — the same direction as the incident ball — is there no such lack of univocal determination.

Petzoldt's ideas about the *Eindeutigkeit* principle soon came to be well known in wide philosophical circles, thanks in no small part to Mach's many flattering citations of Petzoldt on *Eindeutigkeit* in all editions of the *Mechanik* starting with the third (Mach 1897), in the *Wärmelehre* (Mach 1896), in the second and later editions of the *Analyse der Empfindungen* (Mach 1900), and in *Erkenntnis und Irrtum* (Mach 1905).

Einstein no doubt came across these references to Petzoldt when he read the *Mechanik* and the *Wärmelehre*,[5] but more to the point is what he likely read about *Eindeutigkeit* and relativity in various writings that Petzoldt devoted to the topic at the very time that Einstein was struggling with the problems of general relativity. These include the talk that Petzoldt gave to the Deutsche Physikalische Gesellschaft on 8 November 1912, "Die Relativitätstheorie im erkenntnistheoretischen Zusammenhange des relativistischen Positivismus" (Petzoldt 1912b), the much revised second edition of one of his major works in epistemology, *Das Weltproblem vom Standpunkte des relativistischen Positivismus aus, historisch-kritisch dargestellt* (Petzoldt 1912a); or the paper that he published in an early issue of the *Zeitschrift für positivistische Philosophie*, "Die Relativitätstheorie der Physik" (Petzoldt 1914).[6]

At this time Petzoldt was developing a point of view called "relativistic positivism," for which he was to become somewhat famous. He argued that the main lesson of the work of Mach and Avenarius was that all knowledge is knowledge relative to the perspective of a given observer, from which it follows (1) that there can be no talk of a reality independent of some perspective; (2) that the distinction between appearance and reality therefore lapses; and (3) that there is therefore no role left for the traditional conception of substance, which had functioned mainly to ground the appearance-reality distinction. Petzoldt believed that Einstein was arguing for a similar view with his relativization of our physical descriptions of events to different coordinate systems or frames of reference and his consequent remarks about whether or not such phenomena as length contraction and time dilation are real or apparent. Although he never overcame his fundamental confusion between the *physical* idea of relativity to a frame of reference and the *epistemological* idea of relativity to an observer, Petzoldt nevertheless enjoyed some modest repute as a commentator on the philosophical implications of relativity theory, enough so that Einstein would write to Petzoldt in a postcard most probably from late 1914 or early 1915:

> Today I have with great interest read your book in its entirety, and I happily infer from it that I have for a long time been your companion in your way of thinking.—I have told a gentleman who is ill, a friend of mine, about your paper on relativity theory. He has shown a great interest in it. You would give him great pleasure if you were to send him a reprint of it. (EA 19-067)[7]

The book is most probably Petzoldt 1912a, the paper either Petzoldt 1912b or Petzoldt 1914.[8]

If Einstein read Petzoldt's 1912 Deutsche Physikalische Gesellschaft address, he would have found there the following characterization of relativistic perspectivalism:

> The task of physics becomes, thereby, the univocal [*eindeutige*] general representation of events from different standpoints moving relative to one another with constant velocities, and the univocal setting-into-relationship of these representations. Every such representation of whatever totality of events must be univocally mappable onto every other one of these representations of the same* events. The theory of relativity is one such mapping theory. What is essential is that univocal connection. Physical concepts must be bent to fit for its sake. We have theoretical and technical command only of that which is represented univocally by means of concepts.

*Better: representations of events in arbitrarily many of those systems of reference that are univocally mappable onto one another are representations of *"the same"* event. Identity must be *defined,* since it is not given from the outset. (Petzoldt 1912a, 1059)

One could not ask for a clearer statement of the *Eindeutigkeit* principle as it bears upon the way theories may be expected to represent reality via their models: "We have theoretical and technical command only of that which is represented univocally [*eindeutig*] by means of concepts." What does it mean to say that we have such a univocal representation? It means precisely that "every such representation of events must be univocally [*eindeutig*—read 'one-to-one'] mappable onto every other one of these representations of the same events." In other words, it means that each representation, each model, if you will, be isomorphic to all the other models of the theory, which is exactly what we mean today by saying that a theory is categorical, what Carnap is going to define in 1927 as monomorphism.

The footnote to this passage deserves special attention. What do we mean by "same events"? Keep in mind that Petzoldt is using the term "events" in its Minkowskian sense, meaning events as space-time points, points of the space-time manifold. What Petzoldt says is that the identity of events "must be defined, since it is not given at the outset." What is our definition? "Representations of events . . . that are univocally mappable onto one another are representations of 'the same' event." In effect, "events" are defined as equivalence classes under transformations and hence as invariants of the class of transformations assumed by the theory. But precisely this understanding of event individuation is what underlies the move from the hole argument to the point-coincidence argument in Einstein's thinking about the ontological foundations of general relativity.

There are other sources from which Einstein may have or could have learned about the *Eindeutigkeit* principle, including Paul Volkmann, Wilhelm Ostwald, Paul Natorp, and Ernst Cassirer, all of whom participated in one way or another in the turn-of-the-century debate about the merits of and meaning of the principle.[9] But one other contemporary figure bears special mention at this point, namely, Moritz Schlick, who, at about the same time and somewhat independently of Petzoldt, had been developing a related conception in his influential theory of truth as univocal coordination.

Schlick first introduced this idea in his ambitious 1910 essay, "Das Wesen der Wahrheit nach der modernen Logik" (Schlick 1910). Writing from a perspective inspired most likely by Helmholtz's "sign-theory" of knowledge, Schlick argued that judgments are nothing more than signs or symbols (*Zeichen*) coordinated to facts (*Tatsachen*).[10] If this is so, then the most that we can demand of them is that they be coordinated one-to-one with the facts for which we want them to stand. And this, according to Schlick, is precisely what we mean by truth:

> Judgments [are] . . . *symbols* for states of affairs. What conditions do we impose on a symbol? Only a single one, namely, that it be *univocal* [*eindeutig*], or better still, one-to-one in character, i.e., every symbol must correspond to one and only one significatum and every significatum to just one symbol. Since, by its very nature, a symbol can have no other meaning than this, it follows immediately:
>
> A *judgment is true if it univocally designates a definite state of affairs.* (ibid., 466)

This has rightly been faulted as a definition of truth (see, for example, Coffa 1991, 177–79). Thus, if truth is univocal, one-to-one coordination of judgment to state of affairs, falsity can be nothing other than equivocal, one-to-many coordination, which is precisely what Schlick says: A false judgment is "doubly or multiply ambiguous" (*zwei-order mehrdeutig*) (Schlick 1910, 471). The existence of true but ambiguous propositions is sufficient to show how silly is this way of conceiving the nature of falsehood, but we should not jump to the conclusion that the property of our judgment that Schlick was trying to capture with his definition of truth is of no consequence. On the contrary, a comment of Schlick's that was perhaps occasioned by his own dim sense of the oddity of construing ambiguity as falsehood points the way toward the larger significance of his project:

> And it is indeed immediately clear that the only significance of the distinction between true and false statements is, in fact, to preserve the univocalness [*Eindeutigkeit*] of linguistic and conceptual expression, which is the necessary precondition for all understanding, without which every signification and every expression would be altogether meaningless. (ibid., 471)

It is not so much the distinction between truth and falsehood that Schlick is concerned about; it is rather *Eindeutigkeit* itself, the preservation of the univocalness of "linguistic and conceptual expression" being the "only significance" of the distinction, this univocalness being "a necessary precondition for all understanding." This is very much what Petzoldt was arguing about "das Gesetz der Eindeutigkeit," and it is what Einstein assumed about *Eindeutigkeit* in the context of the hole and point-coincidence arguments.

When Schlick turned his attention to the philosophical implications of relativity theory in the late 1910s and early 1920s, he repeated the definition of truth as univocal coordination [*eindeutige Zuordnung*] in virtually every one of his essays on the topic (see Schlick 1915, 1917a, 1917b, 1919, 1920, 1921, 1922a, 1922b). It also figures prominently in his *Allgemeine Erkenntnislehre* (Schlick 1918, §§8, 10; 1925).

Particularly important for its possible bearing on Einstein's deployment of the *Eindeutigkeit* principle, especially at the time of his formulation of the point-coincidence argument in late 1915 or early 1916, is what Schlick wrote in his "Die philosophische Bedeutung des Relativitätsprinzips," which Einstein read (and praised) in late December 1915. There is first a restatement of the definition of truth as univocal coordination:

> The totality of our scientific propositions, in word and formula, is in fact nothing else but a system of symbols *correlated* to the facts of reality; and that is equally certain, whether we declare the real to be a transcendent being or merely the totality and interconnection of the immediately "given." The system of symbols is called "true," however, if the correlation is completely *univocal.* Certain features of this symbol-system are left to our arbitrary choice; we can select them in this way or that without damaging the univocal character of the correlation. It is therefore no contradiction, but lies, rather, in the nature of the matter, that under certain circumstances, several theories may be true at the same time, in that they achieve indeed a different, but each for itself completely univocal, designation of the facts. (Schlick 1915, 149)[11]

One of the many important features of the view adumbrated here is Schlick's rather Duhemian, holistic, underdeterminationist conventionalism, a view that finds its echo in Einstein, even though this specific understanding of the role of conventions in science was later to be repudiated by Schlick himself.[12] The Duhemian, underdeterminationist moment in Schlick's thinking entails that the kind of coordination indicated by the definition of truth is perhaps better characterized as many-to-one, rather than just one-to-one. But whether the coordination is many-to-one or one-to-one, what stands on the right-hand side of this equation, "the facts of reality," are still asserted to be given uniquely. And this latter remains the principal feature of the definition of truth as univocal coordination when it is made the centerpiece of Schlick's subsequent efforts to spell out the philosophical significance of general relativity (the 1915 essay concerns mostly special relativity).

It was in his widely read monograph *Raum und Zeit in den gegen-*

wärtigen Physik (Schlick 1917a, 1917b, 1919, 1920, 1922a) that Schlick undertook to interpret what Einstein meant by his capsule formulation of the chief lesson of point-coincidence argument, the claim that, in general relativity, "time and space thereby lose the last vestige of physical reality." As Schlick himself explained in a letter to Einstein of 4 February 1917: "The essay [is] less a representation of the general theory of relativity itself as it is an exhaustive explanation of the thesis that, in physics, space and time have now forfeited all objectivity" (EA 21-568). Schlick's conclusion is essentially the same as Einstein's. The basic ontology of general relativity is a manifold of point-events individuated as space-time coincidences. Why? Not only because all observation takes the form of noting such coincidences, but more importantly because any arbitrary, continuous, single-valued point transformation "leaves all . . . coincidences totally unaffected" (Schlick 1917a, 241). Schlick adds:

> The whole of physics may be regarded as a quintessence of laws, according to which the occurrence of these space-time-coincidences takes place. Everything else in our world-picture which can *not* be reduced to such coincidences is devoid of physical objectivity. . . . All world-pictures which lead to the same laws for these point-coincidences are, from the point of view of physics, in every way equivalent. (ibid., 241)

Lest there be any doubt about Schlick's commitment to the *reality* of this ontology, he goes on at some length to explain why space-time coincidences should be regarded as real:

> Physics introduces, as its ultimate indefinable conceptions, the coincidences of two *events;* on the other hand, the psycho-genetic analysis of the idea of objective space ends in the conception of the space-time coincidence of two *elements of perception.* Are they to be regarded simply as one and the same thing?
>
> Rigorous positivism, such as that of Mach, affirms them to be so. According to him, the directly experienced elements such as colours, tones, pressures, warmths, etc., are the sole reality, and there are no other actual events beyond the coming and going of these elements. Wherever else in physics other coincidences are mentioned, they are only abbreviated modes of speech, economical working-hypotheses, not realities as perceptions are. Looked at from this point of view, the conception of the physical world in its objective four-dimensional scheme would merely be an abridged statement of the correspondence of the subjective time-space experiences in the realms of the various senses, and *nothing more.*

> This view is, however, not the only possible interpretation of scientific facts. If distinguished investigators in the domain of the exact sciences do not cease to urge that the picture of the world as offered by Mach fails to satisfy them, the ground for it is doubtless to be sought in this, that the quantities which occur in physical laws do not all indicate "elements" in Mach's sense. The coincidences which are expressed by the differential equations of physics are not immediately accessible to experience. They do not directly signify a coincidence of sense-data; they denote non-sensory qualities, such as electric and magnetic field strengths and similar quantities. There is no argument whatsoever to force us to state that only the intuitional elements, colours, tones, etc., exist in the world. We might just as well assume that elements or qualities which cannot be directly experienced also exist. These can likewise be termed "real," whether they be comparable with intuitional ones or not. . . . The picture of the world, as presented by physics, would then be a system of symbols arranged in a four-dimensional scheme, by means of which we get our knowledge of reality; that is *more* than a merely auxiliary conception, allowing us to find our way through given intuitional elements.
>
> The two views stand in opposition; and I believe that there is no rigorous proof of the correctness of the one and the falseness of the other. (ibid., 264–65)

When Einstein read this last remark, he was moved to write to Schlick suggesting an essential clarification about the different kinds of reality represented by Machian "elements of sensation" and space-time events:

> The second point to which I want to refer concerns the reality concept. Your view stands opposed to Mach's according to the following schema:
>
> Mach: Only impressions are real.
>
> Schlick: Impressions and events (of a phys[ical] nature) are real.
>
> Now it appears to me that the word "real" is taken in different senses, according to whether impressions or events, that is to say, states of affairs in the physical sense, are spoken of.
>
> If two different peoples pursue physics independently of one another, they will create systems that certainly agree as regards the impressions ("elements" in Mach's sense). The mental constructions that the two devise for connecting these "elements" can be vastly different. And the two constructions need not agree as regards the "events"; for these surely belong to the conceptual constructions. Cer-

tainly only the "elements," but not the "events," are real in the sense of being "given unavoidably in experience."

But if we designate as "real" that which we arrange in the space-time schema, as you have done in the theory of knowledge, then without doubt the "events," above all, are real.

Now what we designate as "real" in physics is, no doubt, the "spatio-temporally arranged," not the "immediately given." The immediately given can be illusion, the spatio-temporally arranged can be a sterile concept that does not contribute to illuminating the connections between the immediately given. *I would like to recommend a clean conceptual distinction here.* (Einstein to Schlick, 21 May 1917; EA 21-618)

In *Raum und Zeit in den gegenwärtigen Physik,* as in earlier writings, Schlick had argued that while a theory "is *correct* or *true* if the system of judgments indicates the world of facts *univocally,*" there could still be more than one such theory thus univocally coordinated with the facts: "It is, however, possible to indicate identically the *same* set of facts by means of *various* systems of judgments; and consequently there can be various theories in which the criterion of truth is equally well satisfied" (Schlick 1917a, 62). But now Einstein says that such underdetermination is not likely to be evinced as regards the Machian "elements," that it is likely, instead, only with regards to the deep ontology of physics, the ontology of point-events, although, presumably, any one such underdetermined theory will still be univocally coordinated with the deep reality it aims to describe.[13]

What we see emerging here, sometimes with greater clarity, as in the writings of Petzoldt and Einstein, sometimes with lesser clarity, as in the writings of Schlick, is the start of a sophisticated new approach to the problem of the relation between theory and reality. The Schlick of this period was an avowed realist about such things as the ontology of point-events in general relativity, although that was to change within a short time after his move to Vienna in 1922 and his immersion in the Vienna Circle. Einstein was equally explicit about the reality of space-time events individuated as coincidences of world-lines. But in his case, especially, this commitment to an event ontology was tempered by a refusal to characterize such a commitment as part of a broader commitment to realism as a philosophical position. Thus, he could write to Eduard Study on 25 September 1918, this in a kind of a "fan letter" about the latter's *Die realistische Weltansicht und die Lehre vom Raume* (Study 1914):

> "The physical world is real." That is supposed to be the fundamental hypothesis. What does "hypothesis" mean here? For me, a hypothesis is a statement, whose *truth* must be assumed for the moment, *but whose meaning must be beyond all ambiguity.* The above statement appears to me, however, to be, in itself, meaningless, as if one said: "The physical world is cock-a-doodle-doo." It appears to me that the "real" is an intrinsically empty, meaningless category (pigeon hole), whose monstrous importance lies only in the fact that I can do certain things in it and not certain others. This division is, to be sure, not an *arbitrary* one, but instead...
>
> I concede that the natural sciences concern the "real," but I am still not a realist. (EA 22-307)

How can this be? Why would one want to insist upon the reality of univocally determined space-time events and still not want to be called a realist?

One possible reason has to do with the problem of our access to the real. Even if one is granted the existence of a mind-independent reality as that which we aim to describe by our scientific theories, there is still the problem of how, exactly, those theories link up with that reality. This is less a problem at the level of those facets of the real that may be presumed to be more or less directly accessible via the crude and brutal bump of observation, which is perhaps why Einstein cautions Schlick that "two different peoples" who "pursue physics independently of one another" are likely to develop systems that agree about the Machian "elements of sensation." It is a considerably more serious problem at the level of deep theory, where the "two different peoples" are likely to develop "mental constructions" for connecting the elements that may be "vastly different." For at this level, our only access to the real is through our theories themselves.

One would like to be in a position to assert that a true theory is true by virtue of some special relationship to the real, and yet such a special relationship can be guaranteed not for transcendent things-in-themselves, but only for those entities that are introduced (one might say constructed or defined) as elements of the theory's ontology. In a sense, this is the lesson spelled out explicitly in the mid-1930s in the work of Tarski, where truth is defined semantically not as correspondence to a transcendent reality, but instead in terms of a proposition's "satisfaction" by the elements of a model. Since a model for a theory is, at first gander, not the same thing as the reality that the theory putatively aims to describe, it is hard to extract a realism about things-in-themselves from success in the enterprise of discovering and confirming true scientific theories.

Still we think that there is a mind-independent reality, and we think that true theories stand in some privileged relationship with it. A minimum necessary condition for this being the case would therefore seem to be a theory's determining for itself a unique model, or a class of models unique up to isomorphism. If it fails thus to determine a unique model for itself, if it admits a multiplicity of nonisomorphic models, which is to say, in the modern idiom, that it is not categorical, then it would seem to be compatible with the world's being any one of a number of different ways, which would seem, in turn, to undermine any hope of asserting for that theory the envisioned privileged relationship with reality.

If, on the other hand, we can assert that the theory determines for itself a class of models unique up to the point of isomorphism, then it is possible to imagine that one of those models *is* the real world we are trying to finger. But then let us take just one more short step. If we go on to assert that the aspect of the real that scientific theories aim to describe is precisely the *form* or *structure* of the real, then the need for a distinction between model and reality lapses. What the real has in common with the other members of the class of isomorphic models determined by the kind of true theory we are envisioning is nothing more and nothing less than that form or structure. The model, or at least the form of the model, literally *is* the reality that the theory describes. If that is so, then, clearly, *Eindeutigkeit* is a necessary condition on an acceptable theory, for a theory's admitting a multiplicity of nonisomorphic models would mean not only that the theory would be compatible with the world's being any of a number of different ways but also that the theory actually *implies* that the world *is* many different ways. And this latter is surely unacceptable to any philosopher or scientist who considers himself or herself a friend of reality.

The view that I am sketching here has not a little in common with more recent views like Quine's "to be is to be the value of a bound variable" (Quine 1948), Hilary Putnam's "internal realism" (Putnam 1976), and, at least in my opinion, Arthur Fine's "natural ontological attitude" (Fine 1984a, 1984b). Or rather, it has something in common with these views if we confine our attention to the structural aspects of the real. Schlick was, I think, struggling toward more or less the same conception with his famous argument in the *Allgemeine Erkenntnislehre* that all knowledge is knowledge of form, not content. And not unrelated is the attitude toward ontology that accompanied Hilbert's doctrine of implicit definition. But surely the most influential representative of such a view in Einstein's day was Cassirer, to whom we now turn.

3. Natorp and Cassirer on *Eindeutigkeit* and the Problem of Space

Cassirer was the most prominent spokesperson for Marburg neo-Kantianism and critical idealism in the 1910s and 1920s. The student of both Hermann Cohen and Paul Natorp, he began writing about the problem of *Eindeutigkeit* from a neo-Kantian perspective in his influential *Substanzbegriff und Funktionsbegriff* (E. Cassirer 1910). Cassirer's principal aim was to argue that knowledge in general, especially the kind of knowledge epitomized by modern mathematical and scientific theories, is properly understood as concerning itself not with individual substances and their attributes, but instead with patterns of functional relationships: "The concept of function constitutes the general schema and model according to which the modern concept of nature has been molded in its progressive historical development" (ibid., 1910, 27). Such a view of knowledge was anticipated by Leibniz, according to Cassirer (see especially E. Cassirer 1902), but he believed that it had come ever more self-consciously to the fore in the profound foundational investigations in mathematics and the sciences that began to appear in the late nineteenth and early twentieth centuries.

Cassirer's mentor, Natorp, had himself written about the principle of *Eindeutigkeit,* indeed, specifically with regard to whether or not special relativity satisfies the principle, this in his summa, *Die logischen Grundlagen der exakten Wissenschaften* (Natorp 1910). Natorp's approach to the question is shaped by his reading of Kant, for whom, according to Natorp, *Eindeutigkeit* or *Einzigkeit* (singularity — Natorp uses *Einzigkeit* and *Eindeutigkeit* interchangeably) is the characteristic that distinguishes intuition from conception as the avenue through which the real presents itself to us in experience:

> Reality means a determination, such that nothing remains undetermined. Indeterminacy is just mere possibility. In other words: possibility is multifaceted — it always permits a choice; reality is absolutely singular — it is conceived as being determined in a singular fashion, excluding all choice. This singularity plays a great role in Kant's theory of experience; upon its being required rests in particular his distinction of the intuition from the concept; intuition means to him "the representation that can only be given through a singular

> object." Time and space are in this sense "essentially singular" representations, and therefore intuitions; there is only one time, only one space, just as there is only one experience "in which all perceptions are represented as being in thoroughgoing and law-governed connection." Clearly the singularity of time and of space counts necessarily for him as the condition for the singularity of experience; how can the connection of experience be a singular one, without the singular time and the singular space being its foundation? (ibid., 92)

This becomes the basis for Natorp's typically neo-Kantian defense of absolute space and time, his argument being that all empirical determinations of space and time are merely relative and hence fail to evince the necessary uniqueness properties:

> It is certain from the outset . . . that no really achievable foundation for temporal or spatial determination based upon empirical means of determination may claim any absolute security. Nevertheless — indeed, precisely thereby — one makes the fundamental presupposition that there is an absolute time, an absolute space for events in themselves [*des Geschehens an sich*], by means of which, if they were determinable, events themselves would be absolutely determined; by comparison with these, every empirically possible determination could claim only the significance of a useful approximation. Upon what basis do we make this presupposition? It lies originally in the demand, in the final, fundamental supposition that experience is directed toward knowledge of that which is or proceeds "really." *Existence* can be conceived only in a single way, because it really means nothing other than determination in a single way. It is thus an "analytic proposition" that, insofar as existence should be spatio-temporally determined, this determination itself is required to be simply univocal [*eindeutig*], although it is never given as such. Only a simply univocal (and that means: absolute) temporal and spatial determination *would be* the determination of existence itself; that the empirically possible determination can never be simply univocal in no way changes this conditional maxim. (ibid., 328)

This would seem to bode ill for relativity theory. But not so. Natorp argues that while relativity offends against the requirement of *Eindeutigkeit* in its treatment of space and time, it nevertheless achieves a more important univocalness in another domain, something that comes most clearly to the fore in the Minkowskian formulation of the theory:

> Otherwise, the fundamental law of univocal determination loses thereby none of its logical validity; only the possibility of its em-

pirical fulfillment is, in a certain way, restricted. The freedom in the choice of the time parameter, as in that of the spatial coordinates, does not, after all, threaten the unity of our knowledge of nature, because it does not touch the univocalness of the *laws of nature.* Precisely this is perhaps the most important result of Minkowski's investigations: the "invariance" of the laws of nature under all "Lorentz transformations. . . ." A univocalness of empirical temporal and spatial determination more thoroughgoing than that which suffices for the establishment of laws of nature, is indeed not necessary. (ibid., 403–4)

There is, to say the least, a tension here in Natorp's thinking. In the Kantian tradition, a univocal determination of space and time had long been held necessary as securing the ground for the possibility of the univocal determination of empirical objects as presented to us in intuition under the a priori forms of space and time. Natorp argues that such univocalness is lost in special relativity's handling of space and time, and yet he concludes that a univocalness beyond that found in the latter's representation of the laws of nature "is indeed not necessary." How can one be satisfied with a univocalness in the laws of nature if there is no comparable univocalness in the determination of the empirical objects presented in experience to which those laws apply?

Here is where Cassirer picks up the argument in *Substanzbegriff und Funktionsbegriff.* He follows Natorp in asserting the essential ambiguity or *Vieldeutigkeit* of concepts, as opposed to the univocalness or *Eindeutigkeit* of intuition, but he goes beyond Natorp in asking how it can be that inherently ambiguous concepts can nevertheless be employed to construct a complete and univocal representation of a univocal reality.

Cassirer attacks this question by taking up a metaphor deployed by Paul Volkmann in his critique of Petzoldt's *Eindeutigkeit* principle in his *Erkenntnistheoretische Grundzüge der Naturwissenschaften,* the metaphor of vector superposition and "isolation" or, to use the more modern term, decomposition (Volkmann 1896, 80–88). Volkmann's "principle of superposition" held that all phenomena in nature were the result of a superposition of distinguishable elements, but that, like a vector, any such phenomenon could be decomposed into components in any one of a number of different ways, there being no metaphysical or epistemological reason for privileging any one of these decompositions, any more than one set of basis vectors would be inherently superior in vector decomposition, except, perhaps, for reasons of convenience. In this way, Volkmann argued for the essential metaphysical ambiguity of nature.

By contrast, the whole point of constructing a representation by superposition, for Cassirer, is to combine sufficiently many relations as to determine up to the point of isomorphism the reality one is seeking to represent. And it is precisely to the extent that one can hope thus to succeed in specifying a domain of objects univocally by the relations they satisfy that one can claim to have eliminated concepts of substance or "thing" concepts from science in favor of functional relations.

Cassirer's model for this process is Richard Dedekind's definition of "number" by means of the concept of a "progression" (Dedekind 1893), about which Cassirer writes:

> The starting point here seems to be the traditional logical doctrine of a plurality of *things* and the power of the mind to copy [*abzubilden*] them; nevertheless it becomes evident with deeper understanding that the old terms have here gained a new import and meaning. The "things," which are spoken of in the further deduction, are not assumed as independent existences present anterior to any relation, but they gain their whole being, so far as it comes within the scope of the arithmetician, first in and with the relations which are predicated of them. Such "things" are *terms of relations,* and as such can never be "given" in isolation but only in ideal community with each other. (E. Cassirer 1910, 46–47)[14]

What is the significance of what Dedekind has accomplished?

> All the propositions of arithmetic, all the operations that it defines, are related solely to the general properties of a progression; hence they are never directed primarily upon "things" but upon the ordinal relation, which prevails between the elements of certain systematic wholes. . . . The function of "number" is, in its meaning, independent of the factual diversity of the objects which are enumerated; this diversity must therefore be disregarded when we are concerned merely to develop the determinate character of this function. Here abstraction has, in fact, the character of a liberation; it means logical concentration on the relational connection as such with rejection of all psychological circumstances, that may force themselves into the subjective stream of presentations, but which form no actual constitutive aspect of this connection. . . . What is here expressed is just this: that there is a system of ideal objects whose whole content is exhausted in their mutual relations. The "essence" of the numbers is completely expressed in their positions. (ibid., 49–51)

Cassirer believes that Dedekind's approach to defining the numbers can be extended to other sciences:

> Whenever a *system of conditions* is given that can be realized in different contents, there we can hold to the form of the system itself as an *invariant,* undisturbed by the difference of the contents, and develop its laws deductively. In this way we produce a new "objective" form, whose structure is independent of all arbitrariness; but it would be uncritical *naïveté* to confuse the object which thus arises with sensuously real and effective things. We cannot read off its "properties" empirically, nor do we need to, for its form is revealed in all determinateness as soon as we have grasped in its purity the relation from which it develops. (ibid., 52–53)

There are, of course, echoes of Felix Klein's Erlanger program here (see ibid., 153–57), but in Cassirer's estimation, it is Hilbert's approach to the foundations of geometry via the concept of implicit definition that best epitomizes the point of view set forth here. When point and line are defined implicitly as objects satisfying the axioms of geometry, it is not individual points and lines in their particularity that concern us; it is, rather, the structure or form embodied in the system of relations connecting points and lines: "Only this systematic 'complexion' of the elements, and not their particular characters, is taken here as the expression of their essence. In this sense, Hilbert's geometry has been correctly called a pure theory of relations" (ibid., 123).

We can draw a larger lesson about the relation between theories and the "realities" they aim to represent in the mathematical sciences:

> The determination of the *individuality* of the elements is not the beginning but the end of the conceptual development; it is the logical goal, which we approach by the progressive connection of universal relations. The procedure of mathematics here points to the analogous procedure of *theoretical natural science,* for which it contains the key and the justification. (*Cf.* Ch. V.) (ibid., 124)

It is in chapter 5, "The System of Relational Concepts and the Problem of Reality," that Cassirer addresses the problem of *Eindeutigkeit/Vieldeutigkeit* and the problem of reality though the aforementioned critique of Volkmann's doctrine of "isolation" and "superposition."

I shall quote at length from the part of chapter 5 where Cassirer states his major conclusions about the relation between theory and reality:

> While reality presents us a mixture of heterogeneous circumstances seemingly inseparably interwoven and confused, thought demands the separate consideration of each particular moment and the exact determination of the part it plays in the structure of the whole. . . . The complete picture of the total process arises when we connect the par-

> tial system[s] and place them as it were upon each other. This picture of the total process appears not merely as a unitary intuition, but as a differentiated whole, in which the type of dependency between the individual elements is exactly defined. . . . Thus when the principle of "isolation" and "superposition" is explained by the fact that all reality represents only the sum of the manifestations of particular laws of nature, and is to be conceived as *having issued from* these laws, the real epistemological meaning of the thought is concealed. We cannot be concerned here with the origin of things, but only with the origin and character of our insight into things. The "real," as it is grasped in the sensuous impression, is not in and for itself a "sum" of various elements, but first appears to us as an absolutely simply and unanalysed whole. This original "simplicity" [*Einfalt*] of intuition is only transformed into a manifold [*inneren Vielgestaltigkeit*] by the logically analytic work of the concept. The concept is here just as much a source of plurality as it elsewhere appears to be a source of unity. Since we conceive a particular process as successively introduced into different systems, whose general structure can be mathematically deduced, we thereby increase its determinateness, in so far as we define its position in the general plan of our thought with increasing exactness. (ibid. 339–42)

Like Natorp, Cassirer emphasizes the essential "plurality" (*Vielheit*) of concepts, which is contrasted with the essential univocalness (*Eindeutigkeit*) that is the distinguishing mark of intuitions. What is different is Cassirer's pointing out that we can overcome the plurality of concepts, yielding a univocal determination of a functionally or structurally characterized reality, through the accumulation of conceptual relations. To put it somewhat crudely, the idea is that by adding up enough axioms we might eventually constrain the domain of objects we are trying to characterize (numbers, points and lines, and so on) up to the point of uniqueness. As Cassirer says, "[T]he determination of the *individuality* of the elements is not the beginning but the end of the conceptual development."

The goal is complete, univocal determination, as Cassirer makes clear in a discussion of Mach, from which I again quote at length:

> If we represent mathematically a given totality of observations by the "superposition" of several series, we do not indeed thereby increase our knowledge of the absolute and transcendent causes of the process; but we have raised ourselves to a new *type of knowledge*. . . . To describe a group of phenomena, then, means not merely to record receptively the sensuous impressions received from it, but it means to

> transform them intellectually. From among the theoretically known and developed forms of mathematical connection . . . a selection and combination must be made such that the elements given here and now appear as constructively deduced elements in the system, which arises. The logical moment given here cannot be denied even in the theories of empiricism, or under whatever names it may be concealed. The "adjustment of ideas to reality" presupposes the very concept of this reality, and thus a system of intellectual postulates. It is in the principle of the univocalness of natural processes [*Eindeutigkeit des Geschehens* — emphasized with Sperrdruck in the German original] above all, that all these postulates are ultimately combined. "I am convinced," says Mach himself, "that in nature only so much happens as can happen, and that this can happen only in *one* way." All physical process is thus completely determined by the momentarily effective circumstances, and can thus take place in only one way. However, if we analyse the grounds of this conviction, we are implicitly led back to all those fundamental conceptions, which the sensationalistic explanation explicitly denies. The conception of the univocalness [*Eindeutigkeit*] and "stability" of being obviously is not in the content of perception as given in our first immediate experience, but it indicates the goal, which the intellectual labor of science seeks to have this content approximate. This goal can only be reached, if we are able to establish certain permanent relations in the flux of sensations, which differ and have their truth limited to a single moment of time; the rules of these permanent relations we can call to mind independently of change of the momentary material. To the extent that this takes place, the scientific concept of *nature* develops and is confirmed. (ibid., 349–51; translation corrected)

The quote from Mach is taken from one of the places in the *Wärmelehre* (Mach 1896, 393) where Mach himself cited Petzoldt in endorsing the *Eindeutigkeit* principle. Thus, even though Cassirer himself does not cite Petzoldt, there can be no doubt that he was fully aware of the tradition within which he was working.

When Cassirer turned to write about relativity theory in his *Zur Einsteinschen Relativitätstheorie* (E. Cassirer 1921), the broad framework was provided by the point of view on *Eindeutigkeit,* structure, and reality that had first been developed in *Substanzbegriff und Funktionsbegriff.* Cassirer's large aim in 1921 was to defend the Kantian doctrine of the a priori character of space and time as forms of outer and inner intuition. His strategy was to abandon any claim for the a priori status of Euclidean geometry but to insist upon the a priori status of the notions of spatiality and temporality themselves. In what these consist we will see in a moment. But be forewarned that Cassirer's argument will take us right to the heart of the debate we have been tracking about *Eindeutigkeit* and the ontological implications of the hole argument, a circumstance that Cassirer himself highlights in declaring:

> When Einstein characterizes as a fundamental feature of the theory of relativity that it takes from space and time "the last remainder of *physical objectivity*," it is clear that the theory only accomplishes the most definite application and carrying through of the standpoint of critical idealism within empirical science itself. (ibid., 79)

That *Eindeutigkeit* will be the central issue is clear from Cassirer's pointing right away to the larger lessons to be learned from relativity:

> In fact, it can be shown that the general idea of the invariance and univocalness [*Eindeutigkeit*] of certain values, which is given such prominence in the theory of relativity, must recur in some form in *every* theory of nature, because it is part of the fundamental logical and epistemological constitution of such a theory. (ibid., 45; my translation)

He illustrates this point with a discussion of the history of the principle of the conservation of energy, which he presents as the history of a quest for something that can serve as a measure of that which remains invariant and univocally determined throughout natural processes. Once the proper concept of energy is found, this "energy" becomes itself a "determinate object," not in the substantivalist sense preferred by Wilhelm Ostwald, Georg Helm, and their energeticist colleagues, but simply as that which is rendered determinate by the role it plays in the relations that it unifies, for

> the unity and univocalness of the metrical determinations can also be immediately understood and expressed as a unity and univocalness in the determination of the object, precisely because the empirical object means nothing other than a lawlike totality of relations. (ibid., 47; my translation)

Cassirer enlarges upon the idea that the empirical object is nothing more nor less than the totality of a set of relations determined by scientific laws in a discussion of how skeptical doubts about our ability to know the "absolute properties of things" can be overcome by the critical idealist's revision of the concept of truth itself. The skeptical argument draws strength, says Cassirer, from the fact that the skeptic and the dogmatist have conceptions of truth that are "united by a common root," namely, the hypostatization of absolute things and absolute properties:

> As long as thought contents itself with developing, with reference to phenomena and according to demands of its own form, its logical axioms, and truth as a system of pure relations, it moves within its own circle with complete certainty. But when it affirms an absolute, whether of outer or inner experience, it is forced skeptically to annihilate itself with reference to this absolute. (ibid., 54)

Things are different, however, from the point of view of the critical idealist conception of truth:

> The relation is indeed changed when we contrast to both the dogmatic and the skeptical concept of truth . . . the idealistic concept of truth. For the latter does not measure the truth of fundamental cognitions by transcendent objects, but it grounds conversely the meaning of the concept of the object on the meaning of the concept of truth. Only the idealistic concept of truth overcomes finally the conception which makes knowledge a copying, whether of absolute things or of immediately given "impressions." (ibid.)

Such a conception of truth was first clearly indicated by Leibniz in his *Monadology,* this in the form of the common truth expressed by the community of windowless monads:

> This community . . . does not come about by these different pictures of the world being related to each other as copies of a common "original" but by the fact that they correspond functionally to each other in their inner relations and in the general form of their structure. For one fact, according to Leibniz, expresses another when there exists between what can be said of the one and of the other a constant and regular relation. Thus a perspective projection expresses its appropriate geometrical structure, an algebraic equation expresses a definite figure, a drawn model a machine; not as if there existed between them any sort of factual likeness or similarity, but in the sense that the relations of the one structure correspond to those of the other in a conceptually univocal [*eindeutige*] fashion. (ibid., 55; translation corrected)

This Leibnizian, idealistic conception of truth was "taken up and developed" by Kant. But the theory of relativity can be accommodated to it as well, "for, in a general epistemological regard, it is characterized by the fact that in it, more clearly and more consciously than ever before, the advance is made from the copy theory of knowledge to the functional theory" (ibid.).

So the skeptical problem about access to the real is solved, as we saw it already being solved by Einstein, by *defining* the real, the empirical object, in terms of truth. The object of knowledge *is* simply that univocally determined something of which our theories are true. And, as Leibniz saw, that univocally determined something will always turn out to be the form or structure common to the many possible individual instantiations of this truth. To speak again anachronistically, the real empirical object is here defined as the structure common to the many isomorphic models of a theory.

How, more precisely, is relativity theory to be brought into the fold? How can it be assimilated to Cassirer's critical idealism? On the face of it, there is a problem. For Kant, absolute space is characterized by a *logical universality* (not to be confused with the *physical universality* of real extension and space as an all-embracing container) as an ultimate unity of all spatial determinations, as a necessary idea that serves as a rule for considering all motions in it as merely relative. Does relativity theory's repudiation of an absolute physical space violate this requirement? No.

> The logical universality of such an idea does not conflict with the theory of relativity; the latter starts by regarding all motions in space as merely relative because only in this way can it combine them into a definite concept of experience, that unifies all phenomena. On the basis of the demand for the totality of determinations it negates every attempt to make a definite particular system of reference the norm for all the others. The one valid norm is merely the idea of the unity of nature, of univocal [*eindeutige*] determination itself. . . . The "unity of nature" is grounded by the general theory of relativity in a new sense, since it includes under a supreme principle of knowledge along with the phenomena of gravitation, which form the real classical field of the older mechanics, the electrodynamic phenomena. That in order to advance to this "logical universality of the Idea," many trusted presentational pictures must be sacrificed need not disturb us; this can affect the "pure intuition" of Kant only in so far as it is misunderstood as a mere picture and not conceived and estimated as a constructive method.
>
> In fact, the point at which the general theory of relativity must implicitly recognize the methodic presupposition, which Kant calls "pure intuition" can be pointed out exactly. It lies, in fact, in the concept of "coincidence" to which the general theory of relativity ultimately reduces the content and the form of all laws of nature. If we characterize events by their space-time coördinates x_1, x_2, x_3, x_4, x'_1, x'_2, x'_3, x'_4, *etc.*, then, as it emphasizes, everything that physics can teach us of the "essence" of natural processes consists merely in

> assertions concerning the coincidences or meetings of such points. We reach the construction of physical time and of physical space merely in this way; for the whole of the space-time manifold is nothing else than the whole of such coördinations. Here is the point at which the ways of the physicist and of the philosopher definitely part, without their being thereby forced into conflict. What the physicist calls "space" and "time" is for him a concrete measurable manifold, which he gains as the *result* of coördination, according to law, of the particular points; for the philosopher, on the contrary, space and time signify nothing else than the forms and *modi,* and thus the representation of this coördination itself. They do not result for him from the coördination, but they *are* precisely this coördination and its fundamental directions. It is coördination from the standpoint of coexistence and adjacency or from the standpoint of succession, which he understands by space and time as "forms of intuition." (ibid., 83–85; translation corrected)

This long and perhaps difficult passage repays careful scrutiny, for it contains the heart of Cassirer's argument for the large conclusion he wants us to draw, that relativity poses no threat to the Kantian doctrine of space and time as a priori forms of intuition, as long as we understand the latter properly to concern not the Euclidean absolute space and time of Newton, but rather the forms of spatiality and temporality themselves, the forms of coexistence and succession.

In what do these forms consist? Cassirer quotes from Kant's *Inaugural Dissertation:*

> Time is not something objective and real, . . . but it is the subjective condition necessary by the nature of the human mind *for co-ordinating with each other by a fixed law whatsoever things are sensible,* and it is a pure intuition. . . . Space is . . . subjective and ideal and proceeds from the nature of the mind by an unchanging law, *as a schema for co-ordinating with each other absolutely all things externally sensed.* (ibid., 85; emphasis Cassirer's; I follow the translation in Kant 1968, 65–66, 70)

Cassirer then adds: "Whoever recognizes this law and this schema, this possibility of relating point to point and connecting them with each other, has recognized space and time by their 'transcendental meaning'" (ibid.). How is this "pure intuition" of space and time manifested in the context of relativity?

> The most general meaning of this term . . . is merely that of the serial form of coexistence and succession. Nothing is thereby presupposed

> concerning specific metrical relations in the two. . . . When, *e.g.*, in the mathematical foundations of the theory of relativity the formula is deduced for the "distance" of the two infinitely close points $x_1\ x_2\ x_3\ x_4$ and $x_1 + dx_1$, $x_2 + dx_2$, $x_3 + dx_3$, $x_4 + dx_4$, this cannot indeed be conceived as a rigid Euclidean distance in the ordinary sense, since there is involved in it, by the addition of time as a fourth dimension, not a magnitude of space but rather one of motion; but the fundamental form of coexistence and succession and their reciprocal relation and "union" is unmistakably contained in this expression of the general line element. (ibid., 85–86; translation corrected)

But the line element is the fundamental invariant of the general relativistic group, that which, first and foremost, is univocally determined in general relativity. So, in effect, it is the idea of univocal determination generally, and specifically the univocal determination of the general forms of spatiality and temporality, that are here elevated to the status of the a priori, as necessary conditions for the possibility of the presentation of empirical objects in intuition, and, beyond that, for the possibility of scientific knowledge generally.

Cassirer's *Zur Einsteinschen Relativitätstheorie* was well received and widely read. Even Einstein, no close friend of neo-Kantianism, was moved to express his admiration for Cassirer's accomplishment, though he went out of his way to record carefully the specific nature of his most serious reservation, this in a letter to Cassirer of 5 June 1920, written right after Einstein read the book in manuscript:

> I can understand your idealistic way of thinking about space and time, and I even believe that one can thus achieve a consistent point of view. To me, as a non-philosopher, philosophical contrarieties appear more contrarieties of emphasis than contrarieties of a principled kind. What Mach calls *connections,* are for you ideal names, which experience first makes possible. But you emphasize this side of knowledge, whereas Mach wants to make them appear as insignificant as possible. I acknowledge that one must approach the experiences with some sort of conceptual functions, in order for science to be possible; but I do not believe that we are placed under any constraint in the choice of these functions *by virtue of the nature of our intellect.* Conceptual systems appear empty to me, if the manner in which they are to be referred to experience is not established. This appears most essential to me, even if, to our advantage, we often isolate in thought the purely conceptual relations, in order to permit the *logically* secure connections to emerge more purely. (EA 8-386)

Einstein's critique anticipates the reaction shortly to be encountered from thinkers like Schlick and Hans Reichenbach (see Howard 1994). He concedes, in effect, the Kantian point that scientific cognition requires some "conceptual functions" antecedent to experience, but he denies that these conceptual functions are forever fixed and unchangeable, independently of experience. Still, the concession is important. These a priori, if not apodictic, conceptual functions were very soon to be styled by Reichenbach the "constitutive" aspect of the a priori (Reichenbach 1920). And much of the most interesting philosophy of science in the decade of the 1920s was to concern the specific nature of the "constitution" here indicated (see again Howard 1994). It is to Carnap's first steps into that discussion that we now turn.

4. Carnap on Categoricity, Monomorphism, and the Constructionist Project

It is not obvious, at first glance, how, if at all, Carnap's doctoral dissertation, "Der Raum" (Carnap 1921), is related to the later project of the *Aufbau.* The question it asks about the nature of space differs significantly from the constructive project of the *Aufbau,* just as the broadly neo-Kantian philosophical spirit that informs the dissertation differs from the Fregean and Russellian spirit that informs Carnap's work throughout the rest of the 1920s. But absent major psychic disturbance, there are threads of continuity running through every philosophical career, however indistinct they may at some times appear. One of the tasks of the historian is to follow up those threads, not for the purpose of falsifying the career by imposing, in hindsight, a unity not in evidence during the actor's lifetime, but instead for the purpose of understanding better the many stages in that career by setting them in context. In the present case, I hope to understand the project of the *Aufbau* just a little bit better by exhibiting what I think is one of the more important threads of continuity in Carnap's career, a thread that starts with his early engagement with the problem of *Eindeutigkeit* and relativity that we have been tracing here. The reward for our effort will be a clearer understanding of one of the chief constraints under which the *Aufbau*'s project of a *Konstitutionstheorie* was expected to operate, nothing less, namely, than our old friend, the principle of *Eindeutigkeit.*

"Der Raum," which was written under the direction of the neo-Kantian Bruno Bauch, was an ambitious piece of work for a young thinker like Carnap. In it, he undertook to sort through the many confusions and contradictions that had infected the theory of space in the preceding several centuries, confusions and contradictions concerning the source of our knowledge of space, the nature of spatial objects, and the domain of applicability of the theory of space. He proposed to bring order to this domain by recommending a threefold distinction between what he termed "formal space," "the space of intuition" (or "intuitive space" for short), and "physical space." Most important for Carnap was clarifying the grounds for the Kantian claim that space, in some sense, is an a priori form of intuition.

According to Carnap, the three kinds of space stand in a sort of hierarchical relation to one another, determined, after a fashion, by their distance from the empirical. At the top we find formal space, R, comprising the most general structures or forms, with nothing necessarily "spatial" about them, which will then find more distinctively "spatial" instantiations in intuition and physical space. Next comes intuitive space, R', the objects of which are now distinctively spatial, but not with the particularity characteristic of concrete empirical objects. Instead, the objects inhabiting intuitive space are like those "essences" of objects presented in intuition whose comprehension was the goal of Husserl's program of *Wesenserchauung*. Finally, we have physical space, R'', at the bottom, the space of ordinary empirical objects.

Within each of these spaces we can distinguish still more varieties, again arranged hierarchically, but this time with respect not to their distance from the empirical but to their degree of determinateness and generality. Thus, at the top of each of these hierarchies we find n-dimensional topological space, R_{nt}, R_{nt}', R_{nt}'', followed by projective space, R_{np}, R_{np}', R_{np}'', and, finally, metrical space, R_{nm}, R_{nm}', R_{nm}''.

The mathematician is concerned with formal space and its properties. The geometer studies intuitive space and its properties. The physicist is concerned with physical space. When the latter sets about the task of determining which of the many possible Euclidean or non-Euclidean metrical spaces is the space of our physical universe, a problem arises. One can carry out the measurements that alone make possible the determination of the degree of curvature in a certain region of space, and hence the determination of its nature as Euclidean or non-Euclidean, only if one first makes some arbitrary metrical stipulations, conventions if you will, regarding the choice of a physical object or process that will be taken as the norm in our measurement of the metrical properties of physical bodies. But the conventional character of our choice of a measuring rod means that there is a significant degree of indeterminacy in our choice of a particular metrical geometry for the representation of spatial relations in physical space. Choose another measuring rod and you will be led to a different metrical geometry.

Here lies the problem with Kant's argument on behalf of the a priori character of Euclidean geometry as the geometry of the space of pure intuition: Euclidean geometry cannot enjoy this a priori status because it is not univocally determined as the form of physical space. Again, choose a different measuring rod and you get a different geometry.

Carnap then asks whether or not any of the spaces catalogued thus far do evince the kind of determinateness that would alone qualify them for a priori status. His answer is important for the larger story we are following here:

> We have in R_{3t}' the framework of intuitive space that was built up without the concepts of the straight line or the equality of intervals. We can thus build up the corresponding structure, R_{3t}''(three dimensional) topological, physical space, without having to decide about straightness and the metrical relations of the physical line. The only relationships necessary for arranging the empirically given physical-spatial structures in such a framework are the relationships of betweenness (incidence) of points, lines, surfaces, and spatial elements. These relationships for physical-spatial structures can be taken from experience without an agreement about any selected stipulation.... Only *topological* space (by which we always mean only the three-dimensional R_{3t}'') reproduces what is present in experience univocally [*eindeutig*]. By contrast, even projective space, R_{3p}'', is not univocal, because for its construction we must choose and establish a stipulation about straight lines, for which there are several possibilities. Still less are we constrained in the construction of a metrical spatial structure, R_{3m}''; here there are infinitely many different kinds. Which kind of structure we set up depends upon the metrical stipulation that we choose. (Carnap 1921, 37–38)

Why is topological space thus privileged? Carnap says that what we are trying to do here is to divide what is given to us as something integral in experience into two components, deriving from two different sources. The distinction between these two components is similar to, but not identical with, the distinction between content and form (*Stoff und Form*). Instead, he says, we want to draw a distinction within the realm of form between "necessary" and "optional" (*wahlfrei*) form. How do we do this?

> In order to determine about an empirical proposition whether or not it is a factual proposition [*Tatbestandsaussage*], and, in the final analysis, what in it concerns the factual [*Tatbestand*] and what corresponds to the optional form, we must inquire whether or not the empirical proposition remains valid for all possible formations — or, for the

> purposes of our investigation — all possible kinds of spatial structures. Expressed mathematically, this is the case if the content of the empirical proposition remains unchanged (invariant) with respect to the one-to-one, continuous, spatial reformations [*Raumumformungen*] (transformations). This holds for all topological propositions and only for these, i.e., for propositions about the betweenness and connection of spatial structures, and thus for all propositions regarding topological physical space R_{3t}'' and only for this. (ibid., 39)

In other words, topological space and topological relations are privileged because they are univocally determined as that which remains invariant under the transformations of the general relativistic class.

With this we have found our candidate for the status of the a priori form of pure intuition. What earns topological space this status is its relationship with the "factual" [*Tatbestand*]. Carnap writes:

> Insofar as experience is given in a univocal, necessary form, containing no freely chosen stipulations, we have called it the factual [*Tatbestand*]. Accordingly, only those spatial determinations that are contained in the factual can be conditions for the possibility of experience. And they are, as we have seen, only the topological relations, not the projective, and above all not the metrical relations. (ibid., 65)

Carnap notes the relevance of a linguistic analogy. The topological is to the metrical as the meaning of a proposition is to its specific expression in a given language. Just as meaning is invariant under translation, so the topological is that which is invariant under the class of one-to-one, continuous (topological) transformations.

What finally earns for the topological the status of the a priori is precisely its being thus univocally determined as the fundamental invariant in our experience of spatial relations:

> Frequently, in the discussion of this question, it is rightly noted that the "transcendental function" of space can only be ascribed to a univocal spatial form, and that, therefore, non-Euclidean spatial forms cannot be considered for this role. But one cannot draw from this correct assertion the conclusion that only Euclidean space can occupy this position. For this space is on a par with the former, and it can possess just as little or just as much univocalness as any one of the non-Euclidean spaces, such as that one with a uniform curvature of –20. The proper conclusion from that premise can instead go only to topological space, for it alone is set above those others as being completely univocal: The factual in experience cannot appear in several different topological forms. (ibid., 66)

It is worth pausing here to reflect on the place of Carnap's argument in the tradition that we have been tracking regarding *Eindeutigkeit* and relativity. Lest there be any doubt that it does stand squarely within that tradition, consider the sources to which Carnap refers us in his bibliographical note attached to the above-quoted passage from p. 39 of "Der Raum" where Carnap first argues that the factual is the topological because only the topological is determined univocally. The note is headed "Tatbestand" (The factual). For the idea that *"only spatio-temporal intersection* ('coincidence') is physically *determinable,"* Carnap refers us to (among other sources) Einstein 1916a; Petzoldt 1921b; Schlick 1918, 1920; and E. Cassirer 1921. For the idea that "therefore *only topological determinations are univocal,*" Carnap refers us to Schlick 1920. Most instructive is the comparison with Cassirer's position. Cassirer had argued that general relativity acknowledges the basic Kantian point about "pure intuition" exactly "in the concept of 'coincidence' to which the general theory of relativity ultimately reduces the content and the form of all laws of nature" (E. Cassirer 1921, 84), this being the way in which relativity satisfies "the one valid norm . . . the idea of the unity of nature, of univocal determination itself" (ibid.). In the same way Carnap, with a bit more mathematical sophistication, argues that what earns for topological intuitive space, R_{3t}', the status of an a priori form of intuition is precisely that it is the univocal, necessary form in which our experience of spatial relations is given. The only place where Carnap might be held to differ from Cassirer is in his employing the *Tatbestand* vocabulary as a further characterization of univocally determined topological relations, perhaps a reflection of his stronger, if still somewhat incipient, empiricist leanings.

The task of giving a scientific representation of a univocal topological *Tatbestand* had been for some time a central focus of Carnap's work, and it was to continue to be so over the next few years. But over the course of three or four years following the writing of "Der Raum" there are signs of growing despair on Carnap's part over whether or not a suitably univocal construction was possible. The specific idea was to construct an axiom system for a physical theory of space and time using just two primitives, the topological relation of the coincidence, *aKb* (*Koinzidenz*), of two world-points and the relation of temporal succession, *aZb* (*Zeit*), of two genidentical points on a world-line.[15] Carnap first took up the topic as a possible dissertation project in 1919 or 1920, his thinking at that time being recorded in a manuscript outline, "Die logischen Grundlagen der Kinematik" (RCC [Rudolf Carnap Collection] 081-06-01),[16] which he showed to the Jena physicist Max Wien, who thought it too philosophical for a physics dissertation, and to Bauch, who thought it too technical for a philosophy dissertation

(Carnap 1963a, 11). After completing his eventual dissertation, "Der Raum" (1921), and several other papers, Carnap returned to the project of a space-time axiom system in 1924, producing a more thoroughly worked out typescript of one-half of the projected work, "Topologie der Raum-Zeit-Welt" (RCC 081-02-07). This was never published, but its essential content was sketched in a 1925 paper, "Über die Abhängigkeit der Eigenschaften des Raumes von denen der Zeit," and in his 1929 *Abriß der Logistik,* with more detailed expositions being given many years later (Carnap 1954, 1958; see Carnap 1963a, 15).

In the 1920 outline, Carnap declared the goal of the project to be that of a "purely logical construction of kinematics," the latter being understood as a "system of formal relations of the 4-dimensional space-time-manifold." By a "purely logical construction" he intends a "construction on the basis of the basic theorems and concepts of general (deductive) logic (without the addition of specifically quantitative concepts)" (RCC 081-06-01, p. 1).[17] Such a purely logical construction had already been carried through for arithmetic by Frege and Whitehead and Russell. Carnap notes that, in the realm of geometry, such a construction had not yet been carried out for topology, although a start had been made in set theory, whereas Mario Pieri and Russell had already carried out such a construction for projective geometry and Giuseppe Peano, Pieri, Russell, and Hilbert had done the same for metrical geometry, although Hilbert had not produced a purely logical construction because he started from axioms construed as "fundamental facts of our intuition," as opposed to having reduced those axioms to fundamental principles of logic. In other words, what Carnap was proposing to do for his dissertation was to extend the logicist program beyond arithmetic and geometry into the foundations of physics.

In the case of kinematics, the construction should proceed only on the basis of concepts "whose realization can be postulated as being immediately identifiable [*unmittelbar erkennbar*]," avoiding all concepts that either are not "physically identifiable," such as absolute motion, or are reducible to other, simpler basic concepts, as with the example of the concept of mass. What are the admissible concepts for the desired construction of kinematics? There are three: (1) the identity of a physical point in a temporal succession; (2) the relationship of being "temporally earlier" for one and the same physical point; and (3) the spatio-temporal coincidence of two physical points. Inadmissible concepts include: distance; straight line; plane; the relation of spatial betweenness; spatial or temporal continuity; the relation of being "temporally earlier" or simultaneous for two *different* physical points; body; and rigid body (RCC 081-06-01, pp. 2–3).

When Carnap returned to this project in 1924, it was with considerably more sophistication about what would be required from both a technical and a conceptual point of view. The first two of the three basic concepts from 1920 are now collapsed into the one basic concept of the temporal succession of two genidentical points on a world-line, now denoted as *aZb*, the coincidence of two different world-points, now symbolized by *aKb*, remaining as the other basic concept. As before, Carnap is eager to stress the purely logical nature of the construction, which means, among other things, carrying through the construction in a purely coordinate-free manner and without reliance upon any specifically numerical concepts. In Carnap's words: "The topological properties of the physical space-time-world should be deduced axiomatically without the help of coordinates or any numerical measures whatsoever" (RCC 081-02-07, p. 8). And he once again situates the project squarely within the Frege-Russell logicist tradition: "Number-free determinations in the required sense . . . are provided by the *theory of coordinations* [*Zuordnungslehre*] (theory of relations), as the latter has been elaborated by Russell on the basis of the preparatory work of Schröder, Frege, and Peano. It is a special branch of symbolic logic" (RCC 081-02-07, p. 9).

The construction will be an axiomatic one.[18] That has implications for the way we conceive the two primitive concepts of coincidence and genidentical-temporal succession, for these are not to be given explicit definitions; instead, they are defined implicitly through the axioms. In Carnap's words: "These are not defined, their properties are given exclusively through the axioms. . . . The entire axiom system thus constitutes in a certain sense an implicit definition for the coordinations K and Z" (RCC 081-02-07, p. 20). Carnap surely understood from his reading of Hilbert, Schlick, Einstein, and other authors that implicitly defined concepts lack the specificity of explicitly defined ones, for he quickly adds:

> Strictly speaking, that class or coordination which appears everywhere in the defining axioms is only apparently defined by an implicit definition; instead what is defined is the class of those objects (mostly a class of classes or coordinations) that possess the properties indicated by the axioms. Thus, the coordinations K and Z, or the ordered pair K-Z, is not defined here, but rather the class of such pairs that satisfy the axioms. (RCC 081-02-07, p. 20)

At best what is explicitly defined in such a way is the class of ordered pairs of relations (or coordinations, to use Carnap's term) that evince the same structure as *K* and *Z*. In other words, at best the axiomatic approach determines a class of models up to the point of structural isomorphism.

Does Carnap's axiom system yield a suitably univocal determination of the topological space-time structure that he is aiming to characterize? Early in the 1924 manuscript Carnap asserts that the topological space-time structure is completely determined by the *K-Z* relations that are to be defined implicitly by the desired axiom system:

> *Thesis 1a.* From a topological point of view, the spatio-temporal events in the world are completely described [*vollständig beschrieben*] through the specification of the entire set of coordinations K and Z, thus through the specification of the extension of the coincidence- and the genidentical-temporal-relation. (RCC 081-02-07, p. 4)[19]

That the kind of completeness intended here is one sufficient to determine the truth or falsity of every topological proposition is made clear in the gloss on Thesis 1a that Carnap immediately proposes:

> Since every spatio-temporal proposition about the physical world must be present in such a complete description, we can reformulate the theses thus:
>
> *Thesis 1b.* Every spatio-temporal proposition of a topological kind about the physical world can be represented as a mere proposition about the coordinations K and Z. (RCC 081-02-07, p. 4)

What about the completeness of the axiom system itself? The question is not addressed explicitly by Carnap, but it is touched upon ever so briefly in a discussion of the independence of the axioms. Carnap notes that if we discover that an axiom, *Am* (e.g., *A1, A2, . . .*), is not independent, we can either drop *Am* from the axiom system, and replace it wherever it occurs in a proof with the conjunction of some or all of the other axioms from which it is derivable, or we can replace one of the axioms from which it is derivable, say *An,* with a weaker axiom, *An′* in which case the stronger axiom *An* would have to be replaced wherever it occurs in a proof either by the conjunction of *An′* and *Am* or, in some cases, by *An′* alone. In either case, the axiomatic bases for some proofs would change. Nevertheless, as Carnap points out, the validity of all theorems is unaffected: "All valid theorems remain valid, all invalid theorems remain invalid, all undecidable ones undecidable." And then, in a parenthesis, he adds the following about the existence of these undecidable propositions: "(That the latter case arises means that the axiom system is *not definite;* in a system that concerns an extralogical domain of such manifold relations, that can hardly be otherwise)" (RCC 081-02-07).

A proposition is undecidable if and only if neither it nor its negation is a theorem, which is to say that neither it nor its negation is a consequence of the axioms. A formal system that leaves undecidable some of the well-formed formulas of its associated formal language is incomplete in the syntactic sense, and more than likely also in the semantic sense of there being truths under the intended interpretation that are not theorems, assuming, that is, a semantics adequate for a truth definition. Such a system is also noncategorical, for in the normal case one can construct models other than the intended interpretation in which one or more of the undecidable sentences true under the intended interpretation is false, and these models will not all be isomorphic to one another or to the intended interpretation.

How much of this was understood by Carnap in 1924 is unclear, for he says nothing more about the problem of incompleteness or noncategoricity in the manuscript of the "Topologie der Raum-Zeit-Welt." This is somewhat surprising, since it should have been immediately obvious to him that an axiom system with undecidable propositions will not be complete in the way necessary for a properly univocal representation of a putatively univocal topological *Tatbestand.* This fact about axiom systems with undecidable propositions certainly had become clear to him by 1927, as we shall see below when we turn to the discussion of his 1927 essay "Eigentliche und uneigentliche Begriffe." But there is another manuscript from roughly this same time as the *K-Z* system manuscript that gives us a hint that Carnap was worrying about the problem.

The manuscript in question, which is dated "27.12.22–25.1.23," contains the earliest extant development of the method of "quasi-analysis" (*Quasizerlegung*), which was later employed in the *Aufbau* but for a seemingly different purpose than that for which it was first invented. In his "Intellectual Autobiography," Carnap says that he was moved to develop the method of quasi-analysis by his recognition, thanks to the influence of Gestalt psychologists like Max Wertheimer and Wolfgang Köhler, that the world is not presented to us in sense-experience in the form of already realized individual sense-data, but rather in the form of an instantaneous visual field or maybe even an instantaneous total experience. If the constructional project of the *Aufbau* was to mirror, to some extent, actual psychological processes, then these total instantaneous experiences, which Carnap dubs "elementary experiences" (*Elementarerlebnisse*), should form the starting point of the construction, and not the individual sense-data represented by, say, Mach's "elements of sensation." Something resembling sense-data will then be a product of a later stage in the construction, the process for generating them out of holistically conceived elementary experiences being what is designated as "quasi-analysis." In Carnap's words: "I developed a method called 'quasi-analysis,' which leads, on the basis of the similarity-relation among experiences, to the logical construction of those entities which are usually conceived of as components" (Carnap 1963a, 17).

The declared motivation in the *Quasizerlegung* manuscript is quite different. There is no mention of a constructional project like that of the *Aufbau,* nor any mention of any concerns about the genetic, epistemic, or logical priority of one or another kind of element in experience. Instead, the chief concern is with the respective merits of two ways of approaching the understanding of a given domain of objects. The first is "analysis into components" (*Zerlegung in Bestandteile*). It proceeds by specifying for each individual object in the domain all of the properties that characterize it or all of the components of which it consists. The second is a "relational description" (*Beziehungsbeschreibung*). It proceeds by specifying all of the relations that obtain among the objects in the domain.

What chiefly distinguishes the two approaches is that the first, analysis, says something about the individual elements themselves, independently of the other elements, whereas the second, the relational description, concerns only the relation of one element to one or several other elements. A relational description has the advantage of not overstepping the given domain of objects, since it does not decompose the elements of the domain to be described into a set of components themselves not contained in that domain. Carnap says that a relational description is thus an "immanent treatment of the domain." A relational description has the disadvantage, however, of seeming difficulty in the treatment of individual objects, since it says nothing about any individual object without some reference to other objects, which are themselves characterized only by yet further references to still other objects, and so on. The goal of quasi-analysis is to repair this latter defect:

> A procedure will now be explained that makes it possible to modify a given relational description in such a way that it retains the character of an "immanent treatment of the domain," but takes on the formal structure of a Zl [an analysis] and thereby makes possible the individual treatment of the elements. This modification will be designated *Quasi-analysis.* (RCC 089-65-01, p. 2)

The details of the method need not detain us here.[20] Its purpose is what concerns us. As noted, there is no mention in this manuscript of any specifically epistemological aim, such as recovering the constituent units of holistically conceived elementary experiences. The one extended example of the application of the method of quasi-analysis is the same as that given in the *Aufbau* (1928a, §71), namely, musical tones, including complex chords, experienced not as a collection of units but as unanalyzed wholes. But perhaps more telling about Carnap's original motivations is the list of examples of the methods of analysis and relational description that Carnap gives on page 2, this by way of illustrating the strengths and weaknesses of the two approaches:

> *Examples* of *Zl [analysis]:* Description of the set of conic sections through specification of the characteristics of the individual sections; description of a curve through specification of the coordinate equation, thus the ordinate of the point belonging to each individual abscissa; description of a physical state through specification of the values of one (or several) state variables for each position; chemical description of a given substance through specification of its composition out of chemical elements; chronological table of historical persons through specification of birth and death years of each individual.
>
> *Examples* of *relational description:* Description of a geometrical figure consisting of points and lines through specification of its incidence relations; description of a curve through specification of its natural equation, thus the positional relation of each line differential to the set of the foregoing ones; description of a physical change of state through spatio-temporal differential equations, thus specification of the relation of the value of the relevant state variable at any one position and any one time to its values in the neighborhood of the position at the previous time; description of a set of persons through a genealogy, thus through specification of the relations of relatedness that each person bears to the other persons in the set. (RCC 089-65-01, p. 2)

Notice that all but one of the examples in each category is drawn from physics and chemistry. Notice as well that the first and second examples in each category emphasize the same property of a relational description in a geometrical context that Carnap was holding up as the virtue of his axiomatic space-time topology constructed with the tools of the theory of relations, namely, its avoidance of all reliance upon coordinate representations. This suggests that, Carnap's own later recollections notwithstanding, the real motivation behind the method of quasi-analysis was, in fact, to be found in problems encountered in his work on axiomatic space-time topology, not the constructional project of the *Aufbau.* That should not surprise us if we pause to consider the fact that, judging by the evidence of his publications and his unpublished manuscripts from this period, the foundations of physics was the major, indeed, perhaps the exclusive, focus of Carnap's attention prior to the *Aufbau.*

If, however, the motivation for the method of quasi-analysis is to be found in Carnap's work on the foundations of physics, then another question arises: Why would what was earlier and elsewhere being singled out as a virtue of axiomatic constructions via the theory of relations, namely, its avoidance of devices like coordinates that facilitate immediate access to individuals like the points of a manifold, now be regarded as a defect in need of repair? My guess is that it was precisely Carnap's coming to grips with the realization that the axiomatic approach via the theory of relations will inevitably lead to representations that lack the univocal character that Carnap — like Petzoldt, Mach, Schlick, Cassirer, and Einstein before him — thought must be possessed by scientific representations adequate to the univocal reality they aim to describe. Quasi-analysis can be seen, in this light, as an effort to preserve the virtues of a relational description as an "immanent" description, while still making possible an access to specific, individual objects, which, in their specificity and individuality, presumably possess the univocity, the *Eindeutigkeit,* that their implicitly defined counterparts lack.[21]

With this as background, let us, finally, turn now to Carnap's sadly not-well-enough-known 1927 essay, "Eigentliche und uneigentliche Begriffe" (Proper and improper concepts) (Carnap 1927), for it is here that we finally find Carnap making explicit the worry about *Eindeutigkeit* that I have argued was hovering in the background of his struggles with the space-time topology in the early 1920s. The argument of this essay is also crucial to an understanding of Carnap's constructionist project in the *Aufbau.* The latter is an exercise in defining, via construction from elementary experiences, the entire conceptual apparatus of science. But "Eigentliche und uneigentliche Begriffe" is nothing less than a statement of the minimum necessary conditions that must be satisfied by any such definitional enterprise in the sciences. It represents, as it were, the most basic conditions that have to be satisfied by the program of the *Aufbau.* What are these constraints?

Carnap begins by explaining his terminology. Proper concepts are those for which all questions of application can be decided unambiguously or univocally (*eindeutig*). The concept either applies or does not apply; there is no other possibility. Normally, proper concepts that are not themselves fundamental concepts are introduced via explicit definitions of the well-known kind. When such a derivation is given for a proper concept, we say that it is thereby "constituted." And in this way we can arrange all of the concepts in a given domain in a "constitution system." The most important kind of proper concepts is that of "real concepts," concepts that pertain to real objects. It is for their sake that we do science, and they too can be arranged in a constitution system.

Improper concepts are those introduced not via explicit definitions, but instead by means of an axiom system, via implicit definitions. However useful this manner of introducing concepts may prove in mathematics, it suffers from one fundamental disadvantage. It is that many axiom systems fail to constrain the class of their possible models up to the point of isomorphism and so fail to be "monomorphic" (the more common modern expression is "categorical"). They are, instead, polymorphic and so too, by extension, are the "improper concepts" implicitly defined by them.

There is an important connection between monomorphism and decidability. An axiom system, *AS,* is decidable (*entscheidungsdefinit*) if and only if for every sentence, *S,* concerning the concepts defined by *AS,* either *S* or its negation can be derived from *AS* (this is the property now commonly designated as syntactic completeness). An axiom system, so Carnap argues, is decidable if and only if it is monomorphic. Since many axiom systems are not monomorphic and hence not decidable, so too the concepts implicitly defined by those axiom systems fail to be decidable in their application.

Quite apart from its importance in the development of Carnap's thinking, this essay should be appreciated for its being the first place where the modern concept of categoricity, or monomorphism in Carnap's terminology, is clearly defined and its relation to issues of completeness and decidability clearly expounded. Moreover, it was through Carnap's relations with Kurt Gödel and Alfred Tarski that the concept of categoricity later made its way into formal semantics.[22]

In addition to the undecidability that is a consequence of an axiom system's failure to be monomorphic (its being noncategorical), there is the further problem of the indefiniteness or indeterminacy (*Unbestimmtheit*) in the application of concepts defined implicitly by an axiom system. Carnap's favorite example is that of the Peano axioms. There are many possible models for the Peano axioms, ranging from the numbers themselves, to sequences of temporal points, to sequences of spheres of different sizes, and so forth. And while these models may all have in common the same form for the progression of objects contained within them, which means that we can determine definitively whether or not a sequence of objects constitutes such a progression, there is no determining definitively of any one object that may or may not be a member of one of the mentioned sequences, say a particular sphere, whether or not it is a "number" in the Peano sense.

The situation is even worse if we are implicitly defining not a single concept, like "number in the case of the Peano axioms," but a set of concepts, such as the pair "point" and "line" as these are defined implicitly by Hilbert's axioms. For now not only can we not determine definitively about an individual object whether or not it is, say, a point; we cannot even decide about a whole set of such objects whether or not they qualify, collectively, as points, this because of the famous "duality" of Hilbert's axiom system: if one systematically interchanges the terms "point" and "line" everywhere they occur in Hilbert's axioms, one comes up with exactly the same set of axioms that one started with. Thus, just as we have an axiom that states that "two points determine a line," so too we have an axiom that states that "two lines determine a point."

One should be careful to distinguish the problem of indeterminacy in application from the problem of undecidability. Indeed, Carnap can be faulted for having blurred this distinction. It is important to keep the two separate because, as Carnap acknowledges but does not stress, the problem of indeterminacy — if, in fact, it is a problem — is generic to all axiom systems and all implicitly defined concepts, whereas the problem of undecidability is not. There are some monomorphic axioms systems, though they may not be as common as the polymorphic ones. Moreover, Carnap had good reason to suspect that polymorphism may be the norm in axiom systems that will find useful employment in the sciences, for the Lowenheim-Skolem theorems were just then bringing to light the noncategoricity or polymorphism of Zermelo-Fraenkel set theory and its cousins. Still, there was in 1927 as yet no general proof of the necessity of noncategoricity; that was to come only in the 1930s as a corollary to Gödel's first incompleteness theorem. So in 1927, Carnap still had no very good reason for declaring all constitution of concepts via implicit definition to be suspect. The irony, of course, is that the program of constitution via explicit definition will fail, leaving us no other approach but that of Hilbert, with all its many shortcomings from the point of view of one like the Carnap of 1927 who wants the most intimate, point-by-point contact between theory and reality.

Why is such undecidability and indefiniteness unacceptable in a constitution system for the proper real concepts of the sciences? Here is Carnap's answer:

> In the systematic construction of the knowledge of reality, real concepts are constituted step-by-step. As a term in this construction, every real concept possesses an immediate relation to reality. By contrast, improper concepts so-to-say hover in midair. They are introduced through an *AS* that does not refer directly to reality. The axioms of this *AS* and the theorems deduced from them do not properly constitute a theory (since, indeed, they deal with nothing determinate), but only a theory-schema, an open form for possible theories. But if, in the system of knowledge, a real concept appears for which it can be shown empirically that it has the formal makeup indicated in the *AS* for the improper concept, then the *AS* has found a realization: in place of the improper concept, which is just a variable, the real concept can be substituted. Thus, the structures of physical space (points, lines, etc.) evince empirically the makeup that the axioms of geometry express for "points" (in the improper sense), etc.... Through this contact between the real concept and the axioms (the former satisfies the latter), the connection to the entire theory-schema based on the *AS* is accomplished in one stroke. The blood of empirical reality streams in through this one point of contact and flows into the most highly branched veins of the heretofore empty schema, which is thereby transformed into a filled-out theory. (Carnap 1927, 372–73)

It is thus avowedly an epistemological agenda that energizes Carnap's campaign against implicit definition and polymorphic concepts, namely, the desire to secure the most stable possible linkages between theory and experience.

In this respect Carnap's thinking is evolving in a direction similar to that already much in evidence in the work of his logical empiricist colleagues, Reichenbach and Schlick, both of whom worried that concepts implicitly defined by an axiom system were so infected by convention as to be incapable of being brought into a relation with experience secure and determinate enough as to make possible an empiricist answer to neo-Kantian assaults on general relativity (see Howard 1994). Explicit definitions in Carnap's constructional system will then play an epistemic role similar to what Reichenbach and Schlick termed "coordinating definitions," these being the only remaining elements of convention in science, such that, once they are fixed by convention, the truth or falsity of all remaining "empirical" propositions is determined univocally by experience.

When viewed, however, in the context of Carnap's growing worries about the failure of monomorphism, categoricity, or *Eindeutigkeit* in his space-time topology, the argument of the 1927 paper takes on another, somewhat different, significance. For the shift from concepts implicitly defined by an axiom system, as exemplified by the *K* and *Z* relations in the *K-Z* system, to explicitly defined concepts can be seen also as an effort to restore to science a measure of univocity in its representation of a univocal reality or *Tatbestand.*

From an even longer historical point of view, Carnap's turn from implicit to explicit definitions in 1927 represents the triumph of Kant himself over the Marburg neo-Kantians. For the tendency among Marburgers from Cohen to Cassirer had been to exalt the understanding over intuition, with Cassirer arguing that the univocity otherwise characteristic of the real as presented to us in intuition could be recovered at the level of theory by the accumulation of a sufficient number of axioms as to constrain the models for those axioms up to the point of uniqueness or at least isomorphism. What Carnap realizes in the mid-1920s is that, owing to the polymorphism endemic among even moderately powerful scientific theories, that project is doomed to failure. Univocity is to be achieved only by way of more explicit, direct linkages between theory and experience.

The story does not end here, however. For as Carnap himself soon came to realize, the constructionist program of the *Aufbau* also fails, and it fails, ironically, precisely at the point emphasized in "Eigentliche und uneigentliche Begriffe." For one thing, explicit eliminative definitions turn out not to work when it comes to disposition terms, like "soluble," which are employed everywhere in science. And for another thing, the naive faith that explicitly defined concepts could be related to immediate experience in a determinate fashion was to be overturned in the course of the protocol-sentence debate, where Carnap eventually conceded Otto Neurath's point that even at the level of observation sentences it would be impossible to secure determinate, sentence-by-sentence links between individual empirical propositions and an empirical content supposedly associated uniquely with each. Instead, so Neurath argued, a measure of holism, and hence pervasive conventionality, obtains even at the level of observation sentences, and so also, by extension, at all levels of science (see Carnap 1932a, 1932d; and Neurath 1932b). That being so, there will no longer be any epistemological reasons for denying ourselves what all concede to be the advantages of the axiomatic approach, even if that means settling for implicitly defined primitive terms. And once we also accept the lesson of Gödel's first incompleteness theorem, with its corollary that any first-order formal system as powerful as or more powerful than elementary arithmetic will be noncategorical, there will be no avoiding the problem of noncategoricity, the failure of *Eindeutigkeit,* virtually anywhere in science.

The problems of *Eindeutigkeit,* reality, and the foundations of knowledge were very much in the air in 1927. It is instructive to compare Carnap's attitude toward implicit definition, which he regards as suspect even if it succeeds in the univocal determination of the meanings of concepts, with the attitude of Hermann Weyl, as expressed in his *Philosophie der Mathematik und Naturwissenschaft:*

> One might have thought of calling an axiom system complete if in order to fix univocally the meanings of the basic concepts present in them it is sufficient to require that the axioms be valid. But this ideal of uniqueness cannot be realized, for the result of an isomorphic mapping of a concrete interpretation is surely again a concrete interpretation. Hence the final formulation has to be as follows: an axiom system is complete, or *categorical,* if any two concrete interpretations of it are necessarily isomorphic. In this sense the categoricity[23] of Hilbert's axiom system of geometry is guaranteed. Indeed it can easily be shown that a space satisfying these axioms is isomorphic to the algebraic model provided by Descartes' analytic geometry.
>
> A science can only determine its domain of investigation up to an isomorphic mapping. In particular it remains quite indifferent as to the "essence" of its objects. That which distinguishes the real points in space from number triads or other interpretations of geometry one can only *know [kennen]* by immediate intuitive perception. But intuition is not blissful repose never to be broken, it is driven on toward the dialectic and adventure of cognition [*Erkenntnis*]. It would be folly to expect cognition to reveal to intuition some secret essence of things hidden behind what is manifestly given by intuition. The idea of isomorphism demarcates the self-evident insurmountable boundary of cognition. (Weyl 1927, 21–22; translation corrected)

Weyl was writing, of course, from a point of view radically different from Carnap's. Husserl was probably the major intellectual influence on Weyl, who seems to have regarded the axiomatic method as the royal road to Husserlian *Wesenserschauung.* History has judged Weyl the winner.

Notes

I wish to thank the National Science Foundation (Research Grant no. SES-8421040), the American Philosophical Association, the Deutscher Akademischer Austauschdienst, and the University of Kentucky Research Foundation for having provided essential support for part of the research for this essay. My thanks also to John Stachel and to Robert S. Cohen for extending the hospitality of, respectively, the Center for Einstein Studies and the Center for the Philosophy and History of Science, both at Boston University, where some of the research for this essay was carried out. W. G. Heverly, Curator of the Archives of Scientific Philosophy at the University of Pittsburgh, provided essential help in securing copies of materials from the Rudolf Carnap Collection (RCC). Finally, I thank the Hebrew University of Jerusalem, which holds the copyright, for permission to quote from Einstein's unpublished letters; items from the Einstein Archive (EA) are cited by their numbers in the Control Index.

1. For Einstein's version of the argument, see Einstein and Grossmann 1913, 260, and Einstein 1914, 1067. For more recent discussions, see Stachel 1989a and Norton 1984.

2. Einstein probably also benefited from criticisms of the hole argument originating with Hilbert's group in Göttingen, with whose members Einstein had been in fairly regular contact since at least July 1915, the time of a week-long visit by Einstein to Hilbert's Göttingen seminar. For more on the Göttingen critique of the hole argument, see Howard and Norton 1993.

3. That all world-lines are, in fact, geodesics in the underlying manifold is the thesis at issue in what is known as the problem of motion in general relativity. Already in 1915, Einstein had good reasons for assuming this, but proving that world-lines are geodesics turns out to be a nontrivial problem. For a history of the problem of motion in general relativity, see Havas 1989.

4. An example of this reading is found in Friedman 1983b, 22–25; for a detailed critique of this interpretation, see Howard 1992, 212–14 n. 25.

5. For Einstein's reading of the *Mechanik* and the *Wärmelehre,* see Einstein to Mileva Marić, 10 September 1899 (Einstein 1987, doc. 54, pp. 229–31), especially n. 8.

6. The *Zeitschrift für positivistische Philosophie* was the journal of the Gesellschaft für positivistische Philosophie, which had been organized by Petzoldt in 1912, with Einstein being listed as one of its founding members; see Howard 1992, 174–75.

7. The "kranker Herr" was Einstein's friend, Dr. Ludwig Kraft; I thank Robert Schulmann, editor of *The Collected Papers of Albert Einstein,* for this identification.

8. In his *Anhang* to the eighth edition of Mach's *Mechanik* (Mach 1921) — entitled "Das Verhältnis der Machschen Gedankenwelt zur Relativitätstheorie" (Petzoldt 1921a) — Petzoldt writes that Einstein read the 1914 essay "Die Relativitätstheorie der Physik" and expressed to Petzoldt his "agreement" (*Zustimmung*) with it (Petzoldt 1921a, 494).

9. For more on the possible sources for Einstein's conception of *Eindeutigkeit,* see Howard 1992.

10. Helmholtz developed his influential "sign-theory" of knowledge in various of his writings, including, most importantly, his "Die Thatsachen in der Wahrnehmung" (Helmholtz 1878a), which was included in the Hertz/Schlick edition of Helmholtz's epistemological writings (Helmholtz 1921), with extensive annotations by Schlick.

11. Einstein read Schlick's essay in mid-December 1915, immediately after completing his work on the final formulation of general relativity and during the time when, as we saw above, he was explaining the point-coincidence argument to friends like Besso and Ehrenfest. On 14 December, Einstein wrote to Schlick: "Yesterday I received your essay and I have already studied it through completely. It is among the best which have until now been written about relativity. From the philosophical side, nothing appears to have been written on the subject which is nearly so clear. At the same time, you really have complete command of the subject. There is nothing in your exposition with which I can find fault" (EA 21-610).

12. For more on the Duhemian, underdeterminationist moment in Einstein's thinking, see Howard 1984, 1990, 1993, 1994.

13. This point is important enough to bear repeating: Einstein is saying that there may be a multiplicity of empirically equivalent theories that disagree about matters of deep ontology, each telling a different story about, say, the ontology of space-time events. Still, each of these theories, considered by itself, would presumably be expected to satisfy the *Eindeutigkeit* principle, meaning that whatever story it tells about this deep ontology, it tells *only one* such story.

14. Dedekind himself writes: "A thing is completely determined by all that can be affirmed or thought concerning it. A thing *a* is the same as *b* (identical with *b*), and *b* the same as *a*, when all that can be thought concerning *a* can also be thought concerning *b*, and when all that is true of *b* can also be thought of *a*" (Dedekind 1893, 44).

Carnap will argue in 1927 that a fundamental drawback to the constitution of objects through implicit definition via axioms is precisely this circumstance that the things thus defined are given not individually, but only as a community of objects; see below, section 4.

15. Carnap borrowed the concept of genidentity from the psychologist Kurt Lewin (Lewin 1923).

16. The manuscript bears a date in Carnap's hand, "June 1920."

17. This and all following quotes from the Rudolf Carnap Collection (RCC) are quoted by permission of the University of Pittsburgh. All rights reserved.

18. In trying to situate Carnap's project of a space-time topology in its proper historical context, we should not be too quick to assume an inherent opposition between the Hilbert program and the Frege-Russell program, as least not as Carnap understands them here. For while Carnap explicitly employs the formal tools of Russell's theory of relations, he nevertheless also sees his task as being continuous with Hilbert's task in providing an axiomatic foundation for geometry. The only place where he differs significantly from Hilbert concerns the question of the grounding of the axioms, Carnap repudiating what he takes to be Hilbert's grounding of the axioms in intuition. Most important is his clear statement, quoted just below, that an axiom system yields only implicit definitions of its primitives, this idea being a hallmark of the Hilbert approach, one stressed by both Schlick (1918, §7) and Einstein (1921a). Nor was Carnap alone in thinking the Hilbert and the Russell programs compatible, as witness Cassirer's remark, quoted above, to the effect that "Hilbert's geometry has been correctly called a pure theory of relations" (E. Cassirer 1910, 94).

19. Somewhat more problematic than Carnap's Thesis 1a is his Thesis 2a: "From a metrical point of view as well, the spatio-temporal events in the world are completely described through the specification of the entire set of coordinations K and Z" (RCC 081-02-07, p. 4), about which he adds: "Thesis 2a incorporates the idea emphasized in all presentations of the relativity theory since Einstein that out of the entire factual content of experience only the coincidences between two world-points of two world-lines will be utilized in physics" (RCC 081-02-07, p. 5).

20. For an account of the method, see Carnap 1928a, §§69, 71–73.

21. The note of skepticism about the virtues of a relational description in the *Quasizerlegung* manuscript stands in sharp contrast to the staunch advocacy of the relational approach in the manuscript of the *K-Z* system, so much so that one might doubt Carnap's recollection that it was at "about this time [1923 or 1924]" (Carnap 1963a, 15) that he returned to the *K-Z* system after the 1920 outline of the project. The manuscript of the *K-Z* system bears a date of "1924" in pencil on the first page, but that date is most likely not contemporary with the production of the manuscript, since the "1" in "1924" is written not in the German style (with a hook to the top left), but in the American style (with a plain top), a style that Carnap reportedly adopted only after his emigration to the United States

in December 1935. Thus it is possible that the date was added only when Carnap ordered his papers in the early 1960s in preparation for writing his intellectual autobiography for the Library of Living Philosophers volume. If so, the date on the manuscript may be no more accurate than Carnap's forty-year-old memories of the circumstances surrounding its production. Indeed, the differing attitude toward the method of relational description in the 1923 *Quasizerlegung* manuscript and in the *K-Z* system manuscript suggests that the *K-Z* system manuscript was produced prior to the *Quasizerlegung* manuscript. My thanks to W. G. Heverly, curator of the Archives of Scientific Philosophy, for his help with this question.

22. Carnap himself credits Herbert Feigl with inventing the "monomorph-polymorph" terminology; see Carnap 1927, 365.

23. The original German says *vollständigkeit,* meaning completeness.

Einstein Agonists: *Weyl and Reichenbach on Geometry and the General Theory of Relativity*

> These facts alone [the proportionality of inertial and gravitational mass] cannot bring us very far, for their content is still too paltry; one must proceed to speculation, a rather new and uncommon research method in physics . . . [that] Hilbert calls "the axiomatic method."
>
> Gustav Mie 1917

I think it is fair to say that most contemporary readers of Hans Reichenbach's works on the epistemology of geometry have not considered them in the scientific context of their origin, that is to say, against the background of activity in the small but vigorous community of general relativists in the decade or so after the inception of the general theory of relativity (GTR) in November 1915. Among these activities, perhaps the best known to the history of science are the efforts of W. De Sitter, A. Friedmann, G. Lemaître, and others to work out the cosmological implications, including the initial models of an expanding universe, of Einstein's field equations of gravitation. But this was also a period characterized by bold attempts to extend the "geometrization" of gravity, achieved in the GTR, to encompass also the electromagnetic field and thereby to encompass all (known) physical forces within a common physical and geometric framework. In many ways the most significant of these attempts was the "unified theory of gravity and electromagnetism" first put forward in 1918 by the mathematician Hermann Weyl. The theory, of course, was ultimately not successful; Weyl himself abandoned it with the advent of the new quantum mechanics in 1925–26. But it spurred related attempts by A. Eddington, T. Kaluza, and also Einstein, the latter thereby embarking on a search that, in tandem with his nonacceptance of the fundamentally probabilistic character of quantum mechanics, led him into what much of the rest of the theoretical physics community viewed as a scientific wilderness from which he would not again emerge.

My concern here is to show that Weyl's proposed "broadening" of the geometrization of the GTR comprises a particularly relevant context in which to situate the development of Reichenbach's neoconventionalist[1] account of the metric of space-time. In the GTR the "entanglement" of physics and geometry has already occurred in that the metric of space-time is no longer accounted as a rigid structure, fixed for all time, but as dynamically dependent upon (local) matter and energy distributions. But the philosophical questions prompted by the GTR concerning the nature of the geometry of space-time are rendered even more salient in a theoretical context that aspired to bring electromagnetism together with gravitation (and thus in principle all known physical phenomena) into a single unifying representation. In these heady days prior to quantum mechanics, the tantalizing prospect appeared within grasp of a unified theory in which both physics and space-time geometry are constructed *ab initio* within a common geometrical framework. Given the speculative (if not premature) character of these efforts, an inevitable ideological disagreement ensued concerning whether the GTR was a completed theory, and indeed there was even disagreement over what sense of "complete" might here be involved. From one perspective, which corresponds to Einstein's own path of discovery, the GTR is seen as the culmination of a program, initiated by the special theory of relativity (STR), of successively dismantling the concept of a privileged frame of reference. In this light, the GTR is, in principle, to implement a full relativization, eliminating also the privileged inertial frames of the STR. Of course, this was not quite achieved; the GTR retains the conception of "local inertial frame." But from the vantage point of unification, the GTR appears as a necessary yet incomplete step toward a geometric unification of all of physics. Reichenbach and Weyl are exemplars of these quite distinct and rival appraisals of the GTR. What lends heightened interest to the confrontation is that Einstein himself begins this decade as the sole proponent of the former perspective and ends it as virtually the only practitioner of the latter. Thus it is that Reichenbach and Weyl can both appear as champions of Einstein, that is, of different Einsteins.

Weyl and Reichenbach accordingly stand on opposite sides in a resulting debate that ostensibly turns on whether measuring rods and clocks can or should play an epistemologically fundamental role in the new theory. According to Weyl's "broadened relativity theory" (*erweiterte Relativitätstheorie*), the behaviors of such complicated material structures as rods and clocks can only be data to be explained, that is, to be derived from the equations of the "total field," and not stipulatively primitive "facts" licensed in the physical definition of metrical notions.

And it is actually Reichenbach who, early in his career as philosophical savant regarding the theory of relativity, takes up the task of combating the Weyl heresy concerning rods and clocks, for epistemological reasons that are significantly different from those of both Einstein and Pauli, Weyl's leading critics. Reichenbach continues to adhere to the view of physical geometry, rejected by Weyl, that stems from the Helmholtz postulate tying the notion of congruence to the possibility of free mobility of rigid bodies in space (which is possible, of course, only in spaces of constant curvature). Hence, for Reichenbach, the metrical notion of congruence is physically definable (via a "coordinative definition") by the stipulated coincidence of end points of rigid rods, both before and after an intervening spatial separation. The collision of Reichenbach and Weyl occurs precisely here over the meaning of *congruence* in the new context of the GTR and its possible generalization. As we shall see, Einstein occupies a somewhat ambiguous position in this controversy: defending (though increasingly pragmatically) the use of rods and clocks as legitimate macroindicators of the metric interval, while, in his own odyssey of unification, proceeding epistemologically unhindered, even to the extent of suggesting a fundamentally nonmetrical theory as the basis of a unified physics.

In a previous essay (Ryckman 1994), I sought to show that the central component of Reichenbach's neoconventionalism regarding metrical determinations of space-time, an *agreement* regarding the nonexistence of "universal forces," pertains above all to the challenge posed by Weyl's theory to the use of rods and clocks as fundamental posits within the GTR. To be sure, the presentation of this position in his "mature" work of 1928, and especially in the English translation (1958), which omits a crucial appendix, obscures this orientation. This essay will consider the response to Weyl's challenge that is more robustly evident in an earlier work of Reichenbach's, the *Axiomatik der relativistichen Raum-Zeit-Lehre* (1924), an expressly "epistemological" axiomatization of relativity theory and a defense of rods and clocks upon which the later work relies. Here, in treating rods and clocks as empirical postulates of the theory, Reichenbach directly opposes Weyl's proposal for empirically constructing the metric of space-time without them. Along the way, we shall try to illustrate just how inappropriate Reichenbach's approach is to an account of the metric in the GTR.

In the following, a presentation of the relevant parts of Weyl's theory (in §1) is countered with a somewhat detailed consideration and critique (in §3) of the character of the defense of rods and clocks presented in this early work of Reichenbach. Interspersed between is a section tracing Einstein's changing attitude. Our examination of these encounters shows that a profound difference over the very meaning of "geometry" in the new context of the GTR underlies the ostensibly epistemological issue concerning the use of rods and clocks as foundational concepts. In addition, our discussion points beyond to a larger implication concerning the legacy of Reichenbach's method of "coordinative definitions" as taken up within the "received view" of scientific theories.

1. Weyl

1.1 A Hidden History

With an enthusiasm that could scarcely be disguised, Hermann Weyl wrote from Zurich to Einstein in Berlin on 1 March 1918: "This day, as I believe, I have succeeded in deducing electricity and gravitation from a common source." The ingenious character of Weyl's achievement was unmistakable and did not fail to make the corresponding impression on Einstein, who wrote back on 6 April that it was "a stroke of genius of the highest magnitude." Yet within just a few more days, on 15 April, Einstein was to object that despite its seductive mathematical elegance, Weyl's theory was empirically untenable, leading to consequences that did not seem to be in accord with implacable observational fact, the constancy of atomic spectral lines.[2] Einstein's objection was given a rigorous formulation by the twenty-year-old Wolfgang Pauli Jr. in his canonical survey of the theory of relativity published in 1921. And, so the usual story goes, the objection sealed the fate of Weyl's theory, even before the advent of quantum mechanics in the mid-1920s changed forever the classical framework of field-matter interactions. As a result, Weyl's theory is remembered today, if at all, only as the locus of origin of the somewhat misnomered notion of "gauge invariance," which — now taken as involving a factor of phase, and not of length, as Weyl originally thought — has become a major theoretical construct in the physics of elementary interactions.[3]

Like many other chronicles of the history of modern science, this capsule summary of the "received view" of the fate of Weyl's theory achieves its streamlined finality at the rather considerable cost of a good bit of inattention to historical and scientific detail. For one thing, Weyl countered the Einstein-Pauli objection with a second explanatory version of his theory designed to show how material bodies, in particular rods and clocks, "adjust" to the field strengths *where they are,* hence exhibit the congruence behavior which we familiarly attribute to them. For another, this sanitized narrative completely ignores the disquieting fact that the Einstein-Pauli objection — ostensibly concerning a rather straightforward empirical matter — did not succeed in convincing Weyl even long after he had abandoned his theory on other grounds (1949, 288). Moreover, as evidenced by the direction of his own initial work on a "unified theory of fields" in the next six or so years (see §2 below), Einstein did not, apparently, consider that he had delivered a knock-down argument against Weyl's attempt (and the related one, spurred by Weyl, of Eddington) to expand the geometry of space-time to encompass electromagnetism. Following this up, it becomes apparent that Pauli's objections to Weyl turn not so much upon the apparent empirical disconfirmation pointed out by Einstein, but rather upon the failure of Weyl's attempted unified theory (and the later ones of Eddington *and* Einstein) to abide by a rather severe positivist covenant governing what can be said to be physically meaningful. But this is a stricture with which Einstein in his long odyssey in pursuit of a unified theory of fields can hardly be in agreement, at least if we take to heart the injunction expressed in his 1933 Herbert Spencer lecture — to heed what scientists do, and not what they say about what they do.

1.2 Pure Infinitesimal Geometry

Weyl's "unified theory" is perhaps best seen from the conditions of its genesis: as a bold, but ultimately unsuccessful, attempt to remove, from a perspective that could be described as Riemann's own, a "blemish" (*Schönheitsfehler*) from Riemannian geometry, the mathematical framework in whose terms the GTR is cast. In Riemannian geometry, while direction is not path-independent (integrable), length is; transport of a vector around a closed curve, or from one point to a finitely distant other via two different paths, in general changes the orientation, but not the length, of the vector on return to its initial point in the manifold. For this reason, it is in general meaningful to speak of a comparison of lengths but not of directions at points separated by an arbitrary (path independent) distance. Even as the first edition of his classic *Raum-Zeit-Materie* (1918b) was going to press, Weyl conceived what he termed a "pure infinitesimal geometry" (*Reine Infinitesimalgeometrie*), which removed, as he put it, this last "distant-geometry" (*ferngeometrisch*) Euclidean remnant from Riemann's theory of manifolds. According to Weyl, "a genuine local geometry [*wahrhafte Nahegeometrie*] can only be acquainted with a principle of transport of length from one point to another infinitesimally adjacent to it"; from such a perspective, Riemann's geometry is only "a half and inconsistent local geometry" (*eine halbe und inkonsequente Nahegeometrie*) since it assumes the meaningfulness of a path-independent, distant comparison of lengths (Weyl 1918a, 466; 1918c, 385; cf. 1919b, 91; and 1922a, 102). In Weyl's new geometry, both direction (like Riemann) and length (unlike Riemann) are not, in general, path-independent. The epistemological attraction of a purely infinitesimal foundation for differential geometry in physics was, for Weyl, directly evident since it is fully in accord with the Leibniz-Riemann principle that genuine understanding of nature stemmed from formulating its behavior in the infinitesimally small (1923a, 9).

Once the Riemannian assumption of length comparison "at-a-distance" is abandoned, then, as Weyl showed, one begins with a metric geometry of the differentiable manifold that has only a conformal (angle-preserving) structure. Hence, the length of a vector at a given point P is arbitrary up to a choice of scale ("gauge") at P. Until this is done, it is possible to compare only the *relative* lengths of two vectors at P or to determine the angle between them; the *absolute* length of a vector at P remains undetermined. A metrical structure of this kind can be expressed by the "pseudotensor" ("pseudo" since the interval ds is here only *conformally invariant*):

$$ds^2 = \lambda\, g_{ik}\, dx^i dx^k,$$

where the conformal factor λ (>0) is an arbitrary function of position. In this metric, only the ratios and not the absolute values of the g_{ik} have a determinate value. Weyl showed that this conformal geometry corresponds to the light cone structure of space-time in that any two (nonisometric) space-times that are conformally equivalent will have the same light cone structure, noting that Kretschmann (1917) had independently arrived at the same result (Weyl 1918a, 479).

At this point, there are two decisive considerations regarding the conformal geometry. First, such a metric determines only an equivalence class of symmetric (torsion-free) affine connections and not a unique connection as does, according to earlier results of Christoffel and Levi Cività, the metric of Riemannian geometry. This is unsatisfactory because now there is no unique notion of parallel displacement of a vector.[4] So additional structure must be sought to ensure unique compatibility with a metric, hence that the class of affine geodesics and that of metrical geodesics will coincide. Weyl regards this postulate as fundamental; it is "the basic fact [*Grundtatsache*] of infinitesimal geometry" (1923a, 124; 1952 [1922], 124). Second, to determine this metric, additional structure must be added so that it is possible to speak of the same length of a vector that is displaced from P to any point P' in the immediate ("infinitesimal") neighborhood of P. That is, though length comparisons "at-a-distance" have been forsaken, it still must be possible to compare lengths at P with all points lying "infinitely near" P. A large measure of the not inconsiderable aesthetic attractiveness of Weyl's geometry stems from his demonstration that the additional structure required in both cases is limited to a single function of position related to the factor already noted.

To restore the condition of unique determination of an affine connection, Weyl showed that a single connection could be determined, coupled to a given choice of a metric tensor, by incorporating a term containing a function ϕ of position within the definition of the standard Riemannian ("Levi-Cività") connection. Such a connection has the additional property that when the components of ϕ vanish identically at a point, it becomes identical to the "Levi-Cività" connection, as can be seen from the definitions of the two connections in components:

Levi-Cività:

$$\Gamma^r_{ik} = 1/2\, g^{rh}\, [(\partial g_{hi}/\partial x^k) + (\partial g_{kh}/\partial x^i) - (\partial g_{ik}/\partial x^h)],$$

Weyl:

$$\Gamma^r_{ik} = 1/2\, g^{rh}\, [(\partial g_{hi}/\partial x^k) + (\partial g_{kh}/\partial x^i) - (\partial g_{ik}/\partial x^h)] + 1/2\,(g_{ir}\phi k + g_{kr}\phi i - g_{ik}\phi r),$$

where the contravariant tensor g^{ik} is defined by the condition $g^{ik}g_{kj} = \delta_{ij} = 0$, *if* $i \neq j$; 1 *if* $i = j$. Given such a "Weyl connection," it is possible to speak of a manifold with an affine connection where, as in the Riemannian case, there is a unique determination of parallel displacement of a vector at every point.[5]

The next step is to show that there is a metrical concept of "congruent displacement" that already carries with it a compatible concept of infinitesimal parallel displacement, hence that the parallel displacement of a vector leaves vector length unchanged. This is to require that the affinely connected manifold must additionally become a *manifold with a metric connection;* that is, not only must there be a metric defined at every point, but also each point P must be *metrically connected* to the infinitesimal region around it. Thus it must be possible to determine what length at the various points P' in the immediate neighborhood of P corresponds to a given length at P. Analogously to the Riemannian case, it is possible to choose a gauge for the vicinity (tangent space) around P so that congruent displacement of a length from P to any P' in this region does not change this length (this is called a "geodetic gauge"). But in the general case, where the manifold is arbitrarily gauged and l is the length of an arbitrary vector at P, $l + dl$ is the length of the vector arising through congruent displacement to a neighboring point P'. This alteration is given by a linear transformation

$$dl = -l\, d\phi,$$

where the factor $d\phi$ is independent of the transported length, so that ϕ is a function of position. Changing the gauge at P so that $l' = \lambda l$ (λ is the conformal factor noted above) yields $dl' = -l'd\phi'$, where $d\phi' = d\phi - d\lambda/\lambda$. Weyl then showed that a necessary and sufficient condition for dl to vanish at P (which is what is desired) is that ϕ is a linear differential form,

$$\phi = \phi_i\, dx^i, \text{ (with summation over repeated indices).}$$

A Weyl metric then consists of two "fundamental forms," the quadratic form of Riemannian geometry,

$$ds^2 = g_{ik}\, dx^i dx^k, \text{ (summation convention),}$$

and the linear one just defined. These are defined up to the "gauge transformations,"

$$ds'^2 = \lambda\, ds^2, \text{ and}$$
$$\phi' = \phi + d \log \lambda,$$

which will have the physical effect of changing, for example, the lengths of measuring rods and the rates of clocks, by a scalar factor at each point. Besides the requirement of general covariance, that is, of metrical invariance under arbitrary (suitably continuous, and so on) transformations of coordinates, such a geometry satisfies the further requirement of "gauge invariance": there is metrical invariance up to an arbitrary choice of gauge at each point. By the "curl" (or "rot") operation of vector calculus, Weyl then defines from ϕ a covariant, second-rank, antisymmetric tensor that provides a measure of distance curvature, or the extent of nonintegrability, at each point (that is, the departure from Riemannian geometry there). In components it is written

$$F_{ik} = \partial\phi_k/\partial x^i - \partial\phi_i/\partial x^k,$$

where ϕ is the linear differential form already noted, with the consequence that when ϕ is a total differential ($F_{ik} = 0$), length again becomes integrable (path-independent); that is, Riemannian geometry as a special case is recovered.

Thus far we have not left the purely mathematical domain of "pure infinitesimal geometry." Now Weyl's "physical discovery" was that the second-rank tensor just defined is *formally identical* to the electromagnetic field tensor (the so-called Faraday tensor) and thus to the first system of Maxwell's equations (in their tensor formulation) if ϕ_i is seen as the four potential of the electromagnetic field. It is therefore only natural to identify, as Weyl immediately did, these two structures. The startling meaning of this identification is that *all* the phenomena of nature are represented in the fourteen independent coefficients ϕ_i and g_{ik} of the now expanded metric; following Hilbert (1915; see §3.2 below), these in turn are to be deduced from a single universal "world law" (that is, an action principle) of highest mathematical simplicity.[6] Weyl, moreover, argued that just as conservation of energy/momentum corresponds to coordinate invariance, "conservation of electricity" (both the electric potential and the current satisfy the boundary condition $\partial \phi_i / \partial x^i = 0$) corresponds to gauge invariance, as is reflected in the fact that the Maxwell equations are already invariant under conformal transformations, that is, gauge invariant in Weyl's sense (1923b, 308ff.), as Cunningham and Bateman had shown already in 1910 (Weyl 1922c, 286 n. 35). Weyl takes this demonstration to have provided "a strong formal argument in favor of the theory: as origin of the conservation of electric charge one new invariance property [*Invarianzeigenschaft*] of the field laws must be anticipated a priori, involving *one* arbitrary function" (1931, 54).

In setting out to remove a glaring blemish in Riemann's theory of manifolds, Weyl had produced a geometric theory seemingly unifying electricity and gravitation. Since these were the only physical forces recognized in 1918, Weyl was led to triumphantly proclaim the unity of geometry and physics: "Everything real [*Wirkliche*] that transpires in the world is a manifestation of the world-metric: physical concepts are none other than those of geometry" (1918c, 385). For such declarations Weyl was taken to task, by Einstein and Reichenbach among others; the latter as early as his *The Theory of Relativity and A Priori Knowledge* (1920). There Reichenbach wrote, expressing what was probably the consensus of the physics community: "A purely mathematical axiomatization never leads to principles of an *empirical* theory. . . . [I]t is incorrect to conclude, like Weyl and Haas, that mathematics and physics are but one discipline" (1965 [1920], 76). (Inspired by Hilbert's "axiomatic method"—see §3.2 below—and by Weyl's theory, the Leipzig physicist Arthur Haas had written in 1919 a polemical essay entitled "Physics as Geometrical Necessity.") This criticism elicited an indignant response from Weyl.[7] Certainly the purely mathematical road that had led to Weyl's theory would continue to enlist prejudice against it among most physicists. However, what Weyl meant by affirming the unity of physics and geometry was explained in a passage not cited by Reichenbach: it is the task of geometry to investigate the "essence of metric concepts," while physics is concerned to determine "the law" (that is, the integral invariant formulated as an action principle) according to which the actual world is singled out from all other possible four-dimensional metric spaces.[8] For Weyl, the "essence" or "nature" of the metric is given in yet another mathematical result, his group-theoretical treatment of the so-called *Raumproblem,* whose Helmholtz-Lie solution, based as it is on the *ferngeometrisch* postulate of free mobility, neither is adequate to the nonhomogeneously curved spaces permitted by Riemann's theory of manifolds nor survives the scruples of Weyl's stricter "purely infinitesimal" point of view.[9]

1.3 Constructing the Metric without Rods and Clocks

We need now see how the metric of space-time can be empirically constructed according to Weyl. Since for Weyl, the essence of the metric lies in the concept of congruence that is a "purely infinitesimal" concept,[10] the metric is to be viewed solely as a structure of the continuous field whereupon the concepts of vector and "tract" (*Strecke*) employed for its characterization have in themselves "nothing to do" with material measuring rods and clocks (1921d, 473). From this perspective, it is no longer permissible to stipulate that there are measuring rods and clocks to which our chrono-metrical notions are "coordinated," a situation that, to Einstein, robbed the interval *ds* of its empirical foundation (see §2 below). The starting point is a theorem (Weyl 1921b) according to which the conformal and the projective properties of a metric space (in Weyl's sense) univocally determine its metric up to a dilation factor.

This is a purely mathematical result, yet Weyl observes that the conformal and projective structures of a metrical manifold have intuitively clear physical counterparts in the theory of relativity.[11] Thus Weyl provides a so-called geodetic method involving two "directly observable" physical processes that traverse geodesics in space-time: the propagation of light rays (null geodesics) and the inertial trajectories of mass points (timelike geodesics) (1922c, 228–29; 313–14). The conformal properties could be identified by light rays that fix the causal structure of space-time; hence, by observing the arrival of light at points in the immediate neighborhood of a point *O,* the *ratios* of the quantities g_{ik} at *O* may be determined. But the propagation of light determines only the quadratic differential form ds^2 (up to a conformal factor, in Weyl's theory, it is not the g_{ik} themselves but only their ratios that have an empirically ascertainable meaning) while leaving the linear form ϕ unrestricted. This latter may be fixed by considering that the projective properties may be taken to be physically instantiated by the trajectories of freely falling point masses ("geodetic hypothesis"). One may assume that the proper time *s* may be read off from the motion of these unaccelerated "ideal" particles, hence that these projective geodesics carry *affine parameters* and so are *affine geodesics.* There is a compatibility requirement for the conformal and projective structures such that the conformal (null) geodesics $ds^2 = 0$ are included within the class of projective geodesics. Then, through a comparison of two such point masses passing through *O* in different directions, a unit of measure at *O* may be uniquely determined. (One shows that the difference in directional derivatives at *O* is proportional to the gauge at *O*.) In modern terms, the affine structure forces the conformal factor to be a constant (at each point).[12] The physical significance of the theorem is then that the world metric can be fixed *without reliance on measuring rods and clocks*. It is worth noting that this method, as revived by Ehlers, Pirani, and Schild (1972), is of active current interest and that recently Ehlers (1988) has adduced a microsymmetry criterion that provides a nonmetrical means for picking out geodesic paths, thus redeeming the "geodesic hypothesis" against conventionalist leveling arguments such as are found in Grünbaum (1973, 733ff.) or Sklar (1985, 139).

1.4 The Einstein "Prehistory" Objection and Weyl's Rejoinder

Weyl's unified theory, put forward in *two* explanatory versions, received an almost unanimously unfavorable reception from the theoretical physics community, beginning with Einstein himself. Einstein first expressed his disagreement privately in a letter of 15 April 1918 to Weyl and then publicly in a note appended to Weyl's Prussian Academy paper (1918a) announcing his theory; the objection occurs yet again in indirect form in Einstein's widely read "popular" lecture to the Prussian Academy of January 1921, "Geometry and Experience" (1921c; see §2 below). The obstacle appeared to be that Weyl's theory was not in agreement with observation. In fact, Weyl's theory predicted miniscule "second clock" effects that, as Eddington showed, were far below the threshold of observation and, as it later turned out, were within the limits of quantum mechanical modifications.[13] And as Pauli himself had helped to show in one of his first scientific publications, Weyl's theory, though containing field equations of the fourth (and not the second) order, yielded Einstein's field equations as a special case and agreed on two predictions derivable from Einstein's theory: the gravitational red shift and the advance of the perihelion of Mercury. What then was Einstein's objection? We do observe that measuring rods retain their length under transport in electromagnetic fields; prima facie, this is evidence that Riemann's geometry, not Weyl's, is the geometry of space-time. More substantively, Einstein argued as follows: if Weyl's theory is correct, then the spectral lines emitted by atoms would not be very sharp and well-defined as in fact they are observed to be. For if two atoms, say of hydrogen, are together at an initial time in one space-time region and then transported, via different paths to another region of space-time where they are brought together again, then, according to Weyl's theory, we should generally observe a difference in their spectral lines corresponding to their past histories, that is, to the differing values of the electromagnetic fields in the space-time regions they occupied in the interim. Indeed, according to Pauli's calculation, no matter how small the initial difference in the spectral lines of the two atoms posited by Weyl's theory, this difference would "increase indefinitely in the course of time."[14] But astronomical observation tells us that hydrogen atoms everywhere in the heavens exhibit the same spectral signature. So Weyl's theory did not, apparently, correspond to the facts of observation.

Despite his enormous respect for Einstein, Weyl was persuaded neither at the time by this objection nor even three decades later, long after surrendering his theory on grounds of the new quantum mechanics. Instead, he adopted a two-pronged argumentative strategy to counter it, producing in effect a second explanatory version of his theory. On the one hand, he adamantly maintained that the behavior of physical objects such as rods and clocks or, for that matter, atoms, has "as such nothing to do" with the ideal metric notions defined by vector transport: "The functioning of these instruments of measurement is however a physical occurrence whose course is determined through laws of nature and which has as such nothing to do with the ideal process of congruent transplantation of world intervals [*Verpflanzung von Weltstrecken*]" (1919a, 113). To critics like Pauli, this meant that although there was no longer a "direct contradiction with experience," there also no longer existed an "immediate connection" between electromagnetic phenomena and the behavior of measuring rods and clocks; consequently, the connection between electromagnetism and the world metric posited in Weyl's theory is only "purely formal." Eddington similarly (though more sympathetically) interpreted Weyl as giving up any claim to characterize the geometry of the real world and instead as providing only a "graphical representation," i.e., a kind of conventional representation, of "world geometry."[15] Such objections might be summarized: If chronometrical relations are not concerned with measuring rods and clocks, with what *are* they concerned (see Coffa 1979, 283–84)?

Weyl's concern, however, is that this response dodges an explanatory burden that cannot be shirked from the viewpoint of either a pure field theory of matter or a systematic theory of field-matter interactions.[16] For it is just wrongheaded, or as Weyl expressly states, "perverse," to use physical bodies such a rods and clocks, which are *indicators* of the gravitational field, as at the same time instruments to *stipulate* metric relations. To do so is just to treat as a definition ("rigid rod," "clock") what should be explained, that is, *should be derived* from a systematic theory. For such a systematic theory, "Einstein's *definition* of measure determinations in the metrical field with the help of measuring rods and clocks has validity only as a preliminary connection to experience just as does the definition of electrical field strengths as the ponderomotive force on a unit charge." In order, in Weyl's terms, "to close the circle," it is

> necessary, once a suitable action-law has been set up, to *prove* that here, the charged body under the influence of the electromagnetic field, there, the measuring rod under the influence of the metrical field, exhibit, as consequences of the action-laws, that behavior we had originally utilized for the physical definition of the field magnitudes. (1923b, 298)

In the conception of Einstein, the GTR, where admittedly "the conceptual foundations of the theory have no relations with the electromagnetic field" (Einstein 1923b, 448), is certainly not systematic in this sense. Hence it can rely, as it apparently does, on the notion of a "practically rigid rod" that corresponds to "congruence at a distance" (that is, path-independent transport of length). On the other hand, suppose the fundamental metric concept, congruence, is only to be properly conceived as a "purely infinitesimal" concept, as it is in Weyl's elegant generalization of Riemannian geometry. Once this is done, the "practically rigid rod," rather lamely defended in Einstein 1921a (see §2 below), becomes an unprincipled and gratuitous assumption.

Clearly, Weyl had to account for why we do observe the congruence-preserving behaviors of rods and clocks that we do, as well as for the constancy of spectral lines of atoms, despite the requirement of gauge invariance imposed by a "purely infinitesimal" geometry. He does so (and this is the other component of Weyl's explanatory strategy) by invoking a dualism regarding the manner in which physical magnitudes are determined, a distinction that he sees as reprieving his theory from the empirical refutation sketched by Einstein and elaborated by Pauli. Physical quantities are fixed either by a body's following a "tendency of persistence" (*Beharrungtendenz*) or by its "adjustment" (*Einstellung*) to the field strengths where it is, a distinction made concrete by appealing to the different physical behaviors of a spinning top and the magnetic needle of a compass.[17] Whatever its initial orientation, the axial direction of a spinning top is transferred from instant to instant by a tendency of persistence; that is to say, it is governed by the inertial or "guiding" field (*Führungsfeld*). On the other hand, the magnetic needle of a compass *adjusts* to the value of the magnetic field wherever the compass is carried; "adjustment . . . enforces a definite value that is independent of past history and hence reasserts itself after any disturbances and any lapse of time as soon as the old conditions are restored" (1949, 288).

Accordingly, Weyl objects that the Einstein-Pauli "prehistory" criticism of his theory unjustifiably presupposes that measuring-rod lengths and clock periods are altered (or not) through a time-dependent process of *persistence,* whereby the magnitude in question at a given instant is some function of its magnitude at a previous instant. But given the distinction above, this is not at all a *necessary* presupposition, in which case the explanatory burden runs in the other direction. If, for instance, a measuring rod is moved around within a physical field assumed to be *inhomogeneous,* that is, where the field strengths have different magnitudes at different points, an account is surely required as to why we do detect no variable behavior in our measuring rod. One prima facie reason may be that such "deforming" forces at each space-time point are present but are counteracted by electromagnetic forces within the atom that thus produce a state of force equilibrium. This would mean that a massive object, such as a measuring rod, carries with it a determinate magnitude representing the interaction of the gravitational forces of the field and the electromagnetic forces obtaining between the rod's constituent molecules and atoms. If this is so, then a measuring rod may be held to "adjust" to the field strengths where it is *now;* it does not exhibit what was for Weyl a qualitatively distinct, but miniscule (that is, far below the threshold of observation), quantitative difference in pattern of behavior, the tendency for its length to persist, that is, to remain the length it was a moment ago at another point of space-time. The commonsensical objection that of course the length of a measuring rod persists in moving it from one end of a room to another can, presumably, be countered by taking into account the crudity of our everyday experience with middle-sized objects, which are not presumed to interact with the "empty space" in which they (and we!) are located.

What is this field strength? Weyl appears to have been motivated by Einstein's 1919 "scalar-free field equations" (which were intended to provide a more principled footing for the nefarious cosmological constant) to take the relevant field strength here to be the equivalent in his theory to the Riemann scalar of curvature in Einstein's, which has now become a constant, playing the role of keeping electrodynamical forces within electrical "corpuscles" in equilibrium.[18] Analogously to Einstein, Weyl posits a constant "natural gauge," definable in terms of his curvature scalar, which is held to be responsible for maintaining an equilibrium of intra-atomic electrical forces. Hence, we *observe* that measuring rods display the behavior we familiarly think they do: their length — due to "adjustment" to a constant "natural gauge," that is, constant force equilibrium — is unaltered under transport in the gravitational field.[19] Weyl's response to Einstein and Pauli is, then, to say that the alleged "empirical refutation" (prehistory objection) does not touch his theory, since the constancies of "atomic clocks," as also the congruence behaviors of measuring rods, are to be accounted for as arising through *Einstellung* (as indeed they must in a principled "systematic" theory), not through *Beharrung,* as the "prehistory objection" wrongly presupposes.[20]

2. Einstein

As noted above, the Einstein-Pauli "prehistory" (constancy of spectral lines) objection was taken as an authoritatively convincing rejoinder to Weyl's highly speculative theory by a majority of the community of physicists, though perhaps, given what was known at the time, it should not have been.[21] But if we turn to consider a wider range of Einstein's writings and activities in the period 1918–25, we find that, for Einstein, this purported empirical disconfirmation of Weyl's theory was not at all the end of the matter. Taking these into account, it can be seen that Weyl's criticisms of the assumption of rods and clocks as fundamental concepts in the GTR left their mark on Einstein. To these criticisms is probably accountable Einstein's shift from defending the behavior of rods and clocks as evidential requirements of the GTR to a tempered pro tem justification. Besides, the speculative schemes for unification initiated by Weyl, and then modified by Eddington (1921), served as Einstein's own point of departure for his first attempts to formulate a unified field theory. The conclusion then emerges that, far from regarding the "prehistory" objection as decisively undermining Weyl's entire approach, Einstein himself was to become and remain the leading proponent of mathematically speculative schemes for unifying physics, long after Hilbert and Weyl had quit the field. It remains to explore this development, and its grounds, a bit further.

Already at the Bad Nauheim meeting of the German Society of Natural Scientists in late September 1920, a widely publicized confrontation of Einstein with his critics, especially Phillip Lenard, Einstein admitted that at the current stage of the theory's development, it was a "logical weakness" (*Schwache*) of the theory that "it must separately introduce rods and clocks instead of being able to construe them as solutions of the differential equations [i.e., of the field equations]" (1920, 662). And if we turn to the explicit, and "more philosophical," discussions of the status of rods and clocks as foundational concepts in the 1920s, we find this ambivalence in abundance. It is famously voiced in perhaps the most widely read text of Einstein within subsequent logical empiricist philosophy of science, the lecture of January 1921 entitled "Geometry and Experience." Here a thoroughly pragmatic justification of "practical geometry" is somewhat deviously coupled with a fundamental criticism based upon the in-principle difficulties attending the concept of a rigid body. Einstein first makes a case for the validity of the supposition of the existence of Helmholtzian "practically rigid bodies," that is, bodies upon which "two 'tracts' [*Strecke*] found to be equal once and anywhere are equal always and everywhere." This is just to assume, contra Weyl, that congruence relations are path-independent, an assumption for which, alluding to his prehistory objection, "the existence of sharp spectral lines comprises a compelling empirical proof." Nonetheless, Einstein goes on to lodge a criticism of this point of view, which he attributes to Poincaré. There are no actual rigid bodies that correspond to the ideal rigid body of geometry; it is not possible to thus disentangle geometry from physics in this manner. Accordingly, only the whole comprising *G*(eometry) + *P*(hysics) is empirically testable; while *G* and parts of *P* may be chosen arbitrarily, all that matters is that the whole not conflict with experience. This latter viewpoint is correct *sub specie aeterni,* Einstein admits, for reasons quite analogous, if not identical, with whose given by Weyl. Yet even so, for the time being, the former position is to be preferred:

> The concept of the measuring rod and the concept of a clock coordinated with it in the theory of relativity do not find their exact correspondence in the real world. It is also clear that the solid body and the clock do not play the role of irreducible elements in the conceptual structure of physics, but the role of composite structures, which should not play an independent role in the construction of theoretical physics. However, it is my conviction that these concepts, at the present stage of development, still must be introduced as independent concepts. (1921a)[22]

Einstein again adopts a cautious "on the one hand, on the other" mode of presentation in reiterating this conclusion in a little-known "philosophical" essay some four years later. Only now in addition comes the warning that in adopting "the standpoint of the practical physicist,"

> [w]e must however be continually conscious of the fact that the idealization which lies in the fiction of rigid (measuring) bodies [*Körper*] as objects of nature might one day prove unjustified or justified only with respect to certain phenomena of nature. (1925c, 19)

Noting that Riemann had already anticipated this possibility, both in regions that are "not astronomically small" and in the microworld of "electrical elementary quanta" (*elektrischen Elementarquanta*), Einstein observes that Riemann had had the "audacious idea" that "the geometric behavior of bodies might be conditioned by physical realities or forces." Riemann had thus conceived through "pure mathematical speculation" the "unity" of geometry and gravitation attained sixty years later in the GTR. This brief essay concludes with mention of the attempts of Weyl and Eddington to generalize Riemannian geometry so as to "find a place for the laws of electromagnetism in the accordingly expanded conceptual system," endeavors about which Einstein here reserves judgment.[23] In later years Einstein will continue to work both sides of the aisle, sometimes characterizing geometry, without qualification, as "the study of the possible positions and displacements of rigid bodies" (1983 [1928], 164) while also holding that a "*complete* theory of physics" has "no room for the supposition" of rods and clocks (Schilpp 1949, 685–86).

We might now inquire into the reasons underlying the provisional character of Einstein's choice of rods and clocks as the physical correlates of chrono-metrical notions in the context of general relativity where, as he repeatedly points out, coordinate differences do not have an immediate metrical significance in terms of unit rods and clocks (e.g., 1916a, 117). One likely reason seems to be that the assumption of standard measuring instruments played a vital heuristic role in a key Einstein thought-experiment that showed that the space-time geometry of gravitational fields was non-Euclidean. As John Stachel has demonstrated, this thought-experiment — a rigidly rotating disk, the simplest example of a stationary gravitational field — was an essential rite of passage for Einstein in convincing himself, beginning in 1912, that gravitation was incompatible with the flat Minskowski space-times of special relativity and thus of the need to consider the possibility of curved space-time. Essential to its conclusion is that rigid and periodically regular measuring instruments exist with which the geometry of the disk may be determined through actual measurements. Without such intuitive means of disentangling physics from geometry, the thought-experiment would not work. So there is a heuristic and motivational reason for Einstein's retention of rods and clocks in the context of the GTR.

There is a related epistemological attachment. First in his private correspondence with Weyl,[24] and then in his public pronouncements on Weyl's theory, Einstein elaborates upon the "prehistory" objection by contrasting the obvious evidential virtues of the GTR in retaining a path-independent concept of congruence corresponding to "measuring rod geometry." This is a consideration not to be taken lightly, as Einstein wrote to his friend Besso in Zurich in July 1920. Besso, it seems, had been entertaining favorable opinions of Weyl's theory. Einstein responds:

> You think: there is no need to find the invariability of relative extension of a body in the foundation of the theory, that it would be more beautiful if this resulted as a consequence or more acceptable if it had a place in theory as a special hypothesis. However, don't forget that the theory is based on measuring rod geometry [*Maßstabgeometrie]*. Then one accepts that the relative length of measuring rods is a function of its prehistory. It follows that one should find that *actual* measuring rods are relatively invariant. This is why the measuring rods employed in the foundation of [Weyl's] theory are only *imaginary* [*gedachte*] measuring rods which behave otherwise than actual ones. That is horrible [*abscheulich*].[25]

But the most complete statement of this line of thought appears to have been given publicly in Einstein's response to Weyl's presentation at the infamous eighty-sixth meeting of the German Society of Natural Scientists at Bad Nauheim in September 1920:

> In the arrangement of my conceptual system, for me it has become decisive [*massgebend*] to bring elementary experiences into the language of signs [*Zeichensprache*]. Temporal-spatial intervals are physically defined with the help of measuring rods and clocks. If I consider two (such) structures, then their equality is empirically independent of their prehistory. Upon this rests the possibility of coordinating [*zuzuordnen*] a number *ds* to two neighboring world-points. Insofar as the Weyl theory renounces this empirically grounded coordination [*Zuordnung*], it robs the theory [the GTR] of its most solid empirical support and possibilities of confirmation. (1920, 651)

It is expressions such as these that so endeared what was thought to be Einstein's conception of science to certain members of the Vienna Circle such as Philipp Frank (see P. Frank, 1949a). Employing the language of coordination (*Zuordnung*), which lies at the center of the account of cognition in Schlick 1918, a book he had recently read and greatly enjoyed (see Howard, 1984), Einstein here states in no uncertain terms that he considers "norming" the *ds* to ("infinitely small") unit rods and clocks (i.e., rods supply $ds^2 = 1$, clocks $ds^2 = -1$) to be essential to the empirical foundation of his theory. This thought is still operative when, a few months later, in a talk recorded in Vienna in January

1921, he again warns (with presumably Weyl's theory in mind) that unless the *ds* is connected with "the observable facts," a "reality-strange" (*Wirklichkeitsfremden*) theory is the result (Herneck 1976, 103–4).

Even so, in several letters to Weyl during 1918 and 1919, Einstein privately reveals that his opinion of Weyl's theory is far less one-sided than at least his public posture would suggest.[26] More importantly, as evidenced in his scientific work of the period, the initial attempts to construct his own unified theory of gravitation and electricity, it is quite clear that Einstein is not at all constrained by these evidential and heuristic attachments to the validity of rods and clocks. Already on 3 March 1921, he submitted a paper to the Prussian Academy in which he hypothetically considers a relativity theory in which, as in Weyl, the *ds* is only conformally invariant (i.e., only $ds^2 = 0$ is invariant), thus "without making use of the concepts of measuring rods and clocks" (1921b, 262). The next step is yet more radical. By January 1923, in a contribution sent to the academy from the ship *Haruna Maru* en route from Japan, Einstein is prepared to follow Eddington in jettisoning the metrical basis of a combined gravitational and electromagnetic theory altogether. It now appears, "from a purely logical starting point," that "it is much more satisfactory to adopt as the basis of such a theory only [an affine connection] while letting the invariant [ds^2] fall." In fact, from such a *nonmetrical* starting point, Weyl's theory is unsatisfyingly "half-metrical" (1923b, 32). Writing to Bohr from the same ship "near Singapore" on 11 January, Einstein expresses more than hypothetical interest in this option: "I believe that I have finally understood the connection between electricity and gravitation. Eddington has come closer to the truth than Weyl."[27] All thought of unduly jeopardizing the GTR by refusing to give the metric interval *ds* rods and clocks as physical counterparts is surely here abandoned in pursuing this new nonmetrical route to a unified theory. Having embarked upon such a course, Einstein can evidently no longer consider it "decisive" to link the fundamental concept of spatio-temporal intervals with "elementary experiences." Instead, it is in these first strivings at unified field theory that we find incipient indications of a belief in the ability of mathematics to comprehend reality so famously expressed in his Herbert Spencer Lecture at Oxford in 1933. Thus in September 1923, in a report on his new affine theory of the field, Einstein writes that upon choosing a nonmetrical basis for the theory, the following methodological shift ensues: "The search for the mathematical laws which shall correspond to the laws of Nature then resolves itself into the solution of the question: What are the formally most natural conditions that can be imposed upon an affine relation?" (1923a, 448).

As Weyl would point out much later in 1952, there is a certain personal irony in this transformation, for whereas Einstein in 1918–19 had upbraided Weyl for following so purely a speculative approach to physics without "a guiding intuitive physical principle," their roles were soon thereafter reversed. According to Weyl, Einstein came to believe that "the chasm between ideas and experience is so large that only the path of mathematical speculation, whose consequences must naturally be developed and confronted with the facts, has a prospect of success," whereas for Weyl, chastened by difficulties presented to gravitational-electromagnetic unification schemes by quantum mechanics, "my confidence in pure speculation has sunk and a closer connection with quantum mechanical experience seems necessary."[28] Ironies aside, there can be little question that Einstein was far from believing in the total invalidity of the speculative approach of Weyl as the "prehistory" objection might suggest. On the contrary, he thought enough of Weyl's theory and of the related generalization offered by Eddington to adopt them as starting points for what was to be a three-decade-long futile effort to construct a unified theory of fields. And once underway, even as he repeatedly complained that this route led only to exasperating dead-ends, he continued to explore his own variants of the Weyl/Eddington unification schemes, exhausting all the possibilities he deemed reasonable within what he called the "Weyl-Eddington complex of ideas."[29]

Einstein's attitude in this regard is instructively contrasted with the genuine positivism of Pauli. Writing to Eddington just after the publication of Einstein's pure affine theory in September 1923, Pauli stated that he considered such endeavors to be "physically meaningless":

> The most beautiful achievement of the theory of relativity was certainly to have brought the metrical results of measuring rods and clocks, the paths of freely falling mass particles and those of light rays, into a determinate inner bond [*Verbindung*]. . . . [However,] the magnitudes $\Gamma^{\mu}{}_{\nu\alpha}$ [of the affine connection] cannot be directly measured; rather they must be obtained from the directly measured magnitudes first through complicated calculations. No one can empirically determine an affine connection between two vectors in neighboring points if he has not already ascertained the line element. For this reason, I maintain, in opposition to you and Einstein, . . . that to attempt to base a geometry upon an affine connection without a line element is above all meaningless for physics, . . . [for] in that case, we not only have no "natural geometry" but also no "natural theory."[30]

The strictures on meaning in evidence here were already raised in 1919 against Weyl's theory. There, Pauli pointed out, one continually operated with a "meaningless fiction" in supposing a determinate value can be given to field strengths in the interior of an electron, a procedure violating the rule that legitimate quantities in physics must be "observable in principle."[31] That this criterion was not the ground of Einstein's objection to Weyl is clear from an ensuing comment Einstein made in a letter to Born in January 1920: "Pauli's objection is directed not only against Weyl's, but also against anyone else's continuum theory."[32] Naturally, Einstein included himself in the "anyone else" category. Self-admitted epistemological opportunist that he was (see Schilpp 1949, 684), Einstein would not here elevate those twinges of epistemological conscience visible in his objection to Weyl's theory to the status of strictures on the practice of theoretical physics like the positivist one proposed by Pauli. In point of fact, within a few years, Einstein would come into open combat with the "epistemologically soaked orgy" decreed by the new quantum theorists (see Fine 1986b, chap. 6).

3. Reichenbach

3.1 *Die wissenschaftsanalytische Methode*

Discussions of Reichenbach's epistemology of geometry have understandably focused on the "mature" presentation of his views in the *Philosophy of Space and Time* (*Philosophie der Raum-Zeit-Lehre,* 1928; English translation, 1958). To be sure, some recognition has been given to the circumstance that the neoconventionalism there in force differs considerably from the position adopted in his first philosophical monograph on relativity theory published in 1920, which is indeed at variance with his "mature" position. For in 1920, Reichenbach maintains that the metric of space-time is not conventional but "an objective property of the world," the sole subjectivity entering into metrical determinations being restricted to an "arbitrariness" in the choice of a coordinate system. Once a coordinate system is chosen, the components of the metric tensor g_{ik} can be empirically ascertained.[33] But for our story, the greater significance of this early work is that here Reichenbach first outlines a new method for epistemology of science, termed "the method of analysis of science" (*die wissenschaftsanalytische Methode*), which is to replace Kant's "analysis of reason."[34] The idea is that by sharply separating the "subjective" contribution of reason to scientific knowledge from the "objective" contribution provided by the external world (formerly termed "objects of empirical intuition"), one can show thereby how the former is required to constitute the objects of scientific knowledge through "coordination" to the latter.[35] The coordination of mathematical concepts to empirical objects, given in perception, requires "principles of coordination" that coordinate not only "the totality of real things" to "the total system of equations" but *"individual* things" to *"individual* equations" (1965 [1920], 37). This means that in physics, in addition to what Reichenbach terms "axioms of connection," that is, equations expressing empirical physical laws or relations between physical state variables, there is also required a "coordinative system" of principles and axioms whose sole purpose is to enable a univocal coordination of the equations of state to the contents of perception. The question of how such a coordination of concepts to objects can be achieved that is both unique and consistent "belongs in critical philosophy, for it is equivalent to Kant's question: 'How is natural science possible?' " (ibid., 46). By thus "determining the coordination," these principles of coordination "define the individual elements of reality, and in this sense they are *constitutive* of [*sie konstitutiv für*] the real object; in Kant's words: 'because in general only by their means can any object of experience be thought' " (ibid., 53). In performing this constitutive function, these principles retain what is still valid in Kant's conception of a priori elements of knowledge, whereas that sense of the a priori must be dropped according to which such components of knowledge are apodictically certain and unrevisable (ibid., 77).

It is notable that this neo-Kantian analysis of cognition is directed, in the first instance, against unnamed empiricist views that countenance no role for the object-constituting role of certain conceptual elements. But already in 1922 the essential piece of the "mature" conventionalist view of the metric of space-time is in place: the fundamental geometrical concept — congruence — requires, as Helmholtz in particular has shown,[36] a *stipulation* governing transported rigid bodies (measuring rods). The stipulation concerns, of course, that our measuring instruments suffer no nondetectable, hence noncorrectable, deforming forces under transport (*force d'espèce X* [1922b, 34ff.]); "metrical forces" (1969 [1924]; see §3.3 below); or "universal forces" (1958 [1928], §5). Once such a stipulation is made, the thus canonical measuring instruments ("normed" as the unit of the metrical interval *ds*) can be employed for an empirical determination of the metric. As already the example of the rigidly rotating disk shows, in the presence of a (nontransformable away) gravitational field, the resulting metrical determination will be non-Euclidean. In concord with the shift to conventionalism, by 1924 the category of constitutive "coordinative principles" of 1920 has become that of "coordinative definitions" that link purely formal concepts with alleged facts concerning empirically given objects.[37]

A noteworthy aspect of Reichenbach's method of analysis that thus proposes to cleave a physical theory into its empirical and its nonempirical parts (which after the coming "linguistic turn" would be distinguished as its synthetic and its analytic statements) concerns its implied opposition to various contemporary holist views of the relation of physics and geometry in the GTR. For in the contemporary writings of Schlick (1918, 1979 [1922]), Carnap (1921), Einstein (1921a), Eddington (1923, 130), and Weyl (1922c, 62) are credible (even if *sub specie aeterni,* as Einstein has it, see §2 above) arguments to the effect that only the combination of geometry and physics can be tested as a whole, that such a winnowing of nonempirical or conventional chaff from the kernels of pure empirical content, such as Reichenbach envisages, is an epistemological chimera. However, in Reichenbach's hands, the analysis of cognition as coordination gives rise to what is, to all appearances, a verificationist or operationist perspective on scientific theories according to which a theory's empirical foundations or even "cognitive meaning" of its individual terms and expressions require that its conceptual elements be directly identified (via "coordinative definitions") with physical objects to be used in empirical tests of the theory. While it would be futile to deny the apparent similarity of result of viewing scientific theories through the interpretive lenses of Reichenbach's *wissenschaftsanalytische Methode* and as given in, say, the analysis of Bridgman (1927),[38] it is important, I think, to be clear that the philosophical motivation of Reichenbach was, by genesis and intent, sharply different and, in an important respect, conflicting. For the object-constituting role of concepts that Reichenbach's method is designed to emphasize is just the opposite of operationism or verificationism, according to which the entire content of theoretical concepts and entities is reducible to corresponding sets of operations or direct observations. And in fact certainly during the 1920s (thus prior to the verificationist theories of meaning of the Vienna Circle), this type of strict empiricist criterion of meaning is foreign to Reichenbach; for example, he is more than willing to accord theory-ladenness to the "elementary facts" expressed by his axioms (as long as *relativity* theory is not involved [1969 (1924), §1]; as we shall see, this point is moot). Moreover, he firmly opposes positivist readings (by Frank and Petzoldt) of the curious remarks of Einstein concerning "point-coincidences" in the extended exposition of the GTR in 1916. Such coincidences, Reichenbach correctly observes (following Schlick, 1979 [1922], 265), are just the intersections of world-lines; as such the meeting of two elementary particles surely counts as a legitimate "coincidence" in Einstein's sense (Reichenbach 1969 [1924], §4).[39]

3.2 A Constructive Axiomatization

The *wissenschaftsanalytische Methode,* schematically outlined in 1920, received its first detailed application in the *Axiomatik der relativistischen Raum-Zeit-Lehre* published in 1924. This is a work with several levels of interest. It has immediate significance for the history of logical empiricism in that it is the first attempt to give what would become known as a "rational reconstruction" of a physical theory,[40] that is, an exposé of the "logical structure" of a scientific theory wherein empirical and definitional components are clearly distinguished. The proclaimed character of the work as an "epistemological-logical investigation" (*erkenntnistheoretisch-logischen Untersuchung*) (1925, 38) recalls the designation given the *erkenntnislogischen* investigations of the foundations of exact science within the modified transcendental idealism of Marburg neo-Kantians, particularly E. Cassirer (1910; 1921a). As we have seen, Reichenbach's method of *Wissenschaftsanalyse* is similarly concerned to pinpoint the object-constitutive character of certain elements of knowledge in the exact sciences, and it is to this end that axiomatization is to be directed. For as a method of analysis of science, "philosophy is only interested in the logical separation of empirical and logical components" of a theory, and for this purpose, "the value of the axiomatic method" is that "it directly reveals the places where definitions are present; it separates the conceptual components of the theory from experimental content and shows where the discernable problems of physics first begin" (Reichenbach 1927, 130n., 133).[41]

Already in 1921, a preliminary version of his *Axiomatik* had appeared in which Reichenbach announced the epistemological-logical goal of sharply distinguishing the basic empirical assumptions of the theory of relativity, expressed in two groups of axioms governing (1) light signaling and (2) rods and clocks (*Materialaxiome).* These are verifiable by experiment and, as such, distinct from the theory's "conceptual ingredients" (*begriffliche Gehalt*) that are definitional, mere matters of convention. Together, the basic empirical propositions and the conceptual components comprise the theory's "logical structure" upon which depends all of the remaining propositions of the theory of relativity. In this initial published version, a report made to the Congress of German Physicists at Jena in 1921, Reichenbach asserts that the *Materialaxiome* simply affirm the complete identity of measurement results made by employing rods and clocks with those attainable from the *Lichtgeometrie* (light geometry) set out in the five axioms of the first group and that the latter alone suffice for construction of a complete *Raum-Zeit Lehre,* which is his most important result (1921, 684). On the basis of the light axioms alone it can be demonstrated that the theory of relativity is "a valid and complete physical theory."[42] And although no treatment of the metric in general relativity is given, the striking claim is made that the determination of measure within the STR also fixes the metric GTR. This, Reichenbach remarks, is essential: "It is essential that, with the metrical determination of the special theory, that of the general theory is also fixed" (ibid., 686).

These claims are again made on behalf of the 1924 *Axiomatik*'s "physical significance." But now there is actually a purported demonstration of how, in the flat space-times of special relativity, a metrical determination can be made using only light signaling (i.e., the *Lichtgeometrie*).[43] As we shall see (§3.3), this claim will be somewhat qualified, only to be later withdrawn. Furthermore, in maintaining that what is "physically new" (1969 [1924], 76) in Einstein's contention (*Behauptung*) can be summarized in saying that rods and clocks adjust (*einstellen*) not to the classical but to relativistic light geometry,[44] Reichenbach sets the agreement of metrical determinations made by the light geometry and those made with rods and clocks that is at the heart of his *Axiomatik* as also the centerpiece of Einstein's theory; it is "on the basis of our axiomatic representation" that "we can finally pick out *what is affirmed concerning reality [Wirklichkeit] by the relativistic doctrine of space and time.*" Reichenbach will thus respond to the *Axiomatik*'s critics that "Einstein's theory stands or falls with my *Axiomatik,*"[45] a contentious remark, especially in view of the fact that the posited agreement breaks down in gravitational fields. However, the agreement (which is affirmed by Axiom VIII, see below) "holds only in infinitesimal regions for neighboring points," as, not surprisingly, we learn much later (1969 [1924], 167).

Although the *Axiomatik* combines a logical-epistemological orientation with technical discussion requiring familiarity with the calculus of tensors, it does not appear to have been successful in bridging the disparate communities of relativity physicists and scientifically minded philosophers eager to draw out the philosophical significance of the theory of relativity. Due to its explicitly epistemological and "constructive" character, the work did feature a certain novelty. But it was, by Reichenbach's own admission, widely "misunderstood" (1925, 37). Most notably, in its "constructive" concern to partition the theory into an empirical content as distinct from the conceptual structure of the theory, it completely departed from the method of axiomatic treatment of the theory of relativity pioneered by Hilbert. For the GTR, Hilbert had already demonstrated, the starting point is a variational (action) principle whose empirical confirmation or disconfirmation, Reichenbach duly noted (in accord with Pauli), is rather far-removed from actual experiment.[46]

The axiomatization of physics was, of course, one of the famous mathematical "problems" (the sixth) posed by Hilbert at the International Congress of Mathematicians in Paris in 1900; progress on the problem was reviewed in the widely read lecture (1918). Naturally, Hilbert and his school cultivated a "style" of axiomatization that sought to erect physical theories on the most mathematically general grounds.[47] In a contribution on "the foundations of physics" to the Göttingen Academy of Sciences on 20 November 1915, Hilbert (1915) had, independently of and virtually simultaneously with Einstein, succeeded in deriving the fully covariant field equations of gravitation. Hilbert's method of derivation, however, was significantly different: he employed powerful mathematical methods (of the calculus of variations and of integral invariants), and he took as axioms, first, an action function analogous to that of the field theory of matter of Gustav Mie ("Mie's axiom of the world function"),[48] and, second, the principle of general covariance that he called "general invariance" (*allgemeinen Invarianz);* while the latter requirement played the role of a desirable condition of adequacy in Einstein's tortuous route to his field equations, the former was regarded as totally unsupported by Einstein.[49] In so doing, Hilbert claimed to have shown that "electrodynamic phenomena are effects of gravitation."[50] Hilbert's axiomatization prompted Einstein to follow quickly with a derivation of his field equations, sans the Mie theory, from a Hamiltonian principle (Einstein 1916a).[51] However, a very important ideological difference remained. For Hilbert, the axiom of a *Weltfunktion* was not simply a choice representing mathematical elegance but stemmed from a conviction that Einstein's gravitational theory had produced "a completely different attitude toward geometry." Namely, since in the GTR no prior geometry could be assumed, there was a compelling theoretical requirement to develop both physics and the geometry of space-time in one fell swoop.[52] Thus Hilbert initiated a program (see Haas 1919, 1920; Müller 1923) of "axiomatizing physics" by attempting to derive both physics and geometry from a variational principle. The example was instructive to Weyl; even before Weyl conceived of his "unified theory," his initial work in the GTR followed Hilbert's lead in assuming that physics and geometry are subsumed under a variational (action) principle (Weyl 1917, 118ff.); the following year, Weyl would present his "unified theory" in this fashion. But by pointedly proclaiming a different and "constructive" goal for axiomatization, the Reichenbach *Axiomatik* deliberately ventured into philosophical *terra incognita* so far as the relevant scientific community was concerned, and it appears not to have made much impression among theoretical physicists and mathematicians, perhaps the readership Reichenbach most wanted to take notice. But also because of its considerable use of higher mathematics, the Reichenbach *Axiomatik* similarly received little attention from philosophers. One can speculate that the far more straightforwardly philosophical treatment given the subject in his *Philosophie der Raum-Zeit Lehre* of 1928 was, at least in part, a response to the relative neglect of his earlier book (see Schlick 1928).

3.3 Structure

Like its predecessor, the 1924 *Axiomatik* contains two central classes of axioms and, corresponding to these, two classes of definitions: those pertaining to the behavior of light (*Lichtaxiome, Lichtdefinitionen*) and "body axioms" pertaining to the behavior of measuring appliances (*Körperaxiome, Körperdefinitionem*). The axioms express "elementary facts" (*elementare Tatsachen,* §1) on which the theory of relativity is based and that are independent of the theory of relativity, though not of all theory. Each represents an empirically attestable fact in that "each (axiom) signifies (*bedeutet*) an intuitively presentable fact in which nothing further remains that is mysterious or unpresentable."[53] The *Körperaxiome,* translated in 1969 as "matter axioms," implement the notions of rigid rods and natural clocks that are "closed systems," that is, systems that may be considered as isolable from any "physical forces" whose effects may be correctable or considered to be negligibly small, whereas "metrical forces" are disregarded. The axiomatic standing of the congruence behavior of measuring rods and clocks as "elementary facts" rests upon Reichenbach's distinction (definition 21) between "physical" and "metrical" forces (1969 [1924], 88) by virtue of which the "rigid rods" and "natural clocks" figuring in these axioms are *defined.* This distinction recurs in his 1928 book in the more familiar guise of a distinction between "differential" and "universal" forces. But here only by making rods and clocks independent — by stipulation — of "metrical forces" can Reichenbach preserve their standing as "elementary facts," basic empirical postulates of the theory. For the existence of metrical forces ("which depend on the choice of metric") is not independent of general relativity; indeed, the GTR asserts that lengths of rods and periods of clocks are dependent on the surrounding gravitational field strengths. So here the distinction between "physical" and "metrical" forces is a consequence of the *erkenntnislogische* character of the axioms as "elementary facts." Despite their appellation, these "facts" are really idealizations; for instance, it is necessary to specify that rods and clocks are only "infinitesimally closed systems."[54]

The axiomatization takes as its basic concepts only the notions of "real point" (points at which physical objects may be considered to be at rest), "signal" (a physical process propagating between real points), and "simultaneity at a real point." "Earlier" and "later" at a real point are then defined by reference to the departure and return of a signal traversing a closed circuit. The first and larger part of the book concerns the STR. In motivating his approach, Reichenbach provides an imaginative picture of the world as a space filled with mass points similar to the molecules of a gas; these are to be real points at which observers are located who can signal to one another using light rays. The space-time metric within such reference systems ("rigid systems") can be determined solely by the light geometry corresponding to the "light axioms" grouped as I–V. Axioms I and II are topological axioms concerning temporal sequences at a real point. The six axioms grouped under II provide the means for making time comparisons at different points, the most important affirming the existence of "first signals," those signals traversing a closed circuit in least time. Axiom III is a Fermat axiom identifying first signals as directly emitted light signals. The two axioms of group IV, introducing the concepts of "stationary" and "static" systems,[55] have the effect of making the simultaneity relation both symmetric and transitive. The standard $\varepsilon = 1/2$ ("Einstein") simultaneity relation is then defined (definition 8, p. 44). Following a coordinative definition identifying straight lines as light rays (definition 9, p. 51), Reichenbach proceeds to define "spatial straight" and "spatial length" via light rays and return time of light signals; hence "congruence" is defined only by means of the velocity of light. The final light axiom (axiom V) affirms that the light geometry in static systems is Euclidean whereas inertial systems are defined as stationary spatial coordinate systems conforming to the light axioms (axioms I–V).[56] From three definitions (15–17, pp. 71–72) concerning comparison of units in stems moving with respect to one another and two auxiliary theorems, Reichenbach easily derives the result that the Lorentz transformations are mappings preserving the Euclidean character of static systems.

The claim that the thus developed light geometry suffices for the determination of the metric in the flat space-times of special relativity requires some qualification. Reichenbach shows how observers in reference systems situated within a finite distance may use light signals alone to distinguish that their systems are at rest relative to one another. Then he shows that within such "rigid systems" it is possible to define a metric (up to a linear factor) using only light signals. However, the extension of this method of metrical determination to the entire space(time) presents a difficulty. For most general transformations that carry light cones into light cones are not the linear Lorentz transformations but spherical transformations that carry spheres into spheres, the group of so-called Möbius transformations.[57] But in full three-dimensional Euclidean space, or four-dimensional Minkowskian space-time, these contain singularities wherein a point is carried into infinity. So not all real points can be reached with light signals. This means that if light geometry alone is held to be sufficient for the determination of the class of inertial systems, the following problem will present itself. If an observer who believes, after light signaling in his own region of space-time, that he is in an inertial system then signals arbitrarily far outside of this region, he will encounter a singularity, which will lead him to conclude that he is not in an inertial system. For the light axioms hold without singularities only in inertial systems (p. 82).

Reichenbach offers two possible solutions to this difficulty. To uphold the sufficiency of the light axioms for the determination of the metric, a procedure is sketched wherein the observer, assuming his own system S is inertial, can, by constructing another reference system S' relative to his own, calculate the limit at which singularities should occur in S' if it is not an inertial frame. Then, if these singularities are in fact found in S', the observer may conclude that his system S is an inertial system. If the calculated singularities in S' are not found, then S' is inertial and S noninertial, and the observer may then calculate the location of singularities in his own system S.[58] A second route is to adopt the criterion that the metrical determination must be without a singularity at any space-time point. This entails giving up the sufficiency claim, making the reasonable assumption that observational determinations occur anyway only in finite regions. But then it is necessary to invoke "material structures" to determine the class of inertial systems: points at rest relative to one another (as determined by light signaling) and only these may be connected by rigid rods (*Körperaxiom* VII; see immediately below). Hence, the problem of singularities is circumvented by a restriction of metrical determinations to within finite regions and by relying upon the posited existence of rigid bodies (1969 [1924], 13–14, 82).[59]

At this point then come six *Körperaxiome,* grouped as VI–X. The two axioms VI assert path-independent congruence of lengths (as measured by "rigid rods") and intervals (as measured by "natural clocks"); they express "customary presuppositions of measuring with rods and clocks" (p. 93). Axioms VII–X then assert the identity of the geometry of rigid rods and clocks with the light geometry, an identity initially claimed only for inertial systems (p. 89). Axiom VII affirms that only points at rest with respect to one another can be connected by rigid rods. Axioms VIII, IX, and X state the identity of lengths and unit length and time intervals with those of the light geometry. Reichenbach will attempt to extend this identity to noninertial frames later on with the addition of two general relativistic axioms XI (see §3.5 below).

3.4 Countering Weyl

The engagement with Weyl in the *Axiomatik* shows prominently in two strategic places, both involving a general defense of the *Körperaxiome*. The first concerns Weyl's proposal to use the trajectories of force-free mass points in the empirical construction of the metric. Here Reichenbach adamantly states that such a method offers no epistemological advantages to his own proposals involving rods and clocks. This response occurs twice: in the context of the construction of the metric of special relativity, where Weyl would use free mass points to specify the class of inertial systems, and in the context of the construction of the metric of general relativity, where Weyl would use them to provide a determination of the absolute values of the g_{ik}, which have been conformally fixed by the use of light signals only as ratios.[60] In notes in two separate places, the first in the section introducing the *Körperaxiome* (p. 82 n. 27), the second in a section on light geometry in a gravitational field (p. 154), Reichenbach observes that one can, as Weyl does, employ force-free point masses, rather than rods and clocks, as the needed material structures, referring to the treatment in the first appendix to the fourth (1921a) edition of *Raum-Zeit-Materie* (see 1922c, 313–14). But Weyl's approach, it is claimed, faces epistemologically the same problems as the method employing rods and clocks. For just as a rigid body must be *defined* to be one free from the effects of "metrical forces," so can a mass point be said to travel in a straight line (a geodesic) only if it is *defined* as a mass point upon which no net forces act. Weyl's use of mass points offers no epistemological advantages; it is on a par with rods and clocks. That is, it is open to someone to argue that a putatively force-free particle in fact is not traversing a geodesic, but describes a trajectory of a particle with a (nondetectable, "metrical") force acting upon it. This little argument appears to be Reichenbach's first specific application of what will become known as "the method of equivalent descriptions," a conventionalist leveling of the epistemological playing field achieved by pointing out the role of definitional elements in different theoretical proposals. This is the key move in generating a class of "equivalent descriptions": rival but empirically equivalent characterizations of "the facts" of a physical situation, here, pertaining to the metric of space-time. In subsequent works, Reichenbach locates "the philosophical significance" of the theory of relativity precisely in that it enables one to see that definitions are really called for (in the STR, for simultaneity, in the GTR for metrical indicators) where previously empirical cognitions had been sought (1958 [1928], 15). But in this specific instance, one might object that the epistemological comparison is framed too narrowly, in that it is a *law* (admittedly, a law with, in all probability, a vacuous antecedent) that bodies on which no forces are acting travel in uniform rectilinear motion whereas we do not speak of laws of rigid bodies. Appeal to a law here (Weyl calls it the Galilean law of inertia) is a different matter, given the role laws play in the systematization of our physical knowledge, from appeal to the fact of the approximately rigid behavior of rods and periodic behavior of clocks, complicated material structures whose exact behavior remains a task for physical explanation that (many-body quantum theory) will require assumptions far outstripping the law of inertia (cf. Ehlers 1973, p. 34). In fact, it is now commonly held that the Einstein field equation alone actually implies the "geodesic hypothesis" (see §1.3 above) that the world-lines of "test" particles ("small" enough not to be affected by tidal forces) are geodesics of the space-time metric (Wald 1984, 73; however, see Havas 1989). In addition (as was seen in §1.2), Weyl has taken the validity of the inertial law as demonstrating that the world is endowed with an affine structure, that the world is affine in character (Weyl 1922c [1952], 179).

In a second rejoinder to Weyl, Reichenbach observes that one does speak of the "adjustment" (*Einstellung*) of rods and clocks — as seen above, Reichenbach himself refers the "adjustment" of rods and clocks to the "light geometry." Noting that Weyl had first used the term "adjustment" in this connection, Reichenbach states that this characterization must be taken with a grain of salt since it only provides "a statement of the problem" (Reichenbach 1969 [1924], 91). As a characterization of the behavior of material structures, it cannot be taken as explanatory. Of course, if the term is understood literally, it is incompatible with the property of rigidity stipulated in the definition of congruence. Accordingly, Reichenbach maintains that the "situation" regarding the admitted "adjustment" of measuring rods and clocks to the fields in which they are embedded is "formulated rigorously by the matter axioms" without use of the term "adjustment." For Weyl's term merely names, but does not solve, a problem to be resolved by a future theory of matter of which there is, at present, no inkling (ibid.).[61]

3.5 Critique

Up to this point, we have seen how Reichenbach, in defense of his claim that the physical content of the Einstein theory is expressed by the claimed agreement between light and matter axioms, has sought to defend the use of rods and clocks by countering Weyl's proposals to do without them. We now ask: What justification supports Reichenbach's defense of rods and clocks? Reichenbach's epistemological analysis of the theory of relativity has been justly criticized by R. Torretti as "putting the cart before the horse," by which is meant, pride of place is given to special and not to general relativity (Torretti 1983, 241). Nowhere are the grounds for this assessment more clearly displayed than in the *Axiomatik*. In particular, Reichenbach's strategy is to attempt to accommodate, to the extent possible, the rods and clocks that are supposedly licensed in the local inertial frames supported in the STR within the context of gravity (that is, general relativity). To accomplish this, he exploits the limiting process that is at work in the admissible postulate of the "infinitesimal" (local) validity of the STR in the more general setting. For Reichenbach characteristically views this process through, as it were, an inverted lens, seeing metrical determinations in the general case as approximations to what can be established (and justified) locally.

This inverted perspective governs the path taken by Reichenbach in extending his axiomatization of the STR (axioms I–X and definitions 1–21) of part I of the book to the GTR in part 2. For the general theory of relativity, two new *Körperaxiome* are required (axiom XI.1.2 and definitions 23–25, pp. 142–49). From the first general relativistic axiom (XI.1), which asserts the infinitesimal validity of special relativity in every frame ("coordinate system of real points"), Reichenbach takes it to follow that "around every world-point a finite region can be defined in which a (spatial) coordinate system exists" (p. 143). Now a "spatial" coordinate system is essentially a rigid reference system that is both "everywhere present" and "infinitesimally stationary" at every point (definition 23, p. 142); thus we are generally to view measurements made *in finite regions* as if they are in agreement with metrical determinations made *at each point* in such a special coordinate system. Indeed, this is just what is stipulated in definition 24: that in any such coordinate system, "the metrical determination of the world is to be made in such a way that it will become identical at every point with the metrical determination that is locally prescribed by axioms I to X and definitions 1–21 [of the STR] while the same measuring rods and clocks are used everywhere" (142). There would seem to be a rather glaring difficulty with this proposal that has been particularly emphasized by

John Norton (personal communication). For if indeed metrical determinations are generally to be conceived as occurring within such "spatial coordinate systems" (frames of reference) where the STR is valid, then within these regions also the consequences of the STR must obtain. But one of these is that the Riemann curvature tensor vanishes identically in such regions no matter how small since, as a tensor, it is defined at each point. So Reichenbach appears to be attempting to piece together bits of space-time at each of whose points $R_{abcd} = 0$, a condition that, Norton rightly points out, certainly precludes virtually every interesting space-time of general relativity. Hence Reichenbach here appears to rely upon an inconsistent procedure for piecing together the infinitesimal domains of validity of special relativity into finite domains of inhomogeneous gravity.[62]

A related difficulty obtains in his second general relativistic axiom (XI.2). It asserts that accelerating rods and clocks can be considered as "differentially at rest" and hence give the same measurement results as rods and clocks "permanently at rest." Reichenbach states: "Accordingly, axiom XI,2 asserts that every rigid rod l . . . behaves in the same way as a rigid rod l_o . . . which is permanently at rest in an inertial system" (p. 146). Observing that "this result is by no means obvious," Reichenbach goes on to state that it is an assumption of the GTR.[63] However, this assertion, called by Torretti the "rod hypothesis," is problematic. The grounds of the difficulty lie with the notion "differentially at rest," that is, "the momentary inertial rest frame of an accelerating body." This concept cannot be made exact; that is, for an extended body, there is no such thing as *the* momentary inertial rest frame; to the contrary, its various point constituents at any single moment will be "comoving with different inertial frames."[64] Reichenbach is aware of the difficulties consequent upon his "rod hypothesis," but his means of overcoming them involves another problematic application of the inverted limiting principle noted above. For he notes that "the rigid rod and the natural clock are defined as closed systems; but closed systems exist only in inertial systems" (p. 147). Note that this last sentence is still not quite adequate if we take into account the problem of elastic forces, hence that of rigid bodies, already in the STR (e.g., French 1968, 27). But Reichenbach believes he can counter the problem of elastic forces (whereas "metrical forces are to be ignored") by stipulating a limiting process in terms of which these forces can be made to vanish. This is the definition of an "infinitesimally closed" material system (*materielles System*) that is "sufficiently small relative to space-time changes of the gravitational field strength and if the quotient of external physical supporting forces and internal physical forces approaches

0 with the shrinkage [*Verkleinerung*] of the system" (p. 148; translation emended).

Once again, an attempt is made to extrapolate the legitimacy of physical concepts that have a certain restricted domain of validity to more general domains through the inverted application of results obtained from a limiting process. This mode of procedure is in full force in the brief treatment given "the construction of the metric in the general case," where Reichenbach maintains that units of $ds = \pm 1$ can be introduced in any given coordinate system by assuming rigid rods and clocks at rest at each point, thereby (with additional use of light signals) determining the g_{ik} in this system of coordinates (pp. 157–58; p. 157 of the translation omits parts of the German text). However, as Weyl would point out in a critical review, this "fibrillation" of the world has nothing to do with the nature of the metric field.[65]

But independently of these conceptual problems, what remains of the central claim that the geometry of light rays and rods and clocks will always agree? Certainly Reichenbach cannot provide a general demonstration of this claim, for in a generic general-relativistic space-time, there can be no rigid congruences representing rigid bodies (Ehlers 1973, 34). Although he is able, in a section entitled "Light Geometry in a Gravitational Field," to exploit a mathematical result showing that there will be agreement in the special case of the field equations of the vacuum (1969 [1924], 153–54),[66] in general, there is agreement only in the infinitesimal regions compatible with the validity of the special theory. For other gravitational fields, Reichenbach proceeds to show that the metrical determinations of the *Körperaxiome* will *not* agree with light geometrical ones; in a step-by-step discussion, he tracks this failure by considering more and more general types of gravitational fields. Already for static gravitational fields, the metrical determinations of the light geometry and the *Körpergeometrie* no longer coincide (so *Körperaxiom* VIII falls), whereas for stationary fields, such as the rigidly rotating disk (the simplest case; see §2 above), the round-trip light axiom (IV.2) fails; hence determinations of spatial lengths here involve an additional complication. For in order to say that rods everywhere on the disk have the same unit length, a correction factor (corresponding to Lorentz contraction) is required for rods lying tangentially to the motion of the disk. Thus in order to preserve the customary definition of congruence (that is, "no metrical forces") for the full four-dimensional manifold of space-time, it is necessary to invoke "metrical forces" in the definition of congruence for three dimensional rigid rods (p. 178). With respect to even more general "real systems" (restricted only by the condition that it is impossible to transform any two distinct point-events on the same time line onto the same plane of simultaneity), only the topological axioms I and II are held to be valid.[67] For such systems, "the failure of axioms III, IV, and V means that metrical particularities [*Besonderheiten*] no longer exist" (p. 192; translation emended). Finally, in the most fully general case, where the g_{ik} are fully variable in position and time, restricted only by the requirement that the ds^2 be of "inertial index 1" (that is, of signature [1,1,1,–1] with one negative index), the restricted claim is made that the topological order of time, as given by axioms I and II, obtains just in "cut-out" finite domains, while global topological properties are wisely left in abeyance. As the result of these considerations that show that "topological properties turn out to be more constant than metrical ones," the startling admission is that in such general gravitational fields, where there are no rigid congruences, there can no longer be a metric (hence we can no longer really speak of a "geometry"). The rather stunning conclusion of the *Axiomatik* is that if rods and clocks, as the physical correlates of metrical notions, are no longer physically possible, it is necessary to *renounce* the metrical properties of space-time altogether! "[T]he transition from the special theory to the general one represents merely a renunciation of metrical characteristics, while the fundamental topological character of space and time remains the same" (195). This result is, moreover, held to be in harmony with the recognition that topological structure — in particular, the topological distinction of space and time — is, far more than the metrical structure, an object of visualization (*Anschaulichkeit*) in the GTR. Topological structure accordingly affords the means of carrying through "a consistent order of time in terms of causal order."

At this juncture, where the metric is subordinated to the topological structure underlying "the causal order of time," we have reached essentially the same end-point attained in *The Philosophy of Space and Time,* which also recognizes the causal order as "the physical structure into which space-time order can be embedded even when all of the metrical properties of the space-time continuum are destroyed by gravitational fields" (Reichenbach 1958 [1928], 268).[68] This is, it must be said, an astonishing finale. The talk of "renouncing metrical properties" cannot be taken literally, unless metrical properties are *necessarily* associated with rods and clocks, an association that, after all, was inaugurated as a convention! But if the association is a necessary one, then Reichenbach has unwittingly provided a telling (and damning) illustration of *the* fallacy of positivist metascience: an epistemological tail wags the scientific dog (witness the vivid "destruction" of the metric in gravitational fields). But also in the reference to the topological order as "an ultimate fact of nature," we can see just how far Reichenbach's analysis of the metric of space-time has philosophically strayed from Einstein's. For the key step taken by Einstein in arriving at his generally covariant field equations consisted in recognizing that nothing, including its "points," remained of the space-time manifold in the absence of an individuating physical or matter field. This was the lesson of the *Lochbetrachtung,* the so-called "hole argument" (see Stachel 1989a, 1993; and Norton 1984). Without an "individuating field" (such as the metric tensor field), there is no physical way, Einstein reasoned, to individuate the points of the space-time manifold. As Stachel has shown, this is the conclusion underlying the elliptical, and in itself quite puzzling, statement in Einstein (1916a, 117) that the "requirement of general covariance... takes away from space and time the last remnant of physical objectivity." Einstein's considered view, expressed much later, is more explicit: without the metric field, there is not only no residual "empty space" but "absolutely *nothing* and also no 'topological space' " (1952, 155). The implication, surely, is that one cannot (or should not), as does Reichenbach, speak of the primordial objectivity of the topological structure of the space-time manifold. The ontological interdependence of space-time and matter in the GTR forecloses the possibility of asserting the existence of topological structures of space-time in the absence of metrical ones dependent upon surrounding mass-energy distributions. In contrast, Reichenbach's analysis of the concept of congruence through coordinated measuring rods and clocks leads to the peculiar conclusion that where such physical structures are no longer possible, metrical characteristics are to be "renounced" in favor of recognition of the more fundamental claim to objectivity of topological relations. Thus his neoconventionalist solution to the metric of space-time ultimately rests on a grading of "objectivity" that gives pride of place to the topological relations of continuity, connection, incidence, and orientation.

Concluding Remarks

From a contemporary perspective, an amalgam of interrelated philosophical issues are concentrated along the axis Weyl-Einstein-Reichenbach. For one thing, the extraordinary philosophical significance of the GTR in the transformation of our ideas of the geometry of physical space is crystallized along this axis. And attention to its several way stations underscores the difficulty attending accommodation of its revolutionary message within our physical worldview, both then and now. For another, the confrontation between Weyl and Reichenbach continues to echo in current debates in the philosophy of space and time as to whether there is "only a conventional choice" to be made from (*pace* Grünbaum) among "a family of concordant external metrical standards." But, as the context of this confrontation shows, this manner of characterization of the choice involved completely fails to come to terms with the fundamentally conflicting assumptions of the significance of the GTR in the two leading philosophical proponents of differing "external metrical standards." Indeed, the conflict between them can be elided as a "conventional choice" only from one of their perspectives.

There is still a third and much broader consideration. It may not be wrong to see a crucial juncture in the philosophy of physical science lying along this axis. For insofar as logical empiricism was, largely through the interpretive works of Reichenbach on the theory of relativity, able to claim the philosophical mantle of Einstein, a decisive turn was taken for subsequent philosophy of science. However, as we have seen, at the time of appropriation Einstein happened to be in the process of readjusting his public philosophical attire. And it subsequently mattered little that significant philosophical differences would emerge between Einstein and positions held in the Vienna Circle in the early 1930s. What did matter was the apparent imprimatur given logical empiricist orthodoxy concerning the structure of scientific theories and the relation of theory to observation that, of course, was duly transmitted to subsequent generations of philosophers of science. In no small measure, this desirable genealogy enabled the lingering remnant of an adherence to neo-Kantian epistemological imperatives, institutionalized as the method of "coordinative definitions," to become the cornerstone of the "received view" of scientific theories. In its train came a phalanx of attendant philosophical problems and *insolubilia* surrounding the issue of "empirically equivalent descriptions" as well as an overly exulted role accorded to "purely conventional" elements in scientific theories. While this story cannot be pursued here, consideration of the scientific context of Reichenbach's rational reconstruction of the problem of the space-time metric should once again remind that philosophical positions regarding science do not virginally spring Minerva-like from the brow of Jove but are dialectically forged within the contingent scientific *and* philosophical circumstances of the time.

Notes

I am grateful to the Minnesota workshop participants, to Andreas Kamlah, and to the audience at the Center for Philosophy of Science, University of Pittsburgh, where an earlier version of this essay was given. A very special debt of gratitude goes to John D. Norton for stimulating discussions on these matters and for his comments and criticisms. Mate-

1. "Neoconventionalist" in view of his "mature" view that "according to the theory of relativity, the choice of a geometry is arbitrary: but it is *no longer* arbitrary once congruence has been defined by means of rigid bodies" (Reichenbach 1922a, 38).

2. Relevant passages from this Weyl-Einstein exchange are quoted in Straumann 1987.

3. For a usual account, see Pais 1982, 341; for further details, see C. N. Yang 1985 and Straumann 1987.

4. For details, see Torretti 1983, 188–91.

5. Physically, a manifold endowed with an affine connection, based following Levi-Cività on the fundamental idea of infinitesimal parallel displacement of a vector, is interpreted as a world with an inertial-gravitational field, to which Weyl gives the suggestive name *Führungsfeld,* in view of the "undeniable fact that a body which is released in a given world-direction fulfills a uniquely determined natural motion, out of which it can be disturbed only by outside forces" (1923a, 13).

6. "I am audacious enough to believe that the totality of physical phenomena permit deduction out of a single universal world-law [*Weltgesetz*] of highest mathematical simplicity" (Weyl 1918c, 385 n. 4). In a mode of procedure very much in the style of the Mie-Hilbert theory of matter, Weyl would then seek in the next few years, without success, to find a single action principle expressing the force law governing the interaction between the g_{ik} and the ϕ_i; this is to have the form of an integral invariant (Lagrangian density), additively combining a gravitational and an electromagnetic component; see Weyl 1922c (1952), §§35–36, and the later, somewhat chastened, abbreviated approach in Weyl 1923b, §40. The failure to univocally determine a "world-function" played no small part in the critical reception of Weyl's theory. For a critical assessment, see Bergmann 1976 (1943), chap. 16.

7. In a letter to Reichenbach thanking him for sending a copy of his book, apparently the first extant correspondence between them, dated 2 February 1921, Weyl ridicules this criticism: "[Y]ou say that I hold that mathematics (for example the theory of the [Riemann] ζ-function !!) and physics are but one discipline. I hold only that the concepts of *geometry* and of field physics have come together." Weyl's letter is in the Reichenbach archive at Pittsburgh (HR 015–68–04).

8. "The sole distinction between geometry and physics is this: that geometry investigates generally what lies in the essence of metric concepts, while physics determines the law through which the actual world is singled out from among all possible four-dimensional metric spaces of geometry and explores its consequences" (Weyl 1918c, 385).

9. Weyl locates the essence of the metric, hence of congruence, in the idea of the infinitesimal orthogonal group, and he is able to prove, under the assumption that the metric uniquely determines the affine connection, that the group of infinitesimal rotations at a point (for n dimensions) that can be characterized as the set of linear transformations leaving a nondegenerate quadratic differential form invariant is the only group whose mappings of a vector onto itself are "volume true," that is, accord with concept of congruence. This demonstrates the "uniqueness of the Pythagorean determination of measure," hence its apriority, and provides a new solution to the *Raumproblem.* Weyl thus distinguishes the "nature" or "essence" of the metric, which is a priori and the same at each point of space, from its "orientation" from one point to another indefinitely nearby, this being variable and *pace* Riemann and Einstein, locally dependent upon the distribution of matter' (see 1922a and 1923a, lectures 7 and 8). Weyl uses a method that combines Lie's

theory of continuous groups with infinitely small "virtual" variations of the mappings onto itself of an n-dimensional vector body radiating from a point.

10. "The metric depends on the concept of congruence that in any event must be conceived as purely infinitesimal" (Weyl 1922a, 114–15; 1923a, 47).

11. "In the theory of relativity, projective and conformal properties have an immediately intuitive significance. . . . The tendency of persistence [*Beharrungstendenz*] of world direction of a moving material particle, which forces upon it, if it is set loose in a determinate world direction, a determinate 'natural' motion, is a unity of inertia and gravitation. . . . However, the infinitesimal (light) cone realizes in the neighborhood of a world-point the distinction between past and future; the conformal property is the action-connection [*Wirkungszusammenhang*] of the world, through which is determined which world-points stand in possible causal connection with one another. Hence it is also a meaningful fact for physics which comes to expression in the following theorem: *Theorem I. The projective and conformal properties of a metric space univocally determine its metric.* (The proof follows.)" (1921b, 100).

12. This is Norton's (1993, 832) formulation. For a contemporary exposition of Weyl's method, see Ehlers, Pirani, and A. Schild (1972); see also Ehlers 1985.

13. See Eddington 1923, 207. A contemporary text summarizes, "The strength of Einstein's objection seems not as powerful now as at the time when it was raised, since we know the classical physics does not describe atomic phenomena without certain quantum-theoretical modifications" (Adler, Bazin, and Schiffer 1975, 506).

14. Taking the simple case of a static gravitational field, Pauli derives, from Weyl's gauge transformations, the "proper time" equation $\tau = \tau_0\ ^{e\alpha\phi t}$, where α is a factor of proportionality. Pauli explains this equation as follows: "Let two identical clocks C_1, C_2, going at the same rate, be placed at first at the point P_1, at an electrostatic potential ϕ_1. Let the clock C_2, then be taken to point P_2, at potential ϕ_2, for t seconds, and then finally returned to P_1. The result will be that the rate of clock C_2, compared with that of clock C_1, will be increased or decreased, respectively, by a factor $exp[-\alpha(\phi_2 - \phi_1)t]$ (depending on the sign of α and of $\phi_2 - \phi_1$). In particular, this effect should be noticeable in the spectral lines of a given substance, and spectral lines of definite frequencies could not exist at all. For, however small a is chosen, the differences would increase indefinitely in the course of time, according to [this equation]" (1921 [1958]), 196).

15. "The new view entirely alters the status of Weyl's theory. Indeed it is no longer a hypothesis, but a graphical representation of the facts, and its value lies in the insight suggested by this graphical representation" (Eddington 1923, §83, "Natural Geometry and World Geometry"). The antithesis of "natural geometry" versus "graphical representation" will recur in Reichenbach's examination of Weyl's theory (see below). Pauli subsequently also finds this interpretation to characterize the situation very well; see his 1926, 273–74, and also his letter to Eddington of 20 September 1923, quoted in §2 below.

16. Weyl changes his own interpretation of his theory along this axis; for the former, see 1919b; for the latter, 1921b and 1923b, §38.

17. This distinction seems to have been made in print first in 1920, 650; the distinction is discussed more fully in 1921d, 1921b, 1921a, and 1922b. It is remarkable that his distinction is still retained some thirty years later, long after Weyl abandons his own unified theory, in the form of an objection raised from the point of view of a (now future) "systematic theory" to the rigid rods and clocks of Einstein's theory; the objection occurs in a new appendix to 1949 (1927), 288.

18. The motivation is briefly hinted at in 1921c, 474. Einstein (1919) attempts to reinterpret the cosmological term λ, appended to his field equations in 1917 in an effort to satisfy what he would come to call "Mach's principle," as a constant of integration, rather

than as hitherto, "a peculiar universal constant." To do so, Einstein slight modifies his field equations without the cosmological term in a manner that explicitly accords gravitational force a role in holding atoms together. Now the proportionality factor on the left-hand side is changed (from 1/2 to 1/4). But this modification causes the scalar R of the left-hand side (i.e., of $R_{\mu\nu} - 1/4\, g_{\mu\nu}R$) to vanish identically as does the scalar of the right-hand side ($-\chi T_{\mu\nu}$) already when the stress-energy tensor is written in terms of the Maxwell-Lorentz components of the electromagnetic field and the divergence is taken. The modification enables him to derive the result that the scalar of curvature R is constant (1) in all domains in which the current density of electricity vanishes ("empty space") and (2) on every world-line of the motion of electricity (i.e., regarding electricity as a moving charge density). Einstein then gives the following "intuitive" (*anschaulich*) interpretation: "The curvature of scalar R plays the role of a negative pressure which outside of the electrical corpuscles has a constant value R_0. In the interior of each corpuscle there subsists a negative pressure $(R - R_0)$, whose drop maintains the equilibrium of the electrodynamic forces. The minimum of pressure (respectively, the maximum of the curvature scalar) *does not alter with time* in the interior of the corpuscle" (352; emphasis added). Finally, Einstein shows that in regions where gravitational and electrical fields are present, $\lambda = 1/4 R_0$. A. Pais (1982, 257) refers to this paper as "Einstein's first attempt at a unified field theory."

19. See Weyl 1922c, 134ff., for the definition of F (the scalar of curvature in Weyl's generalized metric) in terms analogous to that of the Riemann curvature scalar R; on the basis of F, Weyl develops the notion of a "natural gauge": it is to this "natural gauge" that measuring rods and the frequencies of atomic clocks "adjust." See also the discussion in 1923b, §40, and p. 303: "[M]easuring rod lengths and frequencies of atomic clocks are conserved on the basis of the natural gauge [*natürliche Eichung*], thus in fact are determined through adjustment [*Einstellung*] to the radius of curvature."

20. The physicist Arnold Sommerfeld seems to have thought, at this point, that there was only a minimal difference between Einstein and Weyl, and he urged Einstein to meet Weyl halfway; see his letter to Einstein of 10 August 1921: "I have the feeling as if between you and [Weyl] there is only a really small distinction [*ein ganz kleiner Unterschied*]. [Weyl] would overcome the practical effect of his measuring rod alterations through his [concept of] 'adjustment' [*Einstellung*] and you would restrict the indeterminacy of the world function [*Weltfunktion*] if you would take the g_{ik} as relational magnitudes" (in A. Hermann 1969, 87).

21. Following Pauli, Einstein also views the failure of Weyl's theory to yield solutions corresponding to electrons to be a fundamental flaw; see the letter to Ehrenfest of 4 December 1919 (cited in Seelig 1960, 280) and the letter to Besso of 12 December 1919 (in Speziali 1972, 148). However, this failure would attend all of Einstein's attempts at a unified field theory as well.

22. Translation altered in accord with Stachel 1989a, 94 n. 38.

23. "Whatever the result of these efforts may be, in any case . . . " (Einstein 1925c, 20).

24. Letter to Weyl 15 April 1918: "If one lets go of the connection of the *ds* with rod- and clock-measurements, then relativity theory loses in general its empirical basis" (quoted in Straumann 1987, 416).

25. Letter to Besso of 12 July 1920 (in Speziali 1972, 153).

26. Excerpts from letters in 1918 and 1919 are quoted in Straumann 1987. In particular, in a letter of 16 December 1918, Einstein, referring to a possible visit to Zurich the following February, remarks: "You will see that I am not stubborn [*eigensinnig*] but rather am gladly prepared to enter into any line of thought." Hoffman (1972, 223) also speaks of "Einstein's official argument against Weyl's theory."

27. "Letter from Einstein to Bohr," translated in French 1979, 274.

28. Weyl letter of 19 May 1952 to Carl Seelig in Seelig 1960, 274–75; cf. Sigurdsson 1991, 253.

29. Letter to Besso, 25 December 1925 (in Speziali 1972, 215).

30. Pauli to Eddington, 20 September 1923 (in Hermann 1979, 115–19).

31. "For a physicist this (field strength) is only defined as a force on a test-body and since there are no smaller test bodies than the electron itself, the concept... seems to be an empty, meaningless fiction. One should stick to introducing in physics only those quantities which are observable in principle" (Pauli 1919b; cited in Mehra and Rechenberg 1982, 278).

32. Letter to Born, 27 January 1920 (cited in Mehra and Rechenberg 1982, 279).

33. "[T]he theory of relativity teaches that the metric is subjective only insofar as it is dependent upon the arbitrariness of the choice of coordinates, and that independently of them it describes an objective property of the world" (Reichenbach 1969 [1924], 90). For recognition of the change in Reichenbach's position here, see Friedman 1983, chap. 1.

34. The method (misleadingly translated as "the method of logical analysis") is formulaically characterized as follows: "The process of distinguishing the objective sense of a physical expression from the subjective form of description through transformation formulas, by indirectly characterizing this subjective form, has replaced Kant's analysis of reason" (Reichenbach 1965 [1920], 91–92).

35. Employing the language of "coordination" (*Zuordnung*) in the analysis of scientific cognition, Reichenbach notes (1965 [1920], 85 n. 27), is the centerpiece of Schlick's *Allgemeine Erkenntnislehre* (1918). However, already in his dissertation of 1916 (published in 1916–17) Reichenbach characterizes "the proper task of physics" as "the coordination [*Zuordnung*] of... mathematical propositions of given classes to the objects of empirical intuition" (1916, pt. 3, p. 230); cf. p. 234: "[P]hysical knowledge consists in the coordination to mathematical equations to determinate [*bestimmten*] objects of empirical intuition."

36. Reichenbach's review of the Paul Hertz–Moritz Schlick centenary edition of Helmholtz's *Schriften zur Erkenntnistheorie* (Berlin: Springer, 1921) clearly indicates the source of the transformation in his views, for after an initial sentence describing the contents of the book, it continues: "It is surprising [*überraschend*] here with what certainty is recognized the connection of the congruence axioms with the behavior of rigid bodies; even Poincaré has not expressed conventionalism more clearly" (1922a, 421). Reichenbach refers of course to the well-known essays *"Über den Ursprung und die Bedeutung der geometrischen Axiome"* (1870) and *"Über die Tatsachen, die der Geometrie zugrunde liegen"* (1868) included in this edition. The homage to Helmholtz continues in Reichenbach 1958 (1928); after noting that Helmholtz, not Riemann, "laid the philosophical foundations (of the problem of space)," Reichenbach concludes "Helmholtz' epistemological lectures must therefore be regarded as the source of modern philosophical knowledge of space" (p. 36). The emphasis upon Helmholtz is justly thought misplaced in the context of general relativity where in the generic case of inhomogeneous metric fields (space-times of variable curvature), Helmholtz's solution to *das Raumproblem* based on free mobility of rigid bodies is inappropriate, as Weyl in particular emphasized (see §1.2 above).

37. On Schlick's role in effecting this change of terminology, see Coffa 1991, 201–4.

38. For example, in contrast to some, notably A. Grünbaum, who had seen in Reichenbach's coordinative definitions only a form of "trivial semantic conventionalism," that is, that "the meaning we give to words is arbitrary," Hilary Putnam recognized, thirty years ago, that Reichenbach, in his use of coordinative definitions, "is asserting a quite

special epistemological thesis" that is not trivial. According to Putnam (1963, 121): "That the words [occurring in an empirical law] must *first* be given a meaning by the laying down of *definitions* is not trivial, and indeed, in the opinion of most philosophers of science today, is not true. Yet it is just this that Reichenbach is concerned to assert and in no uncertain terms. He asserts *both* that *before* we can discuss the truth or falsity of any physical law all the relevant theoretical terms must have been *defined* by means of 'coordinating definitions' *and* that the definitions must be *unique,* i.e. must uniquely determine the extensions of the theoretical terms. Such views were quite common when Reichenbach wrote (in 1928)." Putnam is certainly correct in recognizing that coordinative definitions do signal involvement in a significant epistemological thesis. Yet it is not the operationalist and verificationist positions that *were* indeed becoming common by 1928. To see matters thus is to succumb to a perhaps unavoidable anachronism in view of the subsequent trajectory of logical empiricism. But the anachronistic reading is understandable if we slide from speaking about coordination as the method of displaying the object-constitutive role of concepts, as Reichenbach was in 1920–28, to speaking about coordinating empirically verifiable *meanings* to theoretical *terms* ("words" in Putnam's account), precisely as happened following upon "the linguistic turn." In this regard, the analysis of cognition as coordination, in a philosophical context becoming less and less receptive to transcendental idealism, aided and abetted the verificationist transformation.

39. For discussion, see Ryckman 1992.

40. Kamlah 1979, 433; the term "rational reconstruction," however, seems to have first been used with a related sense in Carnap 1928.

41. This paper is dated "Juli 1925."

42. Reichenbach erroneously claims that the Lorentz transformations can be deduced from the light geometry (1921, 685), and light geometry, especially axiom V, renders the metric arbitrary only up to a linear function (686, discussion); see §3.3 below.

43. It is notable that Reichenbach is, in 1924, unaware of the earlier attempts of the Cambridge mathematician A. A. Robb (1914) to axiomatize special relativity using as the only basis concept the signal relation "<" ("after"). Weyl (1923a) had already referred to Robb in this context.

44. The meaning is that the Michelson-Morley experiment concerning the non-detection of an ether drift is taken to show that rods and clocks obey the Lorentz, not the classical Galilean, transformations; for example, Weyl writes: "[W]e must regard *the Michelson-Morley experiment as a proof that the mechanics of rigid bodies must, strictly speaking, be in accordance not with Galileo's Principle of Relativity but with that of Einstein*" (1918b, 136; 1922c, 173–74). By the 3d edition of Weyl's book, the following remark is added (1919b, 149–50): "Since the behavior of rods and clocks remain[s] somewhat problematic for the formation of physical laws, it is of theoretical interest to note that in principle much simpler measuring instruments suffice for fixing the space-time coordinates in an arbitrary reference system, . . . namely light signals and the motions of force-free mass points."

45. This claim is made in Reichenbach 1927, 143; Hentschel (1990, 189) has also found that Reichenbach made the claim in correspondence with a Finnish critic of his axiomatization.

46. "We might think of the content of a theory as summarized in a variational principle; this principle can never be the direct object of experimentation, and yet, depending on the confirmation of its consequences, it may be called true or false with a certain degree of *probability.* In order to avoid this difficulty, it is advantageous to approach the axiomatization in a different fashion. It is possible to start with the observable facts and to end with the abstract conceptualization. . . . Such a *constructive* axiomatization is more

in line with physics than is a *deductive* one, because it serves to carry out the primary aim of physics, the description of the physical world" (Reichenbach 1969 [1924], 4–5; emphasis in original German). Pauli (1958 [1921], 201), already in 1921, expressed skepticism toward what he saw as the too-little empirically oriented Göttingen approach of seeking such a *Weltfunktion:* "[I]t is not at all self-evident from a physical point of view, that physical laws should be derivable from an action-principle. It would, on the contrary, seem far more natural to derive the physical laws from purely physical requirements, as was done in Einstein's theory."

47. See Born 1922 and Mehra 1974, 17–18.

48. Mie's program of a pure field theory of matter, as laid out in three papers in 1912, sought to reduce the problem of the field laws of matter in the four-dimensional world of Minskowski space-time to the determination of a "world function" (*Weltfunktion*) whose integral invariant $\mathcal{L}$, a Lagrangian density, expresses the action contained in an infinitesimal four-dimensional "volume element" of the world ($dw = dx^1dx^2dx^3dx^4 = d^4x$), what Haas (1919) called a "geometrical quantum." Expressed as a Hamiltonian principle,

$$\delta \int \mathcal{L} dw = 0,$$

it states that the change in the total action, for any infinitesimal variation of the field magnitudes that vanish outside a finite region, is zero. For details, problems, and references, see Pauli 1958 (1921), 188–92.

49. See the classic papers of Stachel (1980b) and Norton (1984).

50. From the fundamental variational equation,

$$\delta \int \mathrm{J}\sqrt{g}\, d^4x = 0$$

($\sqrt{g} = \det |g_{ik}|$) derived as a theorem from his first axiom, Hilbert showed that it followed that the electromagnetic four potential ϕ_j depends upon the ten independent gravitational potentials g_{ik}; more generally, that for n unknowns, n variational (Lagrangian) equations can be found such that four are consequences of the remaining $n-4$. On this basis, Hilbert concluded: "[E]lectrodynamical phenomena are effects (*Wirkungen*) of gravitation" (1915, 397). However, Pauli summarized Hilbert's achievement rather differently (and this is the modern view): "It was first shown by Hilbert that the energy-momentum tensor is related to the action function in a simple manner.... *We thus obtain the energy tensor of matter by varying the G(ravitational)-field in the action integral*" (Pauli 1958 [1921], 158).

51. This is not to say that variational methods for his gravitational theory were unknown to, or unused by, Einstein prior to Hilbert's paper of 1915, as Norton (1984) has shown.

52. "The new physics of Einstein's principle of general relativity takes a completely different attitude toward geometry. It presupposes as a basis neither Euclidean nor any other geometry in order to then deduce the particular physical laws; rather, the new physics yields, as I have shown in my first contribution, through one and the same Hamiltonian principle at one stroke the geometrical and the physical laws,... which teach how the measure determination — which at the same time is the mathematical expression of the physical phenomena of gravitation — is connected with the values q_σ of the electrodynamic potentials" (Hilbert 1917, 63–64).

53. Reichenbach 1925, 36. The term *Elementartatbestand* is bit misleading since these "facts," pertaining to the behavior of light rays and rods and clocks, are not elementary in the sense of being theory-independent, but only in the sense of being establishable independently of relativistic theory; thus they rely on, for example, prerelativistic optics and the mechanics of elastic deformations of material bodies.

54. See below; M. Carrier (1990) has given an illuminating discussion of the "ideal" character of facts underlying "constructive" axiomatizations.

55. A stationary spatial system centered about a point *A* consists of all those points *P, P′, P″, …* wherein the time of a light signal from *A* to *P* and back does not alter with time (Reichenbach 1969 [1924], 44). Static systems are those stationary systems in which light signals sent simultaneously from *A* in opposite directions about a closed path simultaneously return to *A* (49).

56. At this point there is a mistake: the proof offered of the uniqueness of the definition of inertial systems contains an error corrected in Reichenbach 1925, 34; the English translation (Reichenbach 1969 [1924]) includes the relevant passage in a footnote on pp. 58–59.

57. In the complex Euclidean plane with the addition of a point at infinity ($z = 0 \rightarrow 1/z = \infty$) to avoid singularities, these are circular transformations that are of the form $z' = az+b/cz+d$, where *a, b, c,* and *d* are complex numbers, and $ad - bc \neq 0$. For details, see Pedoe (1988, chap. 6). For the more involved formulation of Möbius transformations in *n*-dimensional spaces, see Fock 1964, appendix A.

58. This hit-or-miss procedure is abandoned in Reichenbach 1928 (1958), 173, which holds that the determination of the class of inertial systems "is not possible unless we avail ourselves of some physical means other than light signals."

59. Weyl's discussion (1923a, lecture 1; see also 1924) brings clarity to the problem. Beginning from full projective space (since the group of projective transformations is singularity-free), he shows that the Lorentz group, which is the quotient of the Möbius group and the projective group, is the sole singularity-free subgroup of the full Möbius group. He then identifies the trajectories of force-free point masses with the straight lines invariant under projective mappings. A complete treatment is in Weyl 1930.

60. Reichenbach correspondingly assigns matter axioms a different functional role in the STR and the GTR. Hence, the status of matter axioms changes in the two theories. There is "an essential difference" between the STR and the GTR in that the light geometry suffices (with the restrictions noted above) for the construction of the metric in the former, but not in the latter where "material structures" (*materielle Gebilde*) are required to determine the absolute values of the g_{ik} (1969 [1924], 151 and 155); see also the following statement: "The significance of material structures becomes clear: they bring about a comparison of the units at different points. The comparison cannot be achieved by means of light signals" (ibid., 158).

61. Indeed, Reichenbach's response to the challenge posed by Weyl's critique of the status of rigid rods and clocks in general relativity appears nearly verbatim in both his response (1925, 46–47) to critics, which include Weyl, and in his "mature" *Philosophie der Raum-Zeit Lehre* (1958 [1928], 201).

62. For sufficiently strong gravitational fields, this is an illegitimate extrapolation from the principle of "minimal gravitational coupling," that is, that the STR can be expected to hold in a sufficiently small neighborhood of a point *P* where there is an inertial coordinate system.

63. Einstein actually invokes it in his 1922 Princeton lectures (1956, 60n.).

64. Torretti continues: "That is why rods ultimately cannot hold their own as instruments of fundamental measurement in General Relativity" (1983, 315 n. 25). Though not cited in this context by Torretti, already in 1910, von Laue and others had raised fundamental objections to the concept of a "rigid body" in the STR on grounds that rigidity entailed the possibility of transluminal propagation of causal effects; for a recent discussion, see Norton 1992. Von Laue's objection is cited by Weyl four decades later (1951,

75) regarding the fundamental inappropriateness of the notion of a rigid body *already in the STR*.

65. For Weyl (1924, 2127), the "inappropriateness" (*Unsachgemässe*) of Reichenbach's *Ansatz* of rods and clocks here becomes fully evident: "[T]he axiomatic analysis of the metric field is not based on the world-point but on a three-parameter band [*Schar*] of world-lines. . . . [T]his 'fibrillation' certainly has nothing to do with the nature of the field (which) rather would be brought home to us only through the permanence of material bodies and their elementary particles."

66. Reichenbach makes use of a theorem of Schouten and Struik (1921), who in turn have generalized Kastner's (1921) result. The result states that a necessary and sufficient condition for a conformally flat manifold to be mapped upon a Euclidean manifold is that $R_{ab} = 0$, that is, the vanishing of the Ricci tensor. Reichenbach's discussion here appears problematic in that he seems to assume that if $R_{ab} \neq 0$, then light does not travel null geodesics, and "the light geometry cannot be carried through."

67. Reichenbach erroneously infers that in such systems there can be no closed timelike world-lines (1969 [1924], 187); see 1958 (1928), 273, for a correction.

68. See Reichenbach 1958 (1928), 285: "The fact that an ordering of all events is possible within the three dimensions of space and the one dimension of time is the most fundamental aspect of the physical theory of space and time. In comparison, the possibility of a metric seems to be of subordinate importance. It is only the metric, however, which, in the general theory of relativity, has been recognized as an effect of the gravitational field. The essence of space-time order, its topology, remains an ultimate fact of nature, unaffected by these considerations." The talk of "destruction" might lead one to think that Reichenbach is here anticipating later results concerning singularities, but actually these results tell against his causal ordering story. As we understand matters today, general relativistic laws (hence, the metric) break down only at singularities (where, according to the GTR, space-time curvature is infinite); however, if the Penrose-Hawking singularity theorems are on target, here space-time itself breaks down inasmuch as singular points are "cut out" of the manifold, thus taking along the topology that the manifold "normally" supports. However, in my view Reichenbach certainly had no such far-reaching anticipations in mind; rather, his location of "the genuine [*eigentlich,* omitted in the Maria Reichenbach translation] philosophical result of the theory of relativity" in "the causal theory of space and time" (1958 [1928], 269 [303]) stems solely from his narrowly epistemological conception of the metric in the GTR. I'm grateful to Andreas Kamlah for catching the imperfect translation of *Besonderheiten* as "characteristics."

PART III

LOGIC, MATHEMATICS, AND PHILOSOPHY

Warren Goldfarb

The Philosophy of Mathematics in Early Positivism

It is commonly held that the philosophy of logic and foundations of mathematics were of central moment in the development of logical positivism. This common wisdom is, of course, correct. It is also commonly held that there was a single logical-positivist doctrine on the nature of logic and mathematics. Here the common wisdom oversimplifies. For although the major positivist writers were in agreement on the general shape of such a doctrine, differences in their views emerge on closer inspection. A look at the evolution of their views and the differences among them can shed light on the positions at which mature positivism arrived.

It is natural to divide developments into three periods. The first lies prior to the formulation of classical logical positivism, that is, before 1928 or so. Although the roots of many positivist concerns can be seen in the writings of both Schlick and Carnap in this period, it is surprising to note how small a role is played by considerations in philosophy of mathematics. The second period centers on 1930: the main event here is the assimilation and appropriation of the views of Wittgenstein's *Tractatus*. Despite Schlick, Hahn, and Carnap's unanimity (against Wittgenstein) that mathematics as well as logic are tautologous, there are subtle disparities — obscured by the common terminology — in their positions. Finally, the third period is marked by the emergence of Carnap's distinctive position given in *The Logical Syntax of Language*. To my mind, that position marks a profoundly original shift in the conception of the philosophy of mathematics. The earlier views can serve as a contrast that helps to bring out the nature of Carnap's contribution.

1. Protopositivism

Schlick, the founder of the Vienna Circle, sounded many characteristic themes of positivism well before he could properly be called a positivist.

However, his first writing that touches on the philosophy of mathematics displays a rather different bent. In "The Nature of Truth in Modern Logic" (1910), Schlick mentions that the work of Hilbert, Russell, and Louis Couturat on the principles of mathematics could be viewed as casting in doubt the Kantian notion that the certainty of mathematical propositions rests on intuition; but then he winds up taking Poincaré's side against the logicists, with respect to both the ineliminability of mathematical induction by means of a reduction to logic and the justification for induction as lying in the affirmation of powers of mind. "Here too then, we find the 'eternal verities' established through inner or perfect experience" (Schlick 1979a [1910], 85).

Over the next few years, though, Schlick's thinking develops away from such Kantianism. His fundamental theme becomes the *limited* role that intuition or perception plays in knowledge. Knowledge, for him, is conceptual and not grounded in the intuitive. Although there remain some Kantian aspects in Schlick's general system, the "structural realism" he espouses in this period, he explicitly rejects Kant's doctrine of pure intuition. Consequently, of course, he shall have to deny the existence of synthetic a priori truths, and this will raise the question of how mathematics is to be accounted for.

Schlick's answer, in the *Allgemeine Erkenntnislehre* (1918), invokes Hilbert's notion of implicit definition. He calls Hilbert's method "a path that is of the greatest significance for epistemology." That method is "simply to stipulate that the basic or primitive concepts are to be *defined* just by the fact that they satisfied the axioms" (1918, 32; 1975 [1925], 33). Hilbert, of course, devised this notion to apply to geometry. Schlick generalizes it to cover other branches of mathematics — he explicitly mentions number theory and presumably wishes to include analysis — as well as the empirical sciences. He admits that "a system of truths created with the aid of implicit definitions does not at any point rest on the ground of reality" (1918, 35; 1975 [1925], 37); "reality" here means empirical reality. He adds that, for geometry and the empirical sciences, in the end we apply the implicitly defined concepts to the intuitive, although "the moment we carry over a conceptual relation to intuitive examples, we are no longer assured of complete rigor." Schlick is extremely vague, in this book, on the nature of the links between the conceptual and the intuitive or empirical; he talks of the concepts coming in contact with the intuitive, but never says what this comes to.[1]

Schlick's distinction between free-floating, implicitly defined, abstract concepts and a less rigorous application of those concepts to empirical reality seems particularly unsuitable as an account of number theory and analysis. Surprisingly enough, Schlick seems to claim that in "such an abstract science as number theory" there is never any contact with the intuitive, with the real (1918, 35; 1975 [1925], 37). And then, of course, the Kantian question is never answered; the applicability of the a priori sciences of number theory and analysis to the empirical world is never explained; and the need for pure intuition is not eliminated.

A more general difficulty looms here, in any attempt to use implicit definition to replace pure intuition as a grounding for mathematics. Schlick recognizes that we can take an axiom system to provide an acceptable implicit definition only if the system is consistent (1918, 36; 1975 [1925], 38). Now Poincaré had argued over a decade earlier that precisely this need for a consistency proof shows that implicit definitions alone cannot provide the basis for mathematics; if the mathematics is to be legitimized, more must be assumed, namely, whatever principles are used in demonstrating consistency. Thus the recourse to implicit definitions does not, by itself, show that pure intuition can be eliminated (Poincaré 1905, 820). Schlick, despite raising the issue of consistency, simply ignores such an objection, calling the provision of a consistency proof "an internal affair of the theory in question" (1918, 37; 1975 [1925], 39).

In any case, Schlick winds up claiming that mathematics is entirely analytic (1918, 96; 1975 [1925], 115) and for that reason does not lead to new knowledge. Here he means analytic in just Kant's sense (1918, 97; 1975 [1925], 75). This is also odd, one might think. Even if we accept Schlick's employment of implicit definitions, we still need logic to draw the consequences of the axioms, and presumably such logic would go beyond the subject-predicate containment that figures in the Kantian definition. (True enough, in his *Grundlagen der Arithmetik* [1884] Frege called logic "analytic," but in this he was using the word in a redefined sense, not the original Kantian one.) In fact, at this time Schlick does not believe that logic outstrips the Kantian analytic: he believes that the only inferences needed are syllogistic. As he puts it:

> All truths that have precise logical interconnection (that is, that admit of being deduced from one another) can be represented as far as their mutual linkage is concerned by means of syllogisms, specifically in the mood *Barbara.* The Aristotelian theory of inference needs no modification or extension in order to be applicable to modern science. What is necessary is only that the theory of concepts be deepened. (1918, 89; 1975 [1925], 107)

This is astonishing. Thirty-four years after Frege's *Grundlagen* and eight years after volume 1 of *Principia Mathematica,* Schlick thinks himself to have provided an alternative to the Kantian account of mathematics by a breezy invocation of implicit definition and a fallback on syllogisms, and without any attention to details of the logic of the situation. Nor do matters change for ten years. All of the remarks cited above are retained in the 1925 second edition of the *Allgemeine Erkenntnislehre.* Moreover, implicit definition retains its central place in "Erleben, Erkennen, und Metaphysik" (1926). (By this point, though, Schlick no longer uses the "analytic" character of mathematics, as obtained via implicit definitions and syllogisms, as an independent argument against Kant. He exploits a far more direct argument available by then, namely, that general relativity had simply shown Kant to be wrong about geometry and hence about pure intuition.)

There is no further consideration of philosophy of logic or foundations of mathematics in Schlick until he fell under the influence of Wittgenstein's *Tractatus.* This was a period in which, in contrast, his general view of physical science develops considerably. In the *Allgemeine Erkenntnislehre,* the conceptual realm and the empirical world seem to be taken as independent, although with some points of contact. In the 1920s, Schlick starts formulating a different picture, spurred by his understanding of general relativity. In this picture, the conceptual is an overlay on an empirical world already given. In Schlick's account of relativity, there are event-coincidences, and then the conceptual (the implicitly defined, the conventional) includes things like a coordinization and a metric, which simply give us different ways of talking about the same underlying reality.

Schlick's use (or overuse) of implicit definitions is the unstated target of Carnap's earliest writing on philosophy of mathematics and logic, the 1927 paper "Eigentliche und Uneigentliche Begriffe." For Carnap, a proper concept is one built up by explicit definitions from basic concepts; Carnap includes here both real (that is, empirical) concepts and formal (that is, logico-mathematical) ones. An improper concept is one introduced through an axiom system, that is, by implicit definition. The terminology evidently suggests Carnap has doubts about Schlick's notion that implicit definition completely legitimizes the employment of a concept. In the technical part of the paper, Carnap talks about realizations (that is, models) of axiom systems — which is already foreign to the spirit of Schlick's position — and discusses notions like categoricity and syntactic completeness. Finally, he notes two important differences between proper and improper concepts: first, the law of excluded middle does not hold for the latter; and, second, improper concepts are indefinite, in that one cannot say of any given object whether it falls under or does not fall under an implicitly defined concept. He concludes: "These two differences between improper and proper concepts do not get to the heart of the matter; they are only symptoms. The essential difference is this: improper concepts are variables, while proper concepts are constants" (1927, 371). This is just the position Frege had taken against Hilbert twenty years earlier. Hilbert is not providing definitions, Frege said; rather he is using higher level variables, and then the axiom system simply expresses a property of anything the variables indicate (Frege 1984 [1906], 322). Consequently, a sentence using these concepts does not make an assertion; as Carnap puts it, it is a sign for a propositional function. Frege and Carnap both go on to say that such a sentence can be taken as short for a universal statement, to the effect that all concepts that have the properties given by the axiom system also have such-and-such further properties. In the latter statement, Carnap points out, all the variables appear only as bound variables; hence it makes an assertion. Sense is given to assertions "about" the concepts only via these general statements of consequence from the axioms. (The construal of improper concepts as higher-order variables also renders Carnap's treatment of notions like categoricity executable within type theory. The elaboration within the system of *Principia Mathematica* of notions we would take to be metamathematical is carried out at length in an unpublished manuscript of Carnap's from 1930.)

Carnap means to imply that, as implicitly defined concepts are not concepts at all, they cannot be seen as objects of knowledge as Schlick wanted. He concludes that implicitly defined concepts cannot be used to construct a theory, "since they don't concern anything definite." Rather they are used to form a theory-schema, "an empty form for possible theories." The way to give them content is to find proper concepts that can be shown to provide a realization of the axiom system. In an uncharacteristically imagistic phrase, Carnap says, "The blood of empirical reality streams in through this place of contact, into the most ramified veins of the previously empty schema, transforming it into a satisfied theory" (1927, 373). Note that Carnap requires a full realization: that is to say, each implicitly defined concept (each variable) has to be replaced by an explicitly defined one. In this Carnap is denying one view that might be imputed to early Schlick, that the implicitly defined concepts get some real content just by contact with the empirical "at the edges." In the end, for Carnap there is *no* place for implicitly defined concepts in knowledge.

Now Carnap is here speaking particularly of giving empirical content to concepts: the model is geometry, and the procedure he outlines amounts to nothing other than Russell's division of geometry into pure and applied parts. Pure geometry is the statement that certain properties hold of any entities fulfilling the axioms; applied geometry is the investigation of whether explicitly defined spatial concepts constitute a realization of this or that axiom system (Russell 1903, 372ff.).

Once again, though, one can wonder about arithmetic and analysis, where particular physical realizations hardly seem to be the issue. In his discussion of proper formal concepts, Carnap mentions the Whitehead-Russell construction of the objects of mathematics. He contrasts a concept of number as implicitly defined by the Peano axioms (what he calls the Peano-numbers) with those explicitly built up in the logicist manner (the Russell-numbers). Clearly, he would think that definite meaning can be given to the implicitly defined concept only by showing that explicitly defined concepts form a realization of the axioms. Thus, he is fully adopting the logicist agenda: to make the right sense of arithmetic, one has to provide an explicit construction of number. Ultimately, there is no place for implicit definition in the foundations of arithmetic either.

In passing, Carnap also says something about the philosophical status of formal concepts, that is, those used in the logicist construction. This paper was probably written in 1926, and in it there is no explicit reference to the *Tractatus*. Nonetheless, he characterizes the words for logical notions as

> intermediary signs, which themselves do not denote any real concepts. . . . They help to assert something about reality, but they themselves correspond to nothing in reality, they only shape the assertion. Although they have no independent meaning, it is usual to talk of the concepts which are referred to by them; these "logical concepts" or "formal concepts" are, however, . . . of a completely different sort from real concepts. (1927, 358)

The idea expressed here, that logical concepts arise from the representational structure of language but do not themselves represent, is unmistakably Wittgensteinian. This was in all likelihood one of the "interesting and stimulating points" Carnap found in his early, partial, and somewhat cursory reading of the *Tractatus* (see Carnap 1963a, 24). It is clearly an idea to which Carnap was strongly attracted.

The *Aufbau,* published the following year, is for the most part silent on questions in philosophy of logic and foundations of mathematics. There is, however, some evidence of the two positions just canvassed—a commitment to the logicist program and the denial that there are logical entities. Carnap mentions no logic alternative to Russellian type theory; and the whole notion of constitution system underscores the centrality of the logicist reduction: "Even before the introduction of the basic relation(s), we must construct the *logical objects*" (§107). In the final versions of the *Aufbau,* the Tractarian influence was explicit; in particular, the word "tautology" appears as a characterization of logical truths. Even in paragraphs that seem to have an earlier provenance, Carnap indicates that any statement, at any level of the constitution system, is really "about" only the basic elements, despite the amount of class theory that is also involved, thus suggesting that there are no logical objects for the sentences also to be about.

> It follows from the construction on the basis of the same basic objects that statements about all objects are transformable into statements about these basic objects so that... science is concerned with only one domain. (§41)

> In other words... all (scientific) statements can be transformed into statements about my experiences.... [E]ach object which is not itself one of my experiences is a quasi-object; I use its name as a convenient abbreviation in order to speak about my experiences. (§160)

This point, I believe, stems simply from Carnap's way of taking *Principia* to be a "no-class" theory, that is, his agreeing with Russell that classes are logical fictions and signs for classes are incomplete symbols. Even so, the affinity of his thought with doctrines of the *Tractatus* is clear.

There is some such affinity also in Schlick's (much less logically sophisticated) early thinking, in his insistence that mathematical inference does not extend our knowledge and in his view that certain truths are simply conventions, things we agree to take as true as a matter of the language we speak. It is not surprising, then, that when the Vienna Circle studied the *Tractatus* in detail, most quickly found themselves persuaded by his account of logical truth.

2. Wittgenstein's Influence

The Vienna Circle read the *Tractatus* carefully during the academic year 1926–27 and had personal contact with Wittgenstein starting in the summer of 1927. Wittgenstein's characterization of logic as composed of nothing but tautologies, with the consequence that propositions of logic were empty, was within short order adopted by most of the Circle and promulgated in their writings. It appears in Carnap's *Aufbau,* published in 1928, in Hahn's "Empirismus, Mathematik, Logik" (1929), and in

Schlick's 1928 preface (1928) to Waismann's never-published treatise "Logik, Sprache, Philosophie" on Wittgenstein's views.

Two major themes figure in the rhetoric with which Schlick and Hahn surround their explanations of Wittgenstein's notion. The first, prominent particularly in Hahn, is that logic is not about the world but rather about how we talk about the world; it is about transforming one way of speaking to another equivalent way of speaking. "The so-called propositions of logic are directions on how something we have said . . . can also be said in another way, or how a state of affairs we have designated in one way can also be designated in another way" (Hahn 1980 [1929], 41); tautologies "say nothing at all about the objects we want to talk about, but concern only the manner in which we want to speak of them" (Hahn 1980 [1933], 35). Similarly, Schlick says that such propositions "say nothing about existence or about the nature of anything, but rather only exhibit the content of our concepts, that is, the mode and manner in which we employ the words of our language" (Schlick 1979b [1930], 170). In this way Hahn and Schlick express the idea that logical "truth" is not a species of truth: it is a mere artifact of the representational system.

The second theme is that tautologies tell us nothing, have no content. Often this is explained in terms of statements for which there could be no question of verification. "There is no material a priori, i.e., no a priori knowledge of facts; for we cannot know of any observation how it must come out before we have actually made it" (Hahn 1980 [1933], 30). Such a rendering of the idea that the propositions of logic are true come what may—and so for that reason are not really truths at all—gives it a blatantly epistemic tone. This tone is in keeping with a major emphasis in both Hahn and Schlick, namely, that Wittgenstein's "discovery" was essential to justify empiricism and in particular to reconcile empiricism with the apodictic certainty attaching to logic. "Only the elucidation of the place of logic and mathematics (which is of very recent origin) made a consistent empiricism possible" (ibid., 21). Indeed, Hahn actually takes this consequence of the Wittgensteinian view as an argument for it.

Carnap's emphasis when he introduces the notion of tautology hews rather more closely to Wittgenstein. In the *Abriß der Logistik* (1929), "Die alte und die neue Logik" (1930a), and "Die Mathematik als Zweig der Logik" (1930b), whenever he defines the notion of tautology he stresses the truth-table analysis of logical truth and the idea that a tautology rules out no possibilities. (Oddly enough, neither Hahn nor Schlick ever presents a truth-table.) Thus Carnap's explanation does not commit him to notions of "information" or "content" being given by some empirical procedure; it relies more on the Tractarian view that *any* notion of content requires a contrast between what would make the proposition true and what would make it false, which contrast is absent in the case of tautologies.

All three, Schlick, Hahn, and Carnap, differed immediately from Wittgenstein in categorizing not just logic but also mathematics as composed of tautologies. In this difference, Carnap and Hahn here seem most influenced by two factors: their prior commitment to the logicist program and the obscurity of the *Tractatus* in both its criticisms of the logicist reduction and its formulation of an alternative. Hahn writes:

> We have already indicated how pure empiricism is compatible with the existence of logic. The question "How is pure empiricism compatible with the existence of mathematics?" is therefore settled if it can be successfully shown that mathematics is part of logic, and hence, that the propositions of mathematics too do not say anything about the world. . . . Under B. Russell these efforts [to dissolve mathematics into logic] have recently been gathering enormous strength and now seem to be advancing on the road to victory. (Hahn 1980 [1929], 42)

> To be sure, the proof of the tautological character of mathematics is not yet complete in all details. This is a difficult and arduous task. (Hahn 1980 [1933], 35)

Carnap, in "Die Mathematik als Zweig der Logik" (1930b) and "Die logizistiche Grundlegung der Mathematik" (1931), indicates similarly that once the logicist reduction is accepted, the problem remaining is that of showing a sufficiently strong version of the theory of types to be tautologous. In the former paper, Carnap mentions Wittgenstein's view that mathematics is not tautologous but rather is a method of transforming identities. He continues: "Wittgenstein has so far given only a few hints as to the foundations of mathematics; its realization is still unaccomplished" (p. 307). Hahn is more ironic:

> According to Russell, natural numbers are classes of classes; Wittgenstein's view seems to be a very different one; but if we bear in mind that Russell's symbols for classes are incomplete symbols which must be eliminated . . . and if this elimination is carried out according to the rules Russell gives, then we see that the two views are not so different after all. (Hahn 1980 [1930–31], 37)

Nonetheless, only Carnap consistently maintains that the tautologousness of mathematics is a consequence of the logicist reduction ("Mathematics, as a branch of logic, is tautological" [Carnap 1930a, 24; 1959 [1930], 143]). For Hahn also says the following: "I must also say a few things about the place of *mathematics.* Since we have already adopted the view that experience and the logical transformations of logic are our only means of knowledge, . . . the answer has been given in advance: mathematics must likewise have a tautological character" (Hahn 1980 [1930–31], 25). Schlick is even more explicit:

> Although there is at present still considerable disagreement about the ultimate foundations of mathematics, nobody can nowadays hold the opinion anymore that "arithmetical propositions" communicate any knowledge about the real world. . . . Their validity is that of mere tautologies; they are true because they assert nothing of any fact. . . . I repeat: arithmetical rules have tautological character . . . (no matter whether arithmetic is just a part of logic — as Bertrand Russell will have it — or not). (Schlick 1979b [1932], 344–45)

The idea expressed here, that mathematics *must* be tautologous — without so much as a glance at the logicist reduction — is an upshot of the epistemological twist that both Hahn and Schlick give the notion of tautology. "Tautologous" winds up meaning true no matter what the *experiential* facts are, or true but not subject to empirical verification. The underlying picture seems to be this: the empirical world, the world of experience, is given; it is talk of that world that has content. A mathematical proposition does not say anything about this world, does not report particular experiences, is "compatible with *any* observation" (ibid., 346); hence it adds nothing and is just some artifact of how we talk about the experiential facts. In short, as Schlick and (sometimes) Hahn see it, the epistemological status of mathematics is exactly like that of logic, so no further analysis is needed to justify applying the same rubric.

This construal of "tautology," if not completely tendentious, at the least puts serious strain on the notion, forcing it to bear considerable epistemological weight. This can be seen by considering a criticism of Kurt Gödel's, from his manuscript "Is Mathematics Syntax of Language?" (1995), written in the 1950s:

> Mathematical sentences have no content only if the term "content" is taken from the beginning in a sense acceptable only to empiricists and not well founded even from the empirical standpoint. (§5, p. 337)

> The reasoning which leads to the conclusion that no mathematical facts exist is nothing but a *petitio principii,* i.e., "fact" from the beginning is identified with "empirical fact," i.e., "synthetic fact concerning sensations." (§37, p. 351)

For Gödel holds that the notion of content has to include both "conceptual" as well as empirical content, and mathematical propositions do have content of that sort. In the face of this position, if the identification of content with empirical content is not to simply beg the question, some further epistemological property of tautological sentences, beyond the irrelevance to them of experiences, would have to be adduced. The obvious way of attempting to fill this gap would be to find some kind of epistemological transparency that tautologies have, that is, some kind of self-evidence that would correspond to Wittgenstein's notion that "inspection" of the sign suffices to determine that the sentence is true. Given the nonexistence of a decision procedure for first-order logic, it appears that any such attempt will fail.

In contrast, Carnap does not epistemologize the notion of tautology. As I have mentioned, he emphasizes the characterization of "true under all possibilities" and does not invoke empiricism as a premise. To be sure, some realm of fact must be presupposed in order to make sense of the notion of "truth-possibilities," but the characterization does not require it to have any particularly epistemological cast. But then, for Carnap, it will be vital to show that mathematics actually has the property as so characterized. Given the logicist reduction, this comes down to showing that the property is possessed by whatever "logic" is needed for the reduction.

In 1930a, Carnap rehearses Wittgenstein's criticism that the axiom of reducibility is not a tautology. He then discusses Frank Ramsey's division of the paradoxes into mathematical and semantical, his introduction of a simple theory of types, and his argument that that theory — including the axiom of choice and the axiom of infinity (provided it is true) — is indeed tautologous. Carnap, however, calls Ramsey's viewpoint "theological mathematics" (1930a, 307). Finally, he expresses his own view, but without argument: he thinks the simple theory of types is logic, while the axiom of choice and the axiom of infinity should be treated in Russell's manner, as explicit hypotheses of any theorem whose proof requires them.

The heart of the view, then, is that the simple theory of types is tautologous. Carnap thinks that the most serious obstacle to accepting the simple theory lies in alleged difficulties in impredicative definitions, which are used in the logicist constructions of the integers and the real numbers, and so the issue is treated at length in "Die logizistiche Grundlegung der Mathematik" (1931). He presents the standard objection that the application of impredicative definitions involves a vicious circularity. For example, to show that the object identified as the number 3 is an inductive number, one must show, by definition of "inductive," that 3 has every hereditary property 0 has, where P is hereditary iff $(\forall n)(P(n) \rightarrow P(n+1))$. The definiens is a generalization over properties, and of course *inductive* is a property; so it now appears that we are in a circle, implying "it would be impossible to determine whether 3 is an inductive number" (1931, 100; 1983 [1931], 48).

As Ramsey pointed out, if the totality of the properties we generalize over in the impredicative definition exists independently of our specifications, then there is nothing illegitimate in impredicative definitions, and so the simple theory of types is supported. But Carnap rejects this defense of impredicativity as an intrusion of metaphysics into the foundations of mathematics, indeed, as "not far removed from a belief in a platonic realm of ideas." After repeating the rubric "theological mathematics," Carnap asks: "Can we have Ramsey's result without retaining his absolutist conceptions?" (1931, 102–3; 1983 [1931], 50).

To answer the question affirmatively, Carnap argues without invoking ontological considerations that impredicative definitions do not in fact contain a circularity. To show, for example, that 3 is inductive, that is, that 3 has every hereditary property 0 has, does not require subproofs for each property. Rather, we give a general argument, about *any* property: by logical reasoning alone we can show, for an arbitrary, unspecified property P, that if it is hereditary and 0 has it, then, by definition of "hereditary," 1 has it, so that 2 also has it, and so 3 has it. "We do not establish specific generality by running through individual cases but by logically deriving certain properties from certain others." Thus the establishment of a generalization over properties "means nothing more than its logical (more exactly, tautological) validity for an arbitrary property" (1931, 104; 1983 [1931], 51). In the example, when the definitions are unraveled, we are to see that "3 is inductive" is indeed tautological. It is this status that renders Ramsey's metaphysical defense of impredicativity misplaced and otiose, while saving his conclusion.

Carnap's argument here shows how little baggage he wishes to weight the notion of tautology with. However, it seems doubtful that the notion can encompass simple type theory in this nonmetaphysical manner while maintaining its basis in the idea of truth under all possibilities. Carnap's argument denies that generality "refers to objects already given" (1931, 103; 1983 [1931], 51); it is hard to see how this is compatible with Wittgenstein's conception.

3. Logical Syntax of Language

In *The Logical Syntax of Language* (1934b), Carnap drops all talk of "tautology"; the truths of logic and mathematics are now called "analytic." Far more than a terminological change is involved, however, for the setting is entirely different. In particular, "analytic" is a relative, not an absolute, notion; and what it is relative to becomes the center of explanatory attention.

This is Carnap's notion of a language or, as he later calls it, a linguistic framework. A linguistic framework is given by the rules for formation of sentences together with the specification of the logical relations of consequence and contradiction among sentences. The fixing of these logical relations is a precondition for rational inquiry and discourse. There are many alternative frameworks, many different logics of inference and inquiry. Since justification can proceed only grounded in the logical relations of a particular framework, justification is an intraframework notion. Thus there can be no question of justifying one framework over another. Carnap voices this pluralistic standpoint in his Principle of Tolerance: "*In logic there are no morals.* Everyone is at liberty to build up his own logic, i.e., his own form of language, as he wishes. All that is required of him is that . . . he must state his methods clearly, and give syntactical rules instead of philosophical arguments" (1934b, §17). Now Carnap so defines the notion of "analytic" that a sentence is analytic in a linguistic framework if it is a consequence of every sentence. In calling logic and mathematics analytic, Carnap is thus saying that they consist of framework-truths: sentences that any user of the linguistic framework must accept, just by dint of his being a user of that framework. As such, mathematical truths do not describe or reflect any realm of fact; they are simply consequences of the decision to adopt one rather than another linguistic framework.

There were, no doubt, many factors that moved Carnap to this picture. On the technical side, Gödel's incompleteness theorem made it clear that no transparent notion of "tautology" would capture the notion of mathematical truth; at the same time, Carnap took the technique of gödelization, along with the development of metamathematics by the Hilbert school, to show how unobjectionable (syntactic) metalinguistic considerations could be and how misplaced Wittgenstein's scruples on this score were. The pluralism is an official expression of inclinations present earlier on in Carnap's thinking, for example, in his recognition in the *Aufbau* of the legitimacy of different constitution systems. It may also mark both an acknowledgment that the contemporary debate on the foundations of mathematics between classical logicians and intuitionists was not resolvable and a philosophical diagnosis of the way in which the parties in this debate talked at cross-purposes.

The move to a pluralistic conception of linguistic framework, and of logic and mathematics as matters of framework, changes the way Carnap talks of the foundations of mathematics. Earlier, as we have seen, he relies on the logicist reduction to assimilate mathematics to logic and thus to the tautological. In the setting of *Logical Syntax,* there is no need to explain mathematics by way of logic. Consequently, Carnap's logicism becomes merely a commitment to the elaboration of a linguistic framework containing both pure mathematics and the means to apply mathematics to the empirical world. The question of whether mathematical concepts are to be defined in a vocabulary Carnap calls "logical in the narrower sense" — the vocabulary, that is, of *Principia Mathematica* — rather than taken as primitive "is not a question of philosophical significance, but only one of technical expedience" (ibid., §84).

More profound differences between Carnap's position in *Logical Syntax* and the earlier positivist views of foundations of mathematics can emerge from an examination of Gödel's criticisms of positivism in the manuscript mentioned above. Gödel's central argument is based on his second incompleteness theorem. As he puts it, "[A] rule about the truth of sentences can be called *syntactical* only if it is clear from its formulation, or if it somehow can be known beforehand, that it does not imply the truth or falsehood of any 'factual' sentence" (Gödel 1995, §11, p. 339). Evidently, a rule will fulfill this requirement only if it is consistent, since otherwise the rule will imply all sentences, factual and logical alike. The second incompleteness theorem states that mathematics not captured by the rule in question must be used in order to prove the rule consistent. Thus, additional mathematics must be invoked in order to legitimize the rule, and the claim that mathematics is solely a result of rules of syntax is refuted.

This is a powerful argument, but it is important to notice an unspoken presupposition of it. The argument depends on a realm of the "factual" or the "empirical" being available in advance, independently of and prior to the envisaged rules of syntax. As Gödel characterizes the positivist view, first there are empirical sentences, which are true or false by virtue of facts in the world; mathematics is then added, by means of conventional syntactical rules. Gödel's argument is that this addition has to be known not to affect the empirical sentences given at the start, and, by his theorem, to ascertain that requires more mathematics. Hence there is a *petitio.*

As we have seen, a picture of the empirical realm as fixed in advance of the linguistic rules does seem to underlie the ways Schlick and Hahn discuss logic and mathematics and in particular their understanding of the notion of "tautology." Consequently, Gödel's argument is very effective, perhaps even conclusive, against the claim that their notion of tautology could provide a foundation for mathematics. Even Carnap's view of 1930, insofar as his use of "tautology" bespeaks some commitment to a world of fact fixed prior to logic and mathematics, seems prey to this argument. Naturally, if *The Logical Syntax of Language* is taken as continuing this line of thought — and it appears Gödel did so take it — then it will be equally threatened.

However, Carnap's abandonment of the label "tautologous" and, most importantly, the adoption of the Principle of Tolerance show that *Logical Syntax* should be read as presenting a different, more sophisticated position. In this view, there is no notion of "fact" or "empirical world" that is given prior to linguistic frameworks. Sense can be made of such notions only once the rules of a language, and hence mathematics, are in place. That is to say, Gödel's argument, if applied in the setting of *Logical Syntax,* requires a domain of empirical fact conceived as transcending or cutting across different linguistic frameworks. However, as the Principle of Tolerance indicates, it is central to the metaphysics of *Logical Syntax* that any such language-transcendence be rejected. Rather, the notion of empirical fact is given *by way* of the distinction between what follows from the rules of a particular language and what does not, so that different languages establish different domains of fact.[2] In this way, Carnap's view undercuts the very formulation of Gödel's argument.

In fact, consistency proofs are of little interest for Carnap in *Logical Syntax* (see the ends of §§34 and 36). Carnap in no way precludes someone's proposing an inconsistent linguistic framework. Such a framework will certainly not be very useful, but this inexpediency is merely a pragmatic matter. This underscores the radical nature of Carnap's position: the relativity to linguistic frameworks that is claimed for notions like existence and factuality is absolutely thoroughgoing, and the "liberty" in formulating frameworks that is expressed in the Principle of Tolerance is absolutely unrestricted.

As we have seen, Gödel also criticizes the positivist view that propositions of logic and mathematics have no content. Here too, Carnap's pluralism enables him to turn the criticism aside. In *Logical Syntax,* Carnap gives a technical definition, for any given language, of the content of a sentence. His definition has as consequences that mathematical and logical truths have null content and logically equivalent sentences have the same content (§49). Carnap's reply to Gödel's criticism of such a definition of content might well be another remark he includes under the Principle of Tolerance: "It is not our business to set up prohibitions, but to arrive at conventions" (§17), followed by an invitation to Gödel to give a technical definition of his own. For Carnap there would be no meaningful question of which definition is "really" correct. Rather, there are only questions of which definition will reflect to a greater or a lesser extent some of the clearer uses of the informal notion of content that we take ourselves to be reconstructing. These are not yes or no questions; the answers will be pragmatic matters of degree and will be interest-relative.

Clearly, Gödel would reject this line of response. For, given his conceptual realism, what is at stake *is* a yes-or-no question, concerning the actual constitution of a notion of content that does justice to the claims propositions make on an independent reality. But then there is an impasse: agreement is lacking even as to what the argument is about.

Now there is another way in which Carnap's treatment of mathematics can be charged with a *petitio,* without relying on Gödel's argument from the second incompleteness theorem. As has been noted, for Carnap the truths of mathematics are meant to be consequences of the adoption of a linguistic framework. However, this "consequence" cannot be understood in a proof-theoretic sense, that is, as inferability in an axiomatized deductive system, since Gödel's first incompleteness theorem shows that no such deductive system will yield all mathematical truths. Rather, the semantical notion of consequence in the metalanguage must be invoked, and, to define that notion, mathematics of some strength is required.[3] In contrast, in his criticism Gödel formulates the constraint on the syntactic approach that "syntax" has to be finitary. Indeed, Gödel takes the point to be obvious: "The necessity of [this] requirement should be beyond dispute"; for if nonfinitary reasoning is used, the program "is turned into its downright opposite: . . . instead of justifying the mathematical axioms by reducing them to syntactical rules, these axioms (or at least some of them) are necessary in order to justify the syntactical rules" (1995, §§18–19, p. 341).

This elementary point may seem completely convincing; nonetheless, Carnap is oblivious to it. He explicitly notes that his definition of mathematical truth for a language that includes classical mathematics requires an apparatus that outstrips what is formalizable in that language (1934b, §34). Indeed, in general he has no qualms about the introduction of nonfinitary syntactical notions (§45). It should be clear from this that Carnap does not view the reduction of mathematics to syntax as providing a justification for mathematics; the identification of mathematical truths as framework-truths is not meant to legitimize them. Carnap could allow that, while mathematical truths are the result of syntactical rules, our recognition of particular truths, or our trusting any particular formulation of what can be inferred from given syntactical rules, requires more mathematics or different mathematics than that which those rules yield. (Of course, he would also assert, the additional or different mathematics we use is also the upshot of syntactical rules, albeit different ones.) In short, Carnap is not taking the clarification of the status of mathematics contained in *Logical Syntax* to be addressing traditional foundational issues. Those issues are addressed in another way, for they are transformed into questions of what can be done inside various linguistic frameworks or questions of what sort of frameworks are better for one or another purpose. What remains of "foundations of mathematics" is treated by describing, analyzing, and comparing different frameworks. Carnap's discussions of intuitionism (§16), predicativity (§44), and logicism (§84) all exhibit this transformation. Carnap emphasizes the point in his 1934 paper "The Task of the Logic of Science":

> Questions of the logic of science concerning mathematics, or as they are often called, questions about the "foundations of mathematics," are questions of the syntax of the logico-mathematical part of the language of science. The main distinction to be drawn here is whether we are dealing with *assertions* about a given mathematical system... or with *proposals* to set up the language of mathematics in such and such a way... [as to] the dispute about the justification of indefinite and especially impredicate concepts. The question to be asked here is not "Are these concepts meaningful?" but rather formally: "Do we want to incorporate such concepts into our language or not?" (1987b [1934], 65)

Thus, in *Logical Syntax,* Carnap no longer takes there to be any questions about logic and mathematics that are foundational in the traditional sense. He is simply no longer addressing the issues that concerned Kant, Frege, Russell, Hilbert, or even the Carnap of "Logizistiche Grundlegung." Gödel is doubtlessly correct in charging that the positivists, including Carnap in *Logical Syntax,* do not provide a successful epistemological reduction of mathematics to syntax or linguistic conventions. However, on the interpretation just given, this task is simply not one that Carnap takes on.

This leads to a peculiar kind of standoff between Carnap and a "realist" opponent like Gödel. Carnap's position contains a regress: mathematics is obtained from rules of syntax in a sense of "obtained" that can be made out only if mathematics is taken for granted in the metalanguage. As a result, a full exhibition of the syntactical nature of mathematics is never possible. Now, such an account would certainly beg any of the traditional foundational questions. But this does not doom Carnap's view, however, insofar as the structure of that view leaves no place for those foundational questions. Thus, on this score, Carnap's position is entirely coherent. At the same time, the position is not capable of convincing a Gödelian that a faculty of mathematical intuition is unnecessary, for Gödel's view is designed to address — and to support the necessity of addressing — just the philosophical questions that Carnap discards. From Gödel's point of view, Carnap's position *is* viciously circular, or at best philosophically and mathematically empty; while from Carnap's point of view, Gödel's position amounts only to a dogmatic insistence on working within one type of linguistic framework, among the many alternatives.[4]

Notes

1. For a detailed criticism of Schlick on this point, see Coffa 1991, 175ff.
2. For a more detailed treatment of this point, see Ricketts 1994.
3. The technical situation here is elaborated in Goldfarb and Ricketts 1992, 70–71.
4. I am indebted to Thomas Ricketts for many illuminating discussions, and to Geoffrey Hellman for helpful comments.

Thomas Ricketts

Carnap: From Logical Syntax to Semantics

In the mid-1930s Carnap enthusiastically adopted Tarski's technique of truth-definitions to replace purely syntactic treatments of languages in logic. Henceforth, Carnap will develop semantics as the central part of logic to serve the ends previously served by logical syntax. Carnap's move from syntax to semantics may appear to some contemporary eyes to mark a dramatic change of viewpoint. In agreement with Richard Creath, I shall argue that it does not.[1] Given Carnap's very generous conception of syntax and the purposes his syntactic investigations of languages serve inside *Wissenschaftslogik*[2] (the logic of science), the step from syntax to semantics proves to be a small one. Indeed, the step is so small that we should ask why Carnap needed the impetus of Tarski's work to take it. Some commentators have addressed this question by maintaining that before encountering Tarski's work, Carnap, in the grip of verificationism, rejected the notion of truth as a metaphysical pseudoconcept. I agree with Thomas Oberdan that this interpretation is incorrect as regards the antipathy toward truth Carnap voices in *Logical Syntax* (Oberdan 1992, 239). Carnap rejects the notion of truth there because, for good reasons, he believes the notion of truth to be both syntactically intractable and otiose in logic. Carnap's antipathy to truth is thus rooted more in technical than in philosophical considerations.

Carnap's logical syntax program is the topic of section 1 of this essay. This section focuses on the mathematical strength Carnap allows to syntax and on the difficulties this liberal view of syntax seems to pose for Carnap's program. These difficulties, I shall argue, evaporate once the radically deflationary character of Carnap's Wissenschaftslogik is fully appreciated. Section 2 explores Carnap's antipathy toward reference and truth in *Logical Syntax* with special attention to the discussion of truth in §60b and then discusses Carnap's immediate embrace of Tarski's work. The shift to semantics raises — or, as I think, highlights — problems in Carnap's continuing project of Wissenschaftslogik. Section 3 briefly discusses two of these problems: the definition of analyticity and Carnap's view of ontology.

I

In his 1914 paper "On Scientific Method in Philosophy," Bertrand Russell proclaims, "Philosophy... becomes indistinguishable from logic as that word has now come to be used." It concerns itself "with those general statements which can be made concerning everything without mentioning any one thing or predicate or relation" and with the application of this discipline of pure logic to the analysis of the statements of the special sciences (Russell 1914 [1917], 84ff.). Philosophy, so conceived, remains an a priori discipline; but Whitehead and Russell's logic gives it the shared method, goal, and standards of rigor that promise to remake philosophy into a collaborative discipline characterized by piecemeal progress. In a word, philosophy becomes scientific. Russell intimates that philosophical topics that neither fit into this logical analytical mold nor can be turned over to specialists for empirical investigation are best dropped.

Carnap embraces Russell's scientific attitude toward philosophy in his earliest writings. In the early 1930s, Carnap voices this attitude by advocating the view that philosophy is, or should be supplanted by, Wissenschaftslogik. He says in his 1936 paper "Von der Erkenntnistheorie zur Wissenschaftslogik," "It seems to me that *theory of knowledge* [*Erkenntnistheorie*] is in its previous form *an unclear mixture of psychological and logical elements.* That holds as well for the work of our circle, not excluding my own earlier work." Carnap goes on to explain that once empirical psychological questions are factored out, "there remains left as the genuine task of philosophical work the logical analysis of knowledge — of scientific sentences, theories, and methods — hence Wissenschaftslogik" (Carnap 1936a, 36ff.).[3] Wissenschaftslogik provides formal descriptions of languages suitable for the formalization of the various sciences. Reducibility relationships among the sciences can then, Carnap anticipates, be captured in formal terms. In addition, more general, vaguer notions can be made precise via formal explications or replacements. The prime example of a general application of Wissenschaftslogik in this period is Carnap's attempt in "Testability and Meaning" to explicate "empiricist language" and thence "confirmable sentence."[4]

Carnap's conception of Wissenschaftslogik is framed by the logical pluralism set forth in his Principle of Tolerance: "*In logic, there are no morals*. Everyone is at liberty to build up his own logic, i.e. his own form of language, as he wishes. All that is required of him is that, if he wishes to discuss it, he must state his methods clearly, and give syntactical rules instead of philosophical arguments" (Carnap 1937, §17, p. 52).[5] Wissenschaftslogik is thus implemented metatheoretically by the formal description of various languages and by syntactic definitions of various notions. Carnap's shift in the early 1930s to this pluralistic approach to logic is his response to debates in the foundations of mathematics — for example, disputes over the admissibility of impredicative definitions and disputes over the legitimacy of the nonconstructive proofs of classical mathematics. In *Logical Syntax,* Carnap laments the unclarity, dogmatism, and invective that infect these debates, debates that thus resemble those of the pseudoproblems of traditional philosophy (ibid., §§16–17). Carnap finds the source of these fruitless controversies in foundations of mathematics in an ill-defined "striving after 'correctness' " (ibid., xv). Carnap proposes instead that there is no correct or incorrect as regards a language, a logic. Considerations of correctness make sense only in the context of a specific language whose logic affords, so to speak, a language-relative and language-specific notion of correctness. Rather than squabbling, logicians should seek to make their proposals for mathematics precise by formally describing languages that embody them. They are then in a position dispassionately to investigate and compare languages that impose or loosen various constraints on mathematics. These metamathematical investigations owe their clarity and rigor to their formality: they are to employ only syntactic notions. Carnap informally characterizes syntax as "nothing more than *combinatorial analysis,* or, in other words, the *geometry* of finite, discrete, serial structures of a particular kind" (ibid., §2, p. 7). The imprecision that lingers around talk of combinatorial possibilities of spatial elements is, Carnap notes, eliminated when we appreciate how, via gödelization, to interpret syntax in arithmetic: "All the sentences of pure syntax follow from these arithmetical definitions and are thus analytic sentences of elementary arithmetic" (Carnap 1937, §19, p. 57). We shall see momentarily that these two characterizations of syntax are, in their innocence, most misleading: Carnap's conception of syntax is not the contemporary one.[6]

Abstracting from its use, a language is presented syntactically via formation rules and transformation rules (see Carnap 1937, §2, p. 5). Formation rules define sentencehood for the language. Transformation rules define a consequence relation over these sentences. Valid sentences are then consequences of the empty class of sentences; contravalid sentences are those sentences whose consequence class is the universal class of sentences. Much of *Logical Syntax* is devoted to the description of two languages, Language I and Language II. Language I is a language for primitive recursive arithmetic. For Carnap, it represents one way of precisely realizing the constraints urged by constructively oriented mathematicians, including intuitionists. Language II is a version of the simple theory of types plus the axiom of choice; it embodies the perspective of classical mathematics. Carnap envisions that both of these languages may be expanded by the addition of descriptive predicates into languages for empirical science.

Carnap wants the mathematical sentences of these languages to be logically determinate, L-valid (analytic) or L-contravalid (contradictory), as he puts it. This status distinguishes mathematical sentences both from empirical sentences and from metaphysical pseudosentences. Carnap is well aware of Gödel's incompleteness theorems; indeed, he encapsulates the first incompleteness theorem, saying, "Gödel was the first to show that not all analytic sentences are demonstrable" (Carnap 1937, §10, p. 28). Nevertheless, Carnap searches for syntactic definitions of "valid" and "contravalid" over Languages I and II that reproduce the desired true-false dichotomy over the mathematical sentences. Carnap's definition of validity for Language I employs an infinitary omega-rule of inference. The definition of validity for Language II, as Carnap subsequently notes, uses the same resources as a Tarski-style truth-definition (Carnap 1942, 247). From a contemporary vantage point, Carnap allows the full resources of set theory to count as syntactic. Carnap himself is well aware of the strength he allows to syntax: in *Logical Syntax* he remarks that in any consistent language *S* incorporating a minimum of arithmetic, the predicate "analytic in *S*" is indefinable in *S* and definable only in a stronger language, a stronger arithmetic (see Carnap 1937, §34d, p. 113; §60c, p. 219; and §60d, pp. 221–22). Carnap concludes that arithmetic and the syntax it encodes constitute an open-ended mathematical discipline that cannot be exhausted in any single language; and he envisions a hierarchy of languages here: "In other words, *everything mathematical can be formalized, but mathematics cannot be exhausted by one system;* it requires an infinite series of ever richer languages" (ibid., §60d, p. 222).[7]

It is important to bear in mind here exactly what Carnap's reproduction of the true-false distinction in syntactic terms over the mathematical sentences of the object-language comes to. Carnap in *Logical Syntax* rejects reference theoretic notions. We cannot then characterize his goal to be the construction of a definition in the metalanguage whose definiens has a certain extension. Carnap's aim is the construction in a metalanguage of a definition of "valid" ("contravalid") that, for any canonical metalanguage name *s* of what is, informally speaking an object-language mathematical truth (falsehood), has as an L-consequence in the metalanguage the sentence "*s* is valid" ("*s* is contravalid").[8] Of course, Carnap's definition will not mention the notion of metalinguistic L-consequence in defining object-language validity. Metalinguistic L-consequence can only be discussed in a still stronger metametalanguage for stating the logical syntax of the metalanguage. Accordingly, Carnap formulates in his metalanguage a definition of "validity" for the object-language that exploits the L-consequence relation of the metalanguage to yield the desired results. The notion of L-consequence for the metalanguage will then be too strong to be identified with provability in any formal system. It takes the form, as we might put it, of a notion of standard higher-order semantical or set-theoretical or mathematical consequence. In effect then, Carnap characterizes mathematical truth for an object-language in a metalanguage that itself has as L-valid sentences the mathematics it builds into the object-language and then some. I emphasize that Carnap is fully cognizant of the technical situation here.

This open-ended conception of syntax raises questions as to the character, coherence, and interest of Carnap's project in *Logical Syntax*. Michael Friedman has argued that Carnap's use of strong syntax languages is incompatible with the neutrality required by his logical pluralism (Friedman 1992, especially 516–19).[9] The idea is that the pluralism and conventionalism voiced in the Principle of Tolerance will be hollow unless there is a shared perspective from which the adherents of various foundational schools can describe their differences and acknowledge that these differences are linguistic, that is, are the product of a free decision to use one or another language. Such neutrality requires a weak syntax language that is shared by the partisans in foundational disputes — something like primitive recursive arithmetic, Friedman suggests. But a weak syntax language will not syntactically reproduce the true-false distinction over the mathematical sentences of Carnap's sample languages. Ever stronger syntax languages are needed to sustain the view that the mathematics built into any language is analytic. Thus, the requirement of neutrality collides with the status Carnap assigns to mathematics.

Carnap's pluralism in logic does not, I believe, commit him to the requirement of a weak neutral syntax language. The codicil to the Principle of Tolerance states that anyone who wishes to discuss a language should give syntactic rules, not philosophical arguments. If the intuitionist and the classical mathematician wish to compare the full consequence relations for Languages I and II, they will have to use a strong metalanguage to do so. From Carnap's perspective, only dogmatic obduracy would prevent the intuitionist from using a strong, classical metalanguage to this end. Indeed, insistence on a weak universal syntax language is contrary to the Principle of Tolerance, to which Carnap appeals in using strong syntax languages (Carnap 1937, §45, pp. 165–66).[10] Carnap does not put logical syntax forward in order to mediate among disputants in foundations of mathematics. He is no more involved in these disputes than in the metaphysical dispute between realists and idealists. The Principle of Tolerance, despite its name, is not a principle, a thesis. (If it were, it would be stated in a particular Carnapian language with its applicability circumscribed to the languages describable in that language.) Rather, in the Principle of Tolerance, Carnap commends to his audience an attitude that, when imbibed, saps foundational wrangling over correctness and promotes metamathematical investigations. He takes most mathematicians tacitly to share this attitude (ibid., §17, p. 52).

Let us now turn to a second worry, one abetted by this response to the concern about neutrality. I have observed that Carnap's predicates "valid" and "contravalid" for Languages I and II reproduce in Carnap's syntactic terms the true-false distinction over the mathematical sentences of these two languages. Identifying validity with analyticity, Carnap says:

> By means of the concept "analytic," an exact understanding of what is usually designated as "logically valid" or "true on logical grounds" is achieved.... In material interpretation, an analytic sentence is absolutely true whatever the empirical facts may be. Hence it does not state anything about facts.... *Synthetic sentences are the genuine sentences about reality.* (ibid., §14, p. 41)

How does Carnap think that his use of strong syntax languages to define "L-valid" explicates the notion of analyticity informally indicated? The syntactic specification of a language displays how logic and mathematics are built into it. For as noted above, given a mathematical sentence *s* that is an absolute truth under "the material interpretation" of the object-language, the metalanguage syntactic statement "*s* is valid" will be an L-consequence in the metalanguage of the syntactic definition of "valid." The price of this achievement is the use of a strong metalanguage whose mathematics is analytic to build in this way weaker mathematics into the object-language.

The informal notion of analyticity, it might be urged, concerns the grounds for the truth of analytic statements and the basis for our knowledge of them. Carnap's procedure does not appear to explicate any such informal notion of analyticity. Quite the contrary. Carnap's definitions do not show that mathematics is true in virtue of syntactic stipulation, for the mathematics allegedly stipulated must be available in the syntax language in order to frame the stipulation.[11] Gödel gives this point a powerful epistemic twist, arguing that syntactic definitions do not give an alternative to mathematical intuition as an account of mathematical knowledge. Gödel maintains that in order to *justify* the claim that the sentences demarcated by some syntactic rules are true whatever the empirical facts may be, we have to be able to show, in advance of adopting the rules, that the rules are *admissible,* that they "do not themselves imply the truth or falsehood of any proposition expressing an empirical fact" (Gödel 1995, 357). That is, the mathematics and logic of a Carnapian language must be shown to be conservative over the synthetic sentences of the language, if we are to be justified in taking the mathematics and logic to be unconditionally true, that is, empirically contentless truths. Gödel observes that a proof of admissibility is a proof of consistency for the object-language set forth by the rules, and, by the second incompleteness theorem, this proof requires the use of mathematics more powerful than that formalized in the object-language. He concludes that syntactic conventions cannot supplant mathematical intuition as the source of mathematical knowledge, as the latter is required to show the admissibility of the former.

This line of objection underestimates the break Carnap's Wissenschaftslogik makes with philosophy. Correlative questions concerning the grounds of mathematical truth and the nature of mathematical knowledge are precisely the sort of question Carnap eschews as vague and confused. In particular, Carnap does not employ a notion of convention to contrast the status of logic and mathematics with that of empirical science.[12] As regards Gödel's objection, Carnap would deny that *justification* is at stake in admissibility proofs. On the positivist picture Gödel criticizes, we impose the logico-mathematical apparatus of a language on empirical statements. This picture gives rise to a demand for admissibility proofs to justify the alleged empirical vacuity of mathematics. But this picture, while embraced by other members of the Vienna Circle, is not Carnap's.[13] In Carnap's view, there is no clear conception of empirical statement, empirical fact, or empirical possibility, apart from the incorporation of observation predicates into a language with its consequence relation. This is part of the force of the Principle of Tolerance. Carnap, of course, aims to construct languages whose logic and mathematics are conservative over synthetic sentences. A proof of admissibility may then be cited in advocating consideration of a language as a framework for empirical theorizing.

For this purpose, however, there is no vicious circularity in the use of whatever metamathematical resources are required. Carnap's syntactic constructions bear then no explanatory or justificatory weight. From his viewpoint, the hierarchy of ever stronger syntax languages is self-supporting at each level: we use the logical resources of a strong syntax language to characterize the logical resources of an object-language. If there is no explanatory work here, there is no vicious regress in this hierarchy.[14] However, the question remains, How does Carnap intend his formal work to explicate the informal notion of analyticity? I have just discussed how Carnap does *not* conceive of this goal. How *does* he think of it?

I mentioned that Carnap envisions the addition of descriptive predicates to his sample languages. Using this new vocabulary, formalizations of physical laws and even reports of individual observations may be added to the valid sentences of the language by P-rules. Carnap says:

> Whether in the construction of a language S we formulate only L-rules or include also P-rules, and if so, to what extent, is not a logico-philosophical problem, but a matter of convention and hence, at most, a question of expedience. If P-rules are stated, we may frequently be placed in the position of having to alter the language.... But there are no fundamental objections to this.[15] (Carnap 1937, §51, p. 180)

So for Carnap, P-rules as much as L-rules are definitive of a language; a change in P-rules is as much a change of language as a change in L-rules. In the syntactic descriptions of Languages I and II, Carnap just states the L-rules. That is, he just sets forth the rules that fix the logic of the language, leaving for another occasion the specification of P-rules that axiomatize some body of accepted empirical theory. In the context of his discussion of these two languages, the labels "logical rules" and "physical rules" represent an informal classification of syntactic rules, and derivatively of sentences of the language.

Carnap wishes to make precise the distinction informally marked by these labels. In §§50–52 he introduces a general, schematic characterization of L-validity that is applicable to any syntactically given language. Applied to the anticipated expansions of Languages I and II, Carnap's definition more or less tracks the distinction informally marked by talk of analytic as opposed to synthetic sentences. Carnap supposes that we are given the formation and transformation rules of a language, with no separation of transformation rules into L-rules and P-rules. The transformation rules fix the (L- and P-) valid sentences and the contravalid sentences of the language. The union of these two classes comprises the determinate sentences of the language. Carnap now defines the primitive logical vocabulary along the following lines: the primitive logical vocabulary is the largest vocabulary of uncompounded, undefined expressions such that there are sentences constructed solely from that vocabulary and any such sentence is determinate. The other primitive vocabulary is descriptive. An expression that contains a primitive descriptive sign is descriptive. Carnap relies on the distinction between logical and descriptive signs to define L-validity, that is, analyticity. A valid sentence is L-valid if it contains only logical expressions, or if every sentence obtainable from it by the substitution of descriptive signs for its primitive descriptive signs is valid.[16] So we can think of the primitive logical vocabulary as defining a notion of logical form or logical schema for the language. Carnap's basic idea is that a sentence is L-valid just in case it is a substitution-instance of a schema all of whose instances are valid.

There are two crucial points to observe here. First, Carnap's segregation of logical from descriptive signs will yield the desired results only if every mathematical sentence is determinate. Carnap's approach here is thus premised on his open-ended conception of syntax. Second, the intended segregation of logical from descriptive signs, applied to the anticipated enrichments of Languages I and II by descriptive vocabulary and P-rules, requires that there be in the language indeterminate sentences, sentences that are neither valid nor contravalid.

From Carnap's perspective, his syntactic definitions capture what is clear and valuable in the informal distinction between analytic and synthetic sentences. In particular, Carnap's syntactic explicata invoke no notion of something making a sentence true, no notion of truth-in-virtue-of-fact as opposed to truth-in-virtue-of-language or -logic. Carnap's definition of L-validity in *Logical Syntax* thus more displaces than analyzes a traditional notion of analyticity.[17]

II

Before turning to Carnap's adoption of Tarski's semantics in place of logical syntax, we need to consider his attitude toward truth and reference in *Logical Syntax*. It is sometimes thought that Carnap, before Tarski and under the influence of Wittgenstein and verificationism, took truth to be an illegitimate metaphysical notion. Carnap does have something like this attitude toward reference in *Logical Syntax*, but not toward truth.[18] Carnap never claims that truth is a pseudonotion; his attitude is rather that *truth* is both syntactically intractable as well as dispensable in Wissenschaftslogik.[19]

Carnap's hostility to reference in the logical syntax period comes out in his discussion of pseudo-object sentences in part 5 of *Logical Syntax*. These sentences, Carnap tells us, are "formulated as though they refer (either partially or exclusively) to objects, while in reality they refer to syntactic forms and specifically to forms of objects with which they appear to deal." This characterization itself is, Carnap immediately warns, "informal and incorrect" (ibid., §74, p. 285). Indeed, it is itself a pseudo-object statement. Carnap's view is better conveyed by his examples of pseudo-object statements and their syntactic replacements. So, Carnap claims,

> Five is not a thing, but a number,
>
> should be replaced by
>
> "Five" is not a thing word, but a number word. (ibid., §74, p. 286)

And

> The word "luna" in the Latin language designates the moon
>
> should be replaced by
>
> There is an equipollent expressional translation of the Latin into the English language in which the word "moon" is the correlate of the word "luna." (ibid., §75, p. 289)[20]

Pseudo-object sentences then include those that putatively speak of such semantical relations as designation or description between words and things as well as abstract generalizations about entire ontological categories, about things, properties, and facts. I believe that the source for Carnap's aversion to pseudo-object sentences lies in the pluralism in logic that comes with the attitude of tolerance. Pseudo-object sentences suggest the availability of language-transcendent notions of reference, ontology, and fact. They thus tempt us to raise ill-defined questions about the correctness of various languages (see ibid., §78).

Carnap believes that if one attempts to clarify pseudo-object sentences by describing syntactically a language that incorporates them, these sentences prove to be quasi-syntactic sentences. Informally speaking, a property of things is quasi-syntactic if there is a correlated syntactic property of expressions designating things: a thing has the quasi-syntactic property just in case an expression designating the thing has the syntactic property. So the property *positive integer* is quasi-syntactic, as it corresponds to the syntactic property *formal numeral.*

The property *red* is not quasi-syntactic, for there is presumably no syntactic property possessed only by designations of red things. Carnap sketches a syntactic treatment of this idea that jettisons talk of properties and designation: a quasi-syntactic sentence is an object-language sentence that has, in a language that extends an object-language with a syntactic description of it, the same content as a corresponding syntactic sentence.[21]

Thus, when made precise, pseudo-object statements are claims about the syntax of a particular language, accompanied perhaps by a recommendation to use or forswear a language with the syntactic features set forth. Carnap believes that syntactic sentences can serve any clearly conceived purposes in Wissenschaftslogik that pseudo-object sentences serve. So talk of syntactic categories replaces talk of ontological categories. Talk of reference may be supplanted by a discussion of syntactic relationships between two languages.

In Carnap's eyes, truth is more benign than reference. In §60b of *Logical Syntax,* Carnap explains how unrestricted impredicative use of a truth-predicate in standard languages leads to the liar paradox. He concludes from this that, while a truth-predicate for a language cannot be a part of the language, nothing bars us from laying down axioms for a predicate "true-in-*K*" in a metalanguage that properly includes *K*. But Carnap sees little point to this exercise.[22] The metalinguistic truth-predicate is not needed in either pure logic or Wissenschaftslogik. Singular predications of "true" of individual sentences can be replaced by the use of those sentences. Use of "true" in logic for semantic assent can be replaced by syntactic vocabulary. For example, "If a conjunction is true, then so is each of its conjuncts," goes over into, "A conjunction has as L-consequences each of its conjuncts."

Most importantly, however, the truth-predicate Carnap envisions would be a primitive *descriptive* predicate of the metalanguage.[23] Quite correctly, Carnap sees no prospect, even with the resources of open-ended syntax, for syntactically defining a truth-predicate:

> *For truth and falsehood are not proper syntactical properties;* whether a sentence is true or false cannot generally be seen by its design.... (This fact has usually been overlooked by logicians, because, for the most part, they have been dealing not with descriptive but only with logical languages, and in relation to these, certainly, "true" and "false" coincide with "analytic" and "contradictory," respectively, and are thus syntactical terms.) (ibid., §60b, p. 216)

This terse argument needs to be unpacked. Validity and contravalidity are proper syntactical properties that reproduce in suitably strong metalanguages the true-false distinction for the mathematical portions of object-languages. In particular, for each canonical name *s* of such a sentence, exactly one of the metalanguage sentences

s is valid (i.e., true)

or

s is contravalid (i.e., false)

is an L-consequence in the metalanguage of the syntactic definitions of "valid" and "contravalid." In this sense, the syntactic definitions of "valid" and "contravalid" are *complete* over the mathematical sentences of the object-language. Carnap believes that these syntactic definitions cannot be extended to reproduce the true-false distinction over the entire object-language, if the object-language is a descriptive language.

I see in the passage just quoted the following line of thought in support of this conclusion. Via syntactic descriptions in the metalanguage of object-language P-rules, for any canonical name of a determinate object-language sentence *s,* either "*s* is valid," or "*s* is contravalid," will be an L-truth in the metalanguage. But in descriptive languages, there are indeterminate sentences so that the syntactic definitions of validity and contravalidity do not completely characterize object-language truth. Carnap sees no way, using just the logic and mathematics of the metalanguage, to characterize truth completely for nontrivial descriptive object-languages.[24] Moreover, from Carnap's viewpoint, a complete definition of truth for a descriptive object-language, if possible at all, would presumably require the use of descriptive metalanguage predicates and draw on substantive empirical knowledge, that is, the nonlogical resources, the P-rules, of the metalanguage. Only in this way could the definition reproduce the intended true-false dichotomy by yielding for each canonical name *s* of an object-language sentence either "*s* is true" or "*s* is false." But the use of descriptive predicates and the reliance on metalanguage P-rules would destroy the syntactic character of the definition; talk of what the definition *yields* could no longer be understood in terms of L-consequence in the metalanguage. Carnap's instincts in the quoted passage are sound. Nothing in Tarski's procedures shows them to be mistaken. What then accounts for Carnap's ready embrace of Tarski's semantics?[25]

Carnap describes his initial reaction to Tarski's definition of truth in his autobiography in the Schilpp volume:

> When Tarski told me for the first time that he had constructed a definition of truth, I assumed that he had in mind a syntactical definition of logical truth or provability. I was surprised when he said he meant

> truth in the customary sense, including contingent factual truth. Since I was thinking only in terms of a syntactical metalanguage, I wondered how it was possible to state the truth-condition for a simple sentence like "this table is black." Tarski replied: "This is simple; the sentence 'this table is black' is true if and only if this table is black." (Carnap 1963a, 60)

Tarski's insight is his celebrated Convention T, which he states as a material adequacy criterion for a definition of truth. For a definition of truth for a language *L* to be materially adequate, it need do no more than to yield all metalanguage biconditionals of the form:

S is true in *L* if and only *p,*

where *S* is replaced by a canonical name of an object-language sentence and *p* is replaced by that sentence or its metalanguage translation. Given this standard, truth can be trivially characterized for a finite collection of atomic sentences even without knowing which are true. So for the language whose only two sentences are the German sentences "Der Schnee ist weiß" and "Das Gras ist grün," we have:

s is true iff *s* = "Der Schnee ist weiß" and snow is white,
or *s* = "Das Gras ist grün" and grass is green.

Here we have a materially adequate definition of truth for our trivial language. Although "Der Schnee ist weiß" is true, this metalinguistic sentence is not a logical consequence of the definition. Indeed, this definition is available to us apart from any knowledge of the truth or falsity of the two German sentences. Once this insight is in place, it is a straightforward matter to extend the characterization of truth recursively to logically compound sentences and then to convert the recursive characterization into an explicit definition that has as logical consequences, indeed deductive consequences, the substitution instances of Convention T.

Carnap is well aware of the difference between a criterion and a definition of truth. He realizes that his definitions of mathematical truth provide neither a decision procedure nor a search procedure for mathematical truth (Carnap 1937, §34a, pp. 98–102; see also §34d, p. 113).[26] His blind spot lies elsewhere. Carnap's aim in *Logical Syntax,* we have seen, is to develop, in light of Gödel's incompleteness theorems, the means to build mathematics into languages. To this end he requires that the definitions of "valid" and "contravalid" be complete over the mathematical sentences of the language. This requirement gives the syntactic cash value of "building mathematics into languages." Carnap's efforts in *Logical Syntax* are devoted to securing this completeness. He does not see that a definition of truth might be materially adequate but fail to be, in his sense, complete. Tarski is interested in material adequacy, not completeness. For Tarski, it suffices that a truth-definition yield the T-sentences. There is no further requirement that for any or some object-language truths *s,* the truth-definition yield the consequence "*s* is true." In the conversation with Tarski, Carnap comes to grasp the difference between material adequacy and completeness. He accordingly accepts Tarski's definition of truth. Given Carnap's view of syntax, he will not scruple at the mathematical resources Tarski uses. After the description of the meeting with Tarski just quoted, Carnap does observe that Tarski's definition comes at the price of using metalanguage descriptive predicates to define truth for an object-language with descriptive predicates. By syntax period standards, a Tarski's truth-predicate for a descriptive language is itself descriptive; Carnap continues to adhere to this classification of such truth-predicates in his semantics period. However, the T-sentences are logical consequences in the metalanguage of the truth-definition. No use of metalanguage P-rules is presupposed; that is, in the material mode, no knowledge of the extension of the descriptive predicates is presupposed.

III

Carnap readily adopts Tarski's semantics. From now on, canonical formal descriptions of language are provided by formation rules and semantic rules, where the latter include as their core a truth-definition for the language. Given Carnap's open-ended understanding of syntax and his deflationary view of Wissenschaftslogik, his replacement of logical syntax with Tarskian semantics does not represent a major shift in his philosophy. I do wish to discuss briefly two points at which Carnap's adoption of semantics prompts modifications in his views in *Logical Syntax,* namely, the definition of analyticity and the understanding of ontological issues.

I noted how in *Logical Syntax* Carnap develops a language-general definition of analyticity. The definition rests on Carnap's formal segregation of logical from descriptive expressions. Carnap's definition here will yield the desired result in descriptive languages only if there are indeterminate sentences in the language, sentences whose validity or contravalidity are not secured by the transformation rules. With the adoption of semantics, bivalent truth and falsehood replace validity and contravalidity in formal descriptions of the languages in which Carnap is interested. Thus, in *Logical Syntax* the strategy for defining analyticity or L-truth cannot be implemented in the new semantic setting. Carnap recognizes this problem in his first extended presentation of semantics in his 1942 book *Introduction to Semantics.* He says that, for any particular language, we have to rely on an informal understanding of the logical-descriptive distinction and just enumerate the primitive logical expressions. He continues:

> The problem is more difficult in the form it takes in *general semantics.* Here it is the question whether and how "logical" and "descriptive" can be defined on the basis of other semantical terms, e.g. "designation" and "true," so that the application of the general definition to any particular system will lead to a result which is in accordance with the intended distinction. A satisfactory solution is not yet known. (Carnap 1942, 59)

After the move to semantics, Carnap informally glosses F-truths, factual truths, as those truths knowledge of which requires empirical investigation. In contrast, knowledge of L-truths, analytic truths, is based solely on an understanding of the language. Carnap seeks to make this characterization of analyticity precise by eliminating talk of understanding in favor of semantic rules, of truth-definitions (see Carnap 1939, 12–13). Carnap's basic idea is that sentence *s* of an object-language is an object-language L-truth just in case the truth of *s* is a logical consequence in the metalanguage of the truth-definition for the object-language. That is, eliding use-mention niceties, *s* is analytic in the object-language, if the metalanguage sentence

"*s*" is true

is an L-consequence in the metalanguage of the truth-definition and so is itself analytic *in the metalanguage.* Carnap's goal is still to have mathematical sentences be L-determinate. So metalinguistic consequence remains a strong relation not capturable in a deductive system. In *Introduction to Semantics,* Carnap stresses that this characterization of object-language L-truth is not a suitable definition, since it *mentions* metalanguage L-concepts (Carnap 1942, 83–85). To accept it as a definition would be to define "analytic in language *K*" using "analytic in language Meta-*K*." Echoing Tarski's Convention T, Carnap puts forward this characterization of analyticity as an adequacy criterion on any genuine metalinguistic definition of analyticity for the object-language. How then is Carnap to define analyticity, if not generally, then at least on a case-by-case basis for languages that interest him? Carnap's strategy here is in the metalanguage to supplement a Tarski truth-definition for an object-language with a definition of L-truth that meets his adequacy criterion. The semantic rules for a language are then a combination of a truth-definition and a separate L-truth definition. State descriptions are Carnap's device to this end in *Meaning and Necessity.* Later he suggests adding to a truth-definition, the specification of the intended interpretation of a language, a model theoretic definition of L-truth (see Carnap 1963b, 900–903).

Quine, in his attack on analyticity, criticizes Carnap's use of semantic rules to explain analyticity. We can now see more clearly why Quine finds Carnap's technical work here so empty. Quine will not find Carnap's adequacy criterion informative, stated as it is in terms of analyticity in the metalanguage. And Carnap's procedure of simply adding a specification of analytic truths to a truth-definition appears ad hoc, especially with Carnap's use of meaning postulates to secure, for example, the analyticity of the sentences asserting that *warmer than* is a transitive, irreflexive relation (see Carnap 1950a [1956]). There is, I believe, considerable justice in Quine's complaint:

> [W]e might construe an artificial language L outright as an ordered pair whose second component is the class of its analytic statements; and then the analytic statements of L become specifiable simply as the statements in the second component of L. Or better still, we might just stop tugging at our bootstraps altogether. (Quine 1953, §4, pp. 35–36)

Let us now turn to Carnap's postsemantics view of ontology. We saw how Carnap in *Logical Syntax* eschews ontological questions and, with them, questions of reference. Pseudo-object sentences, when made precise, turn into quasi-syntactic sentences that are equivalent to syntactic descriptions of a language. These syntactic descriptions may be cited both in comparing languages and in advocating the use of a particular language. By the Principle of Tolerance, there is no right or wrong in the decision to use one or another language. The transformation of pseudo-object sentences into syntactic sentences both fosters the attitude of tolerance and facilitates the avoidance of pointless squabbling.

With the shift to semantics, Carnap must admit into Wissenschaftslogik the semantic and ontological sentences he had dismissed as pseudo-object sentences. Talk of reference is part of a Tarski truth-definition, and Carnap's formal and informal semantic metalanguages may contain the resources to formulate general ontological theses. The move to semantics may appear then to mark abandonment or serious qualification of the Principle of Tolerance. This is the way A. Church sees matters in his review of *Introduction to Semantics* (Church 1943, 303–4). Nonetheless, adoption of semantics does not seriously affect

Carnap's attitude of tolerance. Carnap now takes the Principle of Tolerance to voice an attitude toward semantical systems and takes it, in this form, to have the same significance for logic and mathematics as previously.[27]

However, semantics does force Carnap to give up the regimentation of pseudo-object sentences by quasi-syntactic sentences as a tool of criticism and clarification. Nor can the classification "quasi-syntactic" be simply replaced by a new classification "quasi-semantic." Every property of things is correlated with a semantic one: for example, every red thing belongs to the extension of the English predicate "red." Moreover, in the extension of a language *K* by a Tarski truth-definition for *K*, each *K*-sentence is L-equivalent to a sentence that predicates truth of it.

The distinction between internal and external questions Carnap explicitly introduces in "Empiricism, Semantics, and Ontology" is supposed to take up the slack here. Internal questions are questions concerning the existence of some kind of entity asked using the language described by some semantical system. These questions are to be resolved either by empirical investigations, whose results are reported using the language, or by logical investigations, that is, on the basis of the logic built into the language by the semantic rules. So in a language incorporating arithmetic, the internal question whether there are numbers is answered by a trivial logical proof. An external question is not asked from within some precisely described language. It purports to concern the ontological presuppositions of a particular language. Are there *really* numbers so that one should adopt a language that incorporates arithmetic? Carnap finds such ontological questions, asked outside of any particular language, irremediably obscure, if taken to admit a correct answer. Putative answers to external questions, sharing this obscurity, are pseudostatements. External questions should rather be understood to ask after the advisability of adopting a language with certain features. Here, adhering to the attitude of tolerance, there is no right or wrong answer, but only noncognitive considerations of expedience.[28]

In *Logical Syntax*, the notion of a quasi-syntactic sentence helps Carnap to enforce a sharp distinction between use of a syntactically described language, syntactic description of a language, and advocacy of the use of a syntactically described language. In particular, the identification and transformation of misleading pseudo-object sentences into syntactic sentences readily seals syntax against metaphysical intrusion. With Carnap's shift to semantics, the distinction between informal semantic claims and metaphysical pseudostatements becomes elusive, almost a matter of emphasis and tone. Early on in *Mean-*

ing and Necessity, Carnap elucidates the vocabulary of his informal metalanguage:

> The term "concept" will be used here as a common designation for properties, relations, and similar entities. . . . For this term it is especially important to stress the fact that it is not to be understood in a mental sense, . . . but rather [as referring] to something objective that is found in nature and that is expressed in language by a designator of nonsentential form. (Carnap 1956, §4, p. 21; cf. Carnap 1937, §34d, p. 114)

He continues on the next page: "I wish to emphasize the fact that the discussions in this book about properties, and similarly about relations, concepts in general, propositions, etc. do not involve a hypostatization." Here the contrast between semantics and metaphysics has become delicate indeed.

Notes

I have benefited from conversations on the topics of this essay with Burton Dreben, Gary Ebbs, Michael Friedman, Alan Richardson, and especially Warren Goldfarb.

1. I agree with Creath (1990, 415) that Carnap's adoption of semantics does not represent a major turning point in his development. I differ somewhat with Creath's characterization of Carnap's "epistemology," his Wissenschaftslogik; and I disagree with Creath's account on p. 411 of Carnap's reasons for rejecting the concept of truth in his syntax period.

2. After this initial use, this term will not be italicized, to avoid any possible confusion with a work by Carnap.

3. See also Carnap 1935, lecture 3, and Carnap 1937, §86, pp. 331–33.

4. Properly speaking, Carnap defines a family of notions of varying strength. Carnap's definitions employ the psychological notion of an observation predicate and are otherwise couched in logical syntactic terms. For further discussion of Carnap's view of empiricism in his syntax period, see Ricketts 1994.

5. It should be noted that Carnap's view of logic and analysis never matches Russell's. For an illuminating comparison of Russell's view in 1914 with Carnap's view in the *Aufbau,* see Richardson 1990.

6. Compare here Quine's distinction between syntax (§53) and protosyntax (§55) in Quine 1951a. Quine, following Carnap's example, allows set membership as a primitive notion of syntax, but not protosyntax. Burton Dreben called Quine's distinction to my attention in this context.

7. I disagree with Oberdan (1992, especially 252 and 255) that Carnap's antipathy toward truth in the syntax period is motivated by a wish to avoid a hierarchy of metalanguages. Carnap (1937) rejects the Tractarian view that one can never discuss the syntax of a language in the language and carefully observes at §18, p. 53: "In every language S, the syntax of any language whatsoever . . . can be formulated to an extent which is limited only by the means of expression in the language S." Carnap is unambiguously

forthright about the hierarchy of metalanguages to which his open-ended conception of syntax commits him.

8. One can in a metametalanguage syntactically define what metalanguage expressions are canonical names of object-language sentences (see Carnap 1937, §63, p. 235).

9. I gave a similar interpretation of neutrality in Ricketts 1982, 122.

10. Equally Carnap should concede the interest of investigating the extent to which the deductive machinery of Languages I and II can be described in a weak syntax language like Language I. He never, to my knowledge, discusses this issue.

11. This objection is simply an application of the core of Quine's criticism (Quine 1976 [1936], §3) of the conventionalist account of elementary logical truth to Carnap's richer view of logic.

12. Carnap explicitly rejects this appeal to convention in Carnap 1963d, 916.

13. For example, this view is articulated in Hahn 1980 (1931) and Hahn 1987 (1933). As Hahn notes, the view grows out of a particular interpretation of Wittgenstein's *Tractatus*. For further discussion, see Goldfarb 1995.

14. This point as well as Friedman's interpretation of *Logical Syntax* are discussed further in Goldfarb and Ricketts 1992.

15. After the shift to semantics, Carnap continues to hold that revision of a synthetic sentence may amount to a change of language (see Carnap 1963d, 921).

16. The general syntax definition of logical expression is in Carnap 1937, §50, pp. 177–78; L-consequence is defined in the next section, on p. 181.

17. For further discussion of how Carnap conceives the explicandum of his general syntax notion of analyticity — more exactly, his general syntax definition of the distinction between logical and descriptive vocabulary — see Ricketts 1994, 187–91.

18. This asymmetrical attitude toward reference and truth goes back to the *Aufbau*. Carnap freely uses the notion of truth to attempt to describe the sense in which constructional systems erected on different bases — for example, a physical as opposed to an autopsychological basis — may have what Carnap calls the same logical value. But he views the notion of reference as bound up with illegitimate metaphysical questions. See especially Carnap 1967 (1928), §161, pp. 256–57: "Strictly speaking, the question should not be phrased as 'What is the nominatum of this object sign?,' but 'Which sentences in which this object sign can occur are true?' *We can make an unambiguous assessment only of the truth or falsity of a sentence, not of the nominatum of a sign, not even of an object sign.*" Furthermore, Carnap (1929, 21), in a discussion of the Berry and Grelling paradoxes, says, "They all contain the pseudorelation of reference [*Bedeutung*], which does not occur at all in a purified language." Alan Richardson brought both of these passages to my attention.

19. Here I disagree with Creath 1990, 411.

20. A syntactical characterization of translation is offered in Carnap 1937, §§61–62.

21. The general syntactic notion of the content of a sentence is defined at Carnap 1937, §49; quasi-syntactic sentences are defined in §73. Carnap identifies pseudo-object sentences as quasi-syntactic in §74, p. 285. In §74, p. 285, Carnap notes that the classification as quasi-syntactic is precise only in the case of syntactically described languages. I thus take the subsequent identification of pseudo-object sentences in everyday language with syntactic claims to be advocacy of an explication for these sentences.

22. Burton Dreben's emphasis is misplaced when he says, "Carnap's syntax period is best viewed as a no-truth period. Analyticity replaced a priori truth and verifiability replaced empirical truth" (Dreben 1990, 86). Carnap's shunning of the concept of truth in *Logical Syntax* is based on the narrower, more technical considerations discussed below.

23. Oberdan (1992, 248–49) claims: "Obviously the assertion that a given sentence is true is quasi-syntactical and since this assertion is equivalent to the given sentence it must be quasi-syntactical too," and he concludes, "[T]he employment of semantical concepts is expressly incompatible with Carnap's syntacticism." This claim is incorrect. At Carnap 1937, §63 p. 236, a passage Oberdan cites, Carnap observes that in a logical language, in which the syntactic definition of analyticity is thereby a definition of truth, the notion of a quasi-syntactic sentence is trivial. This triviality in no way discredits the syntactic definition of truth for these languages. Furthermore, for descriptive languages, the notion of truth is not syntactic. Carnap's conception of pseudo-object sentences is thus no bar to the introduction of a descriptive truth-predicate in a metalanguage.

24. For an example of a trivial case, see the first language discussed in Quine 1976 (1960), §7.

25. Here I differ sharply with Coffa 1991, 303–4. Coffa appreciates neither that Carnap's purposes in *Logical Syntax* require that the definition of "valid" be couched entirely in syntactic terms nor that a merely materially adequate truth-definition would not serve these purposes. Coffa thinks that in the passage from Carnap 1937, §60b (just quoted), Carnap, pulled by verificationist prejudice, slips back into the view that syntactically definable properties must be decidable. This charge mistakes the character of Carnap's commitment to empiricism in his syntax period. The interpretation I present attributes no confusion and inconsistency to Carnap here.

26. I believe that Coffa (1991, 304) is quite mistaken to question Carnap's consistent appreciation of this distinction.

27. See Carnap 1939, 28–29. There Carnap says, "[T]he basis on which logic is constructed, namely, the interpretation of logical signs (e.g., by a determination of truth conditions) can be freely chosen." I believe that the discussion of the Principle of Tolerance at Carnap 1942, 247, should be read in light of the discussion in Carnap 1939. For still later expressions of Carnap's adherence to the Principle of Tolerance, see Carnap 1956 (1950), 247, and Carnap 1963a, 66.

28. The distinction between internal and external questions is presented in Carnap 1956 (1950s), §§2–3, pp. 206–15. My discussion here and in the next paragraph does not do full justice to Carnap's distinction.

Richard Creath

Languages without Logic

In the last few years several philosophers have examined the sense in which Carnap's *The Logical Syntax of Language* (1937 [1934]) (hereafter *LSL*) is really syntax. They have concluded, for the most part, that Carnap's syntax is really semantics or at least that the difference between what Carnap provides and a full-blown semantics would be infinitesimal were there infinitesimals. Here I want to examine the other half of 'logical syntax'. The intent is not to discover whether Carnap's logic is really logic but rather to find out what he means by that notion. More specifically, in the first part of the essay, I shall examine Carnap's definition of 'logical expression' that is found in part 4, "General Syntax," of *LSL*. Surely, this is one of the most serious attempts ever to give a general account of this notion. Unfortunately, Carnap's account seems to fail not merely in detail, but in such a way as to suggest that it is on the wrong track. But if Carnap's strategy does not work, perhaps none will. Since a popular definition of 'analytic' as well as Carnap's own *LSL* definition thereof presuppose a notion of logical truth or logical consequence, it may seem that these difficulties over the logical would likewise vitiate analyticity. In the second part of this essay I shall argue that they do not.

Logic is said to be an old subject, and since 1879 a great one (Quine 1950, vii). Yet even after a prolonged period of greatness, the subject lacks an adequate characterization. And not for want of attention. A century before Frege, Kant had placed logic near the heart of his system. In fact he had two logics, one called pure, for dealing with concepts, and a second called transcendental, which lived in the land of intuitions and which was to secure all those a priori truths not found by purer means. As Coffa has argued (1991), the Kantian story came under attack in the nineteenth century: not only were its basic categories challenged, but philosophers also tried to move the treatment of this or that topic from one Kantian category to another, for example, from transcendental logic to its purer counterpart.

Frege is an illustration. To be sure, his logic is vastly more highly developed than Kant's. But Frege is surprisingly silent on the character of logical truth and on why it would be a good thing from an epistemic point of view to succeed in reducing arithmetic to logic. Indeed, given what he says about geometry, one wonders how seriously Frege took epistemology at all. What he does say suggests motives of avoiding the evils of psychologism and empiricism in arithmetic. But even this neither tells us about the nature of logic nor justifies the reductive program. One is left to suppose that Frege accepted (perhaps uncritically) Kant's conviction of the uniqueness and epistemic importance of pure logic. This remark is *not* an attack on Frege. I am as impressed as anyone with the overpowering clarity of his thought and the magnitude of his achievement. Certainly in the half-century beginning with his *Begriffsschrift* (1967 [1879]) logic was truly a great subject. By the time of *Logical Syntax* it had achieved spectacular results — but still not a definition of logic. Russell had tried, of course, but I shall not pause to explain why I find his account unhelpful. Even after fifty years of so-called greatness we were no nearer a general account of the nature or of the limits of logical truth.

In *LSL* we do get a very ambitious attempt to define the logical, but before looking at the formal mechanics of the definition let us see what the basic informal idea is. Carnap is not looking for a narrow notion of logic that might comprise initially only the predicate calculus and little more. Rather than arguing for or even addressing the issue of logicism, Carnap intends from the very beginning to include within his concept of logic all of what is usually called logic and all of mathematics as well, whether or not the latter is reducible to the former. Moreover, Carnap has a general idea of what separates logic and mathematics from physics: logic and mathematics are not empirical at all but are matters of language instead, while our theories in physics are shaped both by the way we speak and by the specific empirical news we encounter. Not only can Carnap roughly characterize the two sides of the divide, he can also identify the vocabulary to appear on the logical side; it is just the basic vocabulary from the predicate calculus and arithmetic. Sometimes it is strategically wise to begin a definition at the level of basic vocabulary even if one's aim is to define something at the level of sentences or theories. Since Carnap already has a pretty good idea of the vocabulary he wants, this strategy seems reasonable here too. If Carnap can give a general way of picking out a (this) vocabulary as that which gets none of its meaning from any connection with the empirical world, then a logical sentence would be one involving only this vocabulary or in a slightly extended sense, one that uses only this vocabulary essentially.

In short, what Carnap is looking for is something nonempirical down to its most basic conceptual elements.

How this gets translated into a specific strategy for a definition of 'logic' and further into a specific formal definition depends on three factors: (1) Carnap's radical new linguistic conventionalism, (2) his attempt to reconcile the major wings of the Vienna Circle on the issue of the character of scientific laws, and (3) his enthusiasm on discovering how to contain the damage caused by Gödel's incompleteness results.

Concerning the first of these, we must remember that the most central point of *LSL* is that logic (or, more precisely, what has traditionally been called logic) is conventional. This central point in no way depends on logic being special. Other domains might be conventional as well. *LSL* started out modestly enough as an attempt to show, contra Wittgenstein, that it is indeed possible to describe the logical form (that is, syntax) of a language and for a certain fairly weak language to do so within that language itself. This is accomplished by exploiting Gödel's device of arithmetization. While investigating this, Carnap realized that Gödel's shocking and depressing news about incompleteness could be mitigated for a language strong enough to express arithmetic. He had to develop essentially semantic evaluation rules, and the characterization of the consequence relation was no longer definite or carried on inside the language itself. Still, it delivered a form of completeness as well as the rich blessings of classical mathematics. There might be practical reasons for wanting these blessings or practical reasons for preferring the austere beauty of a definite language. In short, there was a choice to be made. And the alternative logics were to be treated as Hilbert had treated the various geometries available at the end of the nineteenth century: the alternatives were definitions, conventions of language. There was thus no uniquely correct language. As Carnap enthused in the foreword to his book:

> The first attempts to cast the ship of logic off from the *terra firma* of the classical forms were certainly bold ones, considered from the historical point of view. But they were hampered by the striving after 'correctness'. Now, however, that impediment has been overcome, and before us lies the boundless ocean of unlimited possibilities. (1937 [1934], xv)

This conventionalism makes the task of isolating the logical much more difficult. With a single language, the usual procedure is to give a list of the logical vocabulary. Because he allows a multiplicity of languages, however, Carnap must "attempt to construct a *syntax for languages in general,* that is to say, a system of definitions of syntactical terms which are so comprehensive as to be applicable to any language whatsoever" (ibid., 167). This attempt at generality is welcome. Quine was eventually to complain about Carnap's later work that a theory of language should include or have available behavioral criteria for its central terms. It should give an account of its chief notions that is applicable to a wide variety of languages. For example, if it speaks of analyticity, the account should be of analytic for *L* for variable *L*. And to give a list without saying how or why the items are on the list is to say virtually nothing except that there are some things on some list. These are all variants on the same complaint, and the general point that Quine is making is legitimate.

It is not beyond the realm of possibility that these demands by Quine were inspired by Carnap's high standard in *LSL*. There Carnap tries to give and account applicable to absolutely any language, an account of what a predicate is, or a variable, or a constant. . . . Not only are the definitions general, but for the primitive terms Carnap uses, 'is a sentence' and 'is a direct consequence of', there is reasonable hope of behavioral criteria: the criterion for 'is a sentence' would be in terms of the utterances that people make or approve. The criterion for 'is a direct consequence of' could be in terms of the arguments that people make or approve. There is no need here to defend or improve these hints about criteria, though I have tried to do that elsewhere.

Carnap's linguistic conventionalism complicated his problem in other ways than guaranteeing a multiplicity of languages to which the definition must apply. There may be practical reasons to choose one convention over another, but at least in principle any sentence can be laid down as following from the null set in virtue of the rules of the language, even sentences that we think of initially as describing the world or our experience. Carnap is explicit that sentences usually thought to express laws of nature or even concrete matters of observable fact can be made into rules of the language (ibid., 180, 316). He has no wish to include their component concepts as logical, so having admitted such sentences as rules, he has to filter these physical concepts back out.

Perhaps this would not be a pressing problem if treating laws as rules of language were merely an abstract possibility. But Carnap saw in such treatment a way to reconcile the two major wings of the Vienna Circle on the issue of the nature of theories, and his wish to reconcile them is thus a second factor shaping his eventual formulation of a definition of the logical. Schlick, apparently under the influence of Wittgenstein's verificationism, insisted that "a law of nature does not even have the logical character of an 'assertion', but represents, rather, a 'prescription for the making of assertions' " (1979b [1931], 188). They are, if anything, rules of inference. Moreover, due to his Kantian orientation, Schlick moves effortlessly from talk of laws to talk of causality and necessity. Neurath, by contrast, is rather countersuggestible where Wittgenstein is concerned, and he has no trouble making assertions of a lawlike sort. But in his way he is as impressed as are the complete verificationists with the fact that all of the observations we could ever have could never logically entail a lawlike claim. This is the so-called Duhemian underdetermination thesis, and Neurath's response to it is to say that since the evidence does not determine the laws we accept, those laws must be conventions. Carnap made similar sounding remarks, both in *LSL* and earlier, but it is important to see how different this "conventionalism" is from the radical conventionalism that Carnap develops in *LSL* and elsewhere at the same time (for example, in 1987a [1932]). Neurath's thesis is that, given all of the observation reports and logical entailment, laws are still not fixed. Carnap's conventionalism agrees to that but insists that what counts as an observation report and what counts as a logical entailment are not themselves given but are determined only by our linguistic conventions.

Carnap's chosen role in the Circle was often that of the conciliator. Now, his radical conventionalism provided the basis for a suggestion that he thought would capture the spirit of each wing's position on laws even if not all of the details. The suggestion is this:

> We may, however, also construct a language with *extra-logical rules of transformation.* The first thing which suggests itself is to include amongst the primitive sentences the so-called laws of nature, i.e. the universal sentences of physics ('physics' here is to be understood in the widest sense).... For the sake of brevity, we shall call all the logico-mathematical transformation rules of S logical or L-*rules;* and all the remainder, physical or P-*rules.* (1937 [1934], 180)

And again:

> If P-rules are desired, they will generally be stated in the form of P-primitive sentences. In the first place, certain most general laws will be formulated as P-primitive sentences; we will call these *primitive laws.*... In the majority of cases, the primitive laws will have the form of a universal sentence of implication or of equivalence. (ibid., 316)

In other words, laws are intelligible as assertions, but what gives them special status as laws is that we have so structured our language that these assertions are or follow from the rules of inference. This likewise accounts for the causal modalities, where causal necessity, for example, is rendered as P-validity. On the other hand, these rules of inference are conventions.

Arguably, this proposal should have made both wings of the Vienna Circle happy, but that it did not do so is not terribly surprising. It is not terribly relevant for my argument either. What is relevant is that (for reasons that were partly academic-political and partly substantive convictions about the nature of science) Carnap thought not only that laws of nature could be represented as linguistic rules but that they were best represented so and that they typically would be so represented. Extralogical inference rules are now not merely abstract possibilities but standard features of languages. But the last thing that Carnap wanted was to put psychological laws of thought on a par with the first-order predicate calculus. So it was urgent to find a way to distinguish the P-rules from the L-rules.

I said a few pages back that one of Carnap's main discoveries in writing *LSL* was a way of restoring a form of completeness to arithmetic in the face of Gödel's famous results. Carnap was certainly excited by his own discovery. This excitement for completeness must be reckoned as a third factor shaping his definition of the logical, for completeness figures in an absolutely pivotal way in the final definition and in the informal discussions leading to it. It is also fair to say that the excitement for completeness bears a portion of the blame when the definitions eventually go awry.

For our purposes the most important terms in Carnap's general discussion are 'logical expression' (as opposed to 'descriptive expression'), 'L-consequence' (as opposed to 'P-consequence'), and 'analytic' (as opposed to 'synthetic' or 'contradictory'). Of these the first is the most basic. The preliminary informal definition that Carnap gives is revealing:

> But if we reflect that all of the connections between logico-mathematical terms are independent of extra-linguistic factors, such as, for instance, empirical observations, and that they must be solely and completely determined by the transformation rules of the language, we find the formally expressible distinguishing peculiarity of logical symbols and expressions to consist in the fact that each sentence constructed solely from them is determinate. (ibid., 177)

What is central here is the tie between logicality and determinateness. (A sentence is determinate just in case [1] it is a consequence of the null set, or [2] every sentence is a consequence of it.) The key to logicality is freedom from extralinguistic factors, and the key to such freedom is completeness of the sort that Carnap had just claimed to find for some fairly strong languages.

While I want to deny the tie between logicality and completeness, it does have some initial plausibility. After all, it would seem, if the rules of the language fix all of the truth-values, *then there is no room left* for empirical interpretation. The point of completeness is to exclude any possibility of any contribution by extralinguistic factors. Carnap concedes that we not only can build laws of nature and even concrete sentences into our language but can go much further: "In the most extreme case we may even so extend the transformation rules of S that every sentence which is momentarily acknowledged (whether by a particular individual or by science in general) is valid in S" (ibid., 180). Even so, the completeness condition does not turn physics into logic. No matter how rich our theories and how vast our current beliefs and even if our beliefs are closed under what is standardly called logical consequence, we do not thereby decide everything that is sayable in the physical language. I suppose that it is just possible that our theories are so strong that only one world in all its detail is consistent therewith. Such Godlike knowledge, or even Godlike opinion, is not a sufficiently serious prospect for us that we need to alter our definitions of logic to accommodate it.

So at this informal level Carnap's definitional strategy has some initial plausibility. But even at this level some doubts naturally arise. Even if completeness could exclude extralinguistic factors that contribute to the meanings of the expressions involved, it does not follow that the failure of completeness guarantees that there *are* such extralinguistic factors. Furthermore, suppose that we do find completeness in some domain. This does not show that the interpretive devices for expressions in this domain are not redundant. That is, the terms here may have extralinguistic interpretations that in fact assign the same truth-values as those assigned by the linguistic rules. And completeness within some domain is no guarantee that when that domain is incorporated into a larger domain, the original terms keep their meanings intact. It is fair to point out, however, that these worries are expressed in a very informal way, and they appeal to notions such as truth and meaning that Carnap officially disavows in *LSL*.

Let us turn then to the precise formal definition that Carnap generates on the basis of the informal one above (where S is a language, $\mathfrak{K}$s are classes, and $\mathfrak{A}$s are expressions):

> Let $\mathfrak{K}_1$, be the product of all expressional classes $\mathfrak{K}_i$ of S, which fulfil the following four conditions. [In the majority of the usual language

> systems, there exists only one class of the kind $\mathfrak{K}_i$; this is then $\mathfrak{K}_1$.] 1. If $\mathfrak{A}_1$ belongs to $\mathfrak{K}_i$, then $\mathfrak{A}_1$ is not empty and there exists a sentence which can be sub-divided into partial expressions in such a way that all belong to $\mathfrak{K}_i$ and one of them is $\mathfrak{A}_1$. 2. Every sentence which can be thus sub-divided into expressions of $\mathfrak{K}_i$ is determinate. 3. The expressions of $\mathfrak{K}_i$ are as small as possible, that is to say, no expression belongs to $\mathfrak{K}_i$ which can be sub-divided into several expressions of $\mathfrak{K}_i$. 4. $\mathfrak{K}_i$ is as comprehensive as possible, that is to say, it is not a proper sub-class of a class which fulfils both (1) and (2). An *expression* is called **logical** ($\mathfrak{A}_l$) if it is capable of being sub-divided into expressions of $\mathfrak{K}_i$; otherwise it is called **descriptive** ($\mathfrak{A}_d$). A *language* is called *logical* if it contains only $\mathfrak{a}_l$; otherwise *descriptive.* (ibid., 177–78)

In other words, the logical vocabulary is the intersection of maximal classes such that everything sayable in them is determinate. The reason for putting the matter this way is clear. If there is more than one class $\mathfrak{K}_i$, then no one of them has more of a claim than any other to be the logical vocabulary because determinateness is the only feature that Carnap indicates for the logical vocabulary and all of these classes have it. But he cannot take the union, for that would allow in indeterminate sentences. So the only reasonable choice left is the intersection.

Carnap is prepared to discover that his definition yields some unexpected results. He himself points out that where the physical laws are included among the transformation rules, the metric tensor $g_{\mu\nu}$ will turn out to be a logical expression in a homogeneous space but not in a nonhomogeneous one. A few minor surprises are perfectly acceptable, but that is not all we find.

If Carnap has the hope of reconciling what are apparently conflicting philosophical views by showing that they are just different languages, then his definition should apply to all languages (as he claims it does), and it absolutely must apply to the most common languages. Suppose therefore that we start with a standard name language. That is to say, suppose we start with a language that uses names for physical objects instead of using sets of quadruples of real numbers for them as Carnap does.[1] No one has ever actually succeeded in delimiting such a set of quadruples for any macroscopic physical object, so we should not wonder that name languages are overwhelmingly more prevalent. Suppose further that we add to our name language quantitative concepts along the lines that Carnap himself suggests and take as part of the definition of the metrical concept of mass the claim that the International Prototype Kilogram (that lump of metal in Paris) has a mass of one kilogram.

If we render 'the International Prototype Kilogram' as the individual constant 'a' and 'has a mass of one kilogram' as the (presumably descriptive) one-place predicate 'F', then 'Fa' is, as part of a definition, determinate. By itself, that is fine, but then {'F', 'a'} is a class of expressions such that everything sayable in it is determinate. What can be added to it in order to make it a maximal class? The answer is, not much; '=' could be added, but thereafter none of the usual machinery of variables and quantifiers could be included (on pain of being able to say that there are two things that are 'F', a claim that is presumably not determinate). Alternatively, some variables and existential quantifiers could be added, but then '=' could not. If everything sayable in the standard logical vocabulary is determinate, we could not add 'F' to such a class without permitting indeterminate sentences. When we take the intersection of all these maximal classes we get next to nothing. Now, 'a' might be in the intersection, but not even that would appear if there were a second quantitative concept, which was defined in part by assigning some metrical value to some individual, for example, assigning the Standard Meter Bar a length of one meter. The empty intersection of the $\mathfrak{K}_i$ is the logical vocabulary according to Carnap's definition. It may be a curious and harmless artifact of the definition of the metric tensor if '$g_{\mu\nu}$' turns out to be a logical expression. But for there to be languages without logic, that is, languages with no logical vocabulary at all, and for these to be standard languages to which Carnap clearly intended his definition to apply, is nothing short of disastrous.

But this is not the only problem. If we want to insist that in the case above 'F' and 'a' are really descriptive (as I suspect Carnap would), then completeness or determinateness in some domain is no guarantee of logicality. The role of completeness was to guarantee, somehow or other, that there was no further room to affect these meanings, at least within this sublanguage. And what happens within the sublanguage would be decisive if one imagines, as a great many philosophers do, that when one adds new concepts to an antecedent language, the old concepts do not change, at least if the old language is a conservative sublanguage of the new. If the logical terms were to get their meaning *fixed* within their own sublanguage, then they might be used to help fix other meanings. Thus, when a logical term is used in a meaning postulate for, say, color (for example, 'Nothing is both red and green all over at the same time'), the meanings of 'nothing' and 'and' are assumed to be completely unaffected. There is something very attractive about this view, but the example of 'Fa' above shows that there is something wrong. Indeed, it seems that the more defensible view would be the holistic one that adding or changing meaning postulates affects the meanings of all the terms of the language. Even a change in the observational rules could result in the putatively logical vocabulary appearing in noninferential, that is, observation, reports. If these terms have an observational role, I see no reason to deny that they are in a straightforward sense descriptive. And all of this leaves completeness in the sublanguage intact. What completeness in a sublanguage does show is that where our concerns are restricted to that sublanguage (and this is often the case), the differences in meaning will not make a difference to whatever issue is at stake in that context.

The idea that the logical terms get their meanings fixed within a sublanguage is analogous to the mistake of thinking that the meanings of observation terms are fixed in ostention and remain fixed as they contribute to the meanings of theoretical terms and even as new meaning postulates are laid down for, say, color words. Carnap saw early and clearly that this was a mistake, but apparently he did not see the analogy to logic.

Not only is completeness or determinateness in some domain no guarantee of logicality, the lack of determinateness is no proof of nonlogicality. Suppose we start with language that we would presume to be completely logical; let us say that it has no sentences that can be used to make observation reports, and it has no correspondence rules or any other device that associates any of its statements with any physical occurrence. Suppose further that a completeness proof is possible for this language either via syntactical devices as in Language I or via the more semantical ones of Language II. Suppose finally that the set of such devices is nonredundant in the sense that no proper subset thereof would yield completeness. If we were to omit even one of those syntactic or semantic devices whereby completeness was established, would we cease to have logic? I find it difficult to imagine that we would. Indeed, in that brief interval between learning of Gödel's incompleteness results and seeing how to repair the damage, did Carnap seriously entertain the idea that *Principia Mathematica* was not logic? I doubt it. I could imagine that before embracing tolerance, Gödel's results would have made Carnap concerned about whether to adopt so powerful a system as *Principia,* but not about whether it was really logic. The fear might have been of skepticism or of mysticism or of whatever, but not of nonlogicality. It is still more implausible to assume that when one loses completeness, empirical interpretation magically appears. Leaving room for empirical interpretation is not the same as providing it.

In sum, then, Carnap's approach puts far too much stress on completeness. He gets languages without logic even in standard cases. Completeness does not guarantee that the vocabulary is free from empirical interpretation, and the lack of completeness does not guarantee that the vocabulary has such an empirical interpretation. Obviously, Carnap's general definition of logical expression does not work in its present form. More than that, if my speculations about what may have led Carnap to his definition are correct, then it is unlikely that any technical modification would work better in serving those ends. Even setting Carnap's specific motivation aside, I cannot see any general way of delimiting the logical truths, either by isolating the logical vocabulary or in any other way. Plainly, I have not proved that no definition will work, and I have no wish to deny that there are truths in what is historically called logic. But I do doubt that they constitute any special kind of truth or that they contain any special kind of vocabulary. We are still free to list some terms in a particular language and call those the logical terms of that language. We are also free to list terms of a second language and call those the logical terms of the second language. No doubt we will continue to do so whenever the utility of the second list is sufficiently analogous for a given purpose to the first. Where I am pessimistic, however, is over the prospect of finding a general definition of 'logical' applicable to terms from languages in general. Having forsworn the general project, we must remember how little we learn from a mere list. The arguments are different, but my conclusion *so far* is in essentials in harmony with that of Tarski (in 1936 [1956]) and Quine (in 1960).[2]

But perhaps we need not draw their ultimate conclusion, namely, that analyticity is likewise infirm. It is to this issue that the last part of this essay is devoted. The first task here is to discover what happens in *LSL*. Then it will be possible to compare that with Carnap's later views.

Because Carnap allows P-rules as noted, he needs (or wants) to distinguish the *logical consequences* (L-consequences) from the physical ones. The former are those that, once defined terms are replaced with primitive ones, involve only logical expressions essentially. An *analytic* claim is, in turn, an L-consequence of the null set.

It might be thought that difficulties over 'logical expression' irremediably infect 'analyticity'. Certainly they do, given how that term is defined in *LSL*. And the situation there is even worse than that. The notion of logical consequence (and hence of analyticity) presupposes a notion of definition, and there are severe problems with this notion as it is presented in *LSL*. This is because the definition of 'definition' is both narrower and wider than we would expect. A symbol (elementary expression), say, a_1, can be defined in terms of symbols $a_2, \ldots, a_n$ just in case for every sentence S_1 in which a_1 occurs, there is another sentence S_2 such that S_2 contains only $a_1, \ldots, a_n$, and such that S_1 and S_2

are derivable from each other. Note that Carnap does not here distinguish between L-derivability and P-derivability (indeed, his chart [1937 (1934), 185] falsely suggests that the demonstrable, that is, that which is derivable from the null set, will be a subset of the L-valid), so that the only thing that matters for the mutual derivability of S_1 and S_2 is getting from one to the other in a finite number of steps. But this rules out cases in which S_1 and S_2 are logical consequences of one another but not mutually derivable. Why the restriction? Carnap gives us no motivation at all.

Some will find the definition of 'definition' too wide as well. This is directly tied to the inclusion of P-derivability within derivability. Carnap may have fully intended to do just that (cf. 1967 [1928], ix; and Schilpp 1963, 945–56), but it does have some unexpected consequences. Suppose we intend to adopt as P-primitive sentences the usual laws expressed as $F=MA$, $E=MC^2$, and $v=kt/p$. Since what flanks the '=' is in each case mutually derivable (that is, any full sentence involving what appears on the left is mutually derivable with a sentence that results from replacing that expression throughout with what appears on the right), what appears on the left of the '=' can be defined by what appears on the right. When we make the replacements in this way we get an L-valid claim, so the original laws are L-valid as well. What we (and I suppose Carnap) expect is that they should turn out to be P-valid. Such surprising results are not confined to cases where the putative laws themselves have a single symbol on one side of an equation, at least if the other expressive resources of the language are suitable. Suppose we adopt as rules of the language (expressed in the usual pidgin English):

(1) Creature with a heart ≡ creature with a kidney

(2) Hearted ≡ creature with a heart

(3) Kidneyed ≡ creature with a kidney

Ordinarily we would take (1) as P-valid and (2) and (3) as definitions and hence as L-valid. Likewise we would take as P-valid their consequence:

(4) Hearted ≡ kidneyed

But, according to the definition of 'definition', 'hearted' can be defined as 'kidneyed', and so (4) turns out to be L-valid against our expectations.

The foregoing have been difficulties that plague chiefly a term leading up to 'analytic'. Apart from these, Carnap's treatment of analyticity in *LSL* denies that feature to claims involved in *partial* definitions. Geometrical terms can be partially (and implicitly) defined by a set of axioms without reducing geometry to arithmetic via Cartesian coordinates. Color words are partially defined by such claims as 'Nothing is both red and green all over at the same time'. Theoretical terms are only partially defined via implicit definitions that relate one theoretical term with another or that specify in observational terms such things as test procedures. Examples include our earlier sentence, 'The International Prototype Kilogram has a mass of one kilogram'. All this is sad and ironic because Hilbert's treatment of geometrical axioms as implicit definitions is the very model around which *LSL* is built. Carnap's thesis of the conventionality of logic (and philosophy) is a generalization of Hilbert's view of geometry. Carnap recognized fairly quickly that an account of partial definition would be needed. Indeed, the classical discussion of it occurs in 1936 in his "Testability and Meaning" (1936–37).

By the time that the English edition of *LSL* was published (1937) the situation that had motivated Carnap's various definitions had changed substantially. By that time Schlick had died, and Wittgenstein had not had direct contact with the Vienna Circle for a number of years. Neurath had not changed, but it was already plain that he was not going to accept any account of truth. And it was such semantical concerns that Carnap was pursuing. In any case the notion of confirmation now seemed more promising as a way to deal with the gap between evidence and laws than did treating laws as conventions. Laws could be thought of simply as universally quantified sentences about the physical world. Carnap had also moved to America, where his primary interlocutors were Church and Quine on semantics. Except for the last few pages of *Foundations of Logic and Mathematics* (1939), he basically turned questions of the interpretation of theoretical laws over to Reichenbach and Hempel, at least for a good many years following 1937.

Even though there were these changes in motivation, there was not a wholesale discontinuity with Carnap's move into semantics. His later philosophy is in many ways a development out of *LSL*. Conventionality is still a central issue (perhaps the central issue), and analyticity is supposed to be its indicator. Carnap still employs (1942) a distinction between logical and descriptive signs, but he no longer offers a general account of it. Instead, the distinction is drawn only for particular languages, and this is done by means of lists of vocabulary. There is still reference to L-concepts, but the meaning of "L-" has changed. For example, L-validity is better understood as validity in virtue of the language rather than as validity in virtue of the more narrowly logical fragment of it. Analyticity is given a general treatment, but in the absence of a general account of the logical the two are not tied. Carnap typically renders 'analytic' as 'follows from the semantical rules alone'. As a result, the real ancestor of analyticity in the later work is 'valid' in *LSL* rather than 'analytic' in *LSL.*

These changes are important, for they allow Carnap to treat partial definition, and they avoid the other difficulties discussed earlier. The changes, however, are somewhat disguised by continuing to use the "L-" prefix, even though it is used now for different notions. The changes are also disguised by Quine's continuing formulations of analyticity. Remember that most of us learned about analyticity from Quine, not Carnap, and Quine often renders 'analytic', as 'can be turned into a logical truth by putting synonyms for synonyms' (1951b, 23). This bears a very strong family resemblance to Carnap's *LSL* talk of replacing defined terms with their definitions. There are, however, a couple of reasons, at least from Carnap's point of view, for resisting putting the matter in just this way. First, it would make the claim that what is historically called logic is analytic utterly unilluminating. Second, Quine's phrasing might mislead the unwary (and a great many philosophers have been misled) into thinking that 'analytic' will not apply to those cases recently discussed under the heading of partial definition. For example, consider

(5) Nothing is both red and green all over at the same time.

It would seem that (5) is necessarily true and guaranteed by the meaning of its component terms. But none of these terms can be explicitly defined in any ordinary way, so the replacements of synonym for synonym on a word-by-word basis seems to lead nowhere. Quine is not misled. His phrase does work but only (as he is also aware) because of what might be called a "trick." In Carnap's view from the 1940s onward, all of the analytic truths, including (5) as well as all of the logical truths however narrowly or broadly defined, will be semantic consequences of the null set and hence of each other. They will thus, in a weak sense, be synonyms. Therefore, (5) becomes a logical truth by replacing an expression, namely, the whole of (5) itself, with a synonym, namely any logical truth you like. Actually, the same trick can be made to work on a word-by-word basis. Carnap would hold that 'red' and 'red and not green' are mutually derivable and hence (weak) synonyms. Replacing the former in (5) by the latter yields a logical truth. It works, but it is still a trick. While Quine's phrase is technically adequate, many will miss the "trick" and be misled into thinking both that nothing has changed in Carnap's treatment of analyticity and that his later formulations cannot deal with partial definition.

I am not trying to convince readers that Carnap's formulations of analyticity in his later work are ideally perspicuous; we all know that they are not. Nor will I try to convince readers here that his notion of analyticity is intelligible and does useful work in epistemology. Rather I have confined myself to emphasizing that Carnap's mature treatment of analyticity does not face the problems of isolating the logical/mathematical expressions, or the problems of his earlier notion of definition, or, finally, the problem of being unable to deal with partial definition. These problems with the original formulations are real and persistent, and together they provide a convincing case for Carnap to have made the changes he did.

I have not argued that no general definition of 'logical expression' will work, but the prospects do not look good. If there are languages without logic, that is, without a special class of logical expressions or of logical truths, perhaps it is an idea that we can get along without. If the notion of analyticity is intelligible and epistemologically attractive, as I have elsewhere tried to suggest, then our strategy should be to illuminate the traditional field of logic by appeal to analyticity, rather than vice versa. In this way if someone were later to come along and give a plausible characterization of some narrower class of logical truths, well and good. If not or until that time we as logicians and as epistemologists can proceed without that characterization as we make our way across Carnap's "boundless ocean of unlimited possibilities."

Notes

1. The initial problem that I am describing does not arise for Carnap's Language I and Language II in *LSL*. This is because these are position languages, in these cases languages whose basic individual constants are in effect the natural numerals expressed as 0, 0′, 0″, and so on. Thus, there is only one individual constant that is not composed out of more basic expressions. In this respect the languages are unusual. And they are no more economical in basic expressions than, say, Russell's in which '0', '1', and '+' are taken as primitive expressions rather than '0', '′', and '+' as Carnap does. Moreover, Carnap's languages are still open to some of the objections raised in the text about linking completeness and logicality. Even if for his own languages Carnap's definition of the logical worked perfectly, the definition would still be a failure in that it would not apply with the generality he both wants and needs. It is simply not enough for there to be some language(s) where the logical-descriptive distinction can be drawn.

2. Note that my argument is that the notion of logicality fails at the level of expressions or sentences within a larger descriptive language. This still leaves room to talk of a logical language or sublanguage. But here completeness has little or nothing to do with successfully delimiting logical languages. What makes the language logical, if anything, would be the absence of correspondence rules, observational rules, or other empirical interpretive devices. Completeness is neither necessary nor sufficient.

PART IV

EXPERIENCE, EMPIRICAL KNOWLEDGE, AND EMPIRICISM

Thomas Oberdan

Postscript to Protocols: Reflections on Empiricism

The perennial rivalry between foundationalist and coherentist epistemologies has frequently led philosophers to recall the protocol-sentence controversy in the Vienna Circle as a seminal episode in the development of analytic thought. According to the conventional wisdom current just a few years ago, the dispute arose when the physicalists (including Otto Neurath, Rudolf Carnap, and Carl Hempel) repudiated the Circle's traditional foundationalist epistemology. Foundationalism, as conceived in the early days of positivism, was based on the idea that, loosely speaking, there is a select class of statements that are noninferentially warranted and that provide, via inferential relations, the epistemic warrant for all other statements. This doctrine was offered in support of the correspondence conception of truth, which explicates truth as a property of statements that correspond to facts in the world. Then foundational statements, the noninferentially warranted ones, are themselves accepted because, when they are "compared" to the facts, they are found to "correspond" to the way things are. But the early Vienna Circle thinkers could not conceive how the correspondence notion, and its foundational explication, could be defended without the attendant doctrine that foundational statements are certain. To oppose these conceptions, the physicalists proposed the fallibilist idea that all beliefs, even foundational ones, are corrigible and may be abandoned in the face of contrary evidence or conflicts with other accepted beliefs. They also disavowed the associated understanding of truth, denying that observation provides the means for "comparing" statements and facts or, for that matter, that there is any sense to the correspondence conception of truth at all. Thus it appears that what separated the two sides in the clash over protocols concerns the difference in their reactions to the doctrine of certainty: the physicalists abandoned it whereas Schlick and his wing (Friedrich Waismann, Ludwig Wittgenstein, and Edgar Zilsel) argued strenuously to retain it.

In recent years several philosophers have noted how profoundly misleading the received wisdom concerning the protocol dispute is (Coffa 1991; Uebel 1992b; Oberdan 1993). Thus what was, only a few years ago, the received account of the controversy is no longer as well received as it once was. Its primary flaw is its focus on subsidiary issues, without seeing how they are unified in the chief problem the Circle members were attacking, the question of how language relates to experience. Once this insight is recognized, others soon follow. Thus it immediately becomes obvious that Neurath's condemnations are directed toward *any* philosophical conception of truth and not just the specific doctrine of correspondence. Although it may be possible and even useful to describe how scientific practitioners come to accept certain beliefs, any philosophical explication of the nature of truth is implicitly a prioristic and hence antithetical to Neurath's own emerging naturalism.[1] Encouraged by Neurath, Carnap also repudiated truth-doctrines, though his reasons derived from the syntacticist sentiment that fueled his early 1930s philosophy. Nor, for that matter, are the motivations behind the early physicalists' rejection of truth any longer acceptable today, for naturalism is not incompatible with correspondence, and Carnap's early syntacticism was inherently flawed (Oberdan 1992, §5). But if the early physicalists' reasons for repudiating truth are recognized as the fruits of excessive positivist ardor, they may be quietly overlooked, and the rehabilitation of Neurath's and Carnap's Circle-era philosophies as cogent and tenable epistemologies may be launched with some prospect of success.

The remaining villain in the piece, then, is Schlick, whose otherwise admirable defense of correspondence was founded on his silly ideas about certainty. These notions, summed up in the idea that foundational statements are guaranteed by the feelings experienced during the act of verifying them, supposedly constituted a central dogma of Schlick's later epistemology. Throughout his career, his treatments of scientific knowledge were always torn between a holistic, formal analysis and an atomistic, naive empiricist one. By relying on certainty to bridge the gap, Schlick attempted to unify the formal with the intuitive and thereby overcome the tensions inherent in his earlier epistemological thought. But Schlick's views on the source of certainty in individual experience were implausible — if not incredible — and his attempt to overcome his earlier philosophic schizophrenia must be deemed a failure. Unfortunately, this new attempt to comprehend the protocol-sentence controversy persists in misconstruing Schlick's role in the historical debate as much as the old received view did. Like its predecessor, the new line of interpretation focuses on the doctrine of certainty as the central problem in Schlick's protocol-era epistemology, while neglecting the fact that the physicalists' objections to Schlick's defense of foundationalism were primarily directed at the conception of language underlying his epistemology, and thus failing to recognize that Schlick's major epistemological contentions were founded on his understanding of language. Yet for Schlick, more than the other participants in the controversy, the new line of interpretation promises to yield deeper insight than has hitherto been glimpsed.

This theme will be pursued by exploring the major objections the physicalists raised to Schlick's defenses of foundationalism and correspondence. Specifically, it will be argued, first, that the physicalists' primary objection to Schlick's account is that it entails a dogmatism contrary to the conventionalism of Carnap's *Logical Syntax,* as expressed in his famous Principle of Tolerance. Second, the physicalists thought the logico-epistemic link posited by Schlick to connect statements and observations is vitiated by his emphasis on the privacy of experience. Finally, Schlick's attempt to explain his views, by means of foundational "affirmations" (*Konstatierungen*), was regarded as contentious, since it relied on an obscure (and apparently unintelligible) understanding of the semantics of demonstrative expressions. The argument against these criticisms requires showing that Schlick explicitly endorsed a variety of conventionalism like the one expressed in Carnap's Principle of Tolerance and featured it in his approach to the dissolution of philosophical pseudoproblems. Then it emerges that the chief difference between Schlick's conception of language and the physicalists' was whether or not statements can be related to experience to determine their truth or falsity. This was, of course, the crucial issue for Schlick's defense of empiricism, one totally obscured by the physicalists' discussions of protocols. After explaining how Schlick explicitly provided for the application of expressions in his conception of grammar, it will be argued that his account of language issues in a tenable account of the determination of the truth of empirical statements, which does not depend on the doctrine of certainty. So if Schlick's silliness about certainty is charitably ignored, the resulting epistemology represents a viable alternative to the physicalists' conception — one that has thus far been wholly ignored by the emerging account of the protocol-sentence controversy. Moreover, it will be argued in conclusion that the chief difference between Schlick and the physicalists in the protocol-sentence controversy concerns the importance of empiricism, relative to other doctrines, in their respective philosophies. Indeed, in Schlick's "final" philosophy, empiricism is just as important as conventionalism, and in this respect his ultimate views recommend themselves more strongly to empirically minded conventionalists than those of Neurath, Carnap, or Hempel.

1. Foundations and Truth

The analysis of observation developed by Carnap proceeded from the premise that the syntactic analysis of protocols exhaustively reveals their salient epistemic features (Carnap 1932a, 91, 38–39). This assumption impedes any alternative analyses of protocols that deploy nonsyntactic methods, while at the same time relegating to the natural sciences (especially behaviorist psychology) questions of the origins and grounds of the protocols themselves (Carnap 1932b, 178, 182). Then the logical analysis of the relation between protocols and the observed situations that prompt them would be expressly prohibited, and questions concerning the warrant or truth of protocols would lie beyond the realm of proper philosophical investigation. To counter this view, Schlick defended traditional correspondentism by explaining how affirmations mediate the relation between physicalistic protocols and immediate experience. In his conception, the grounds for accepting a protocol sentence like "A white spot appeared on the screen at such and such a time and place" would be that an observer had registered, at the time and place mentioned, the affirmation "Here now white." Thus conceived, the locus of affirmations in the epistemic scheme falls somewhere in the nether region between proper scientific expressions and immediate experience. As devices introduced to bridge the chasm separating language and experience, affirmations cannot be conceived as belonging wholly to one side of the divide or the other. But if affirmations are neither expressions of scientific language (strictly speaking) nor genuine elements of the stream of experience, then what — precisely — are they?

Schlick's equivocal answer to this question plagued his conception from the start, for when he first introduced affirmations, he characterized them as both mental phenomena ("inner episodes") and genuine linguistic expressions (Schlick 1979b [1934], 381). This led Otto Neurath to smirk that affirmations "can sometimes be treated as statements, sometimes as non-statements" (Neurath 1983 [1934], 159). Carnap saw that Schlick's characterization of affirmations could only be disambiguated by ignoring his remarks about their status as "mental acts," focusing exclusively on their function as linguistic expressions. Then affirmations are to be understood as protocols of an independent phenomenal language, distinct from the language of the scientific system, along the lines of the first method Carnap described in his essay "On Protocol

Sentences" (Carnap 1987a [1932d], 458–63). Moreover, as Carnap repeatedly emphasized, affirmations thus construed must be translatable into sentences of the system language in order to provide evidence for scientific assertions (Carnap 1932a, 81). Thus Carnap interpreted affirmations as extrasystematic protocols fully translatable into singular physicalistic claims. Then it would follow, as he wrote Schlick on 17 May 1934, that their conceptions really differed very little (ASP 029-28-14, p. 1). Carnap referred to the passage in his essay "On Protocol Sentences," where he said that all testing ultimately rests on the experiences of an individual, including his experiences of the linguistic behavior of other persons. "This fact, that testing rests on the perceptions of the tester, forms the legitimate kernel of truth in 'methodological solipsism' " (Carnap 1987a [1932d], 469). And Carnap granted that this fundamental point of consensus may not have been sufficiently emphasized in his own essays on protocols (ASP 029-28-14, p. 1).

Carnap further noted that if Schlick rejected the assimilation of affirmations to protocols outside the system language, then his conception would violate the thesis of physicalism that all scientific statements are translatable into the intersubjective language of physics (ibid.). Schlick responded that he had affirmed physicalism as early as his *General Theory of Knowledge,* and his present aim was only to deny that all statements are hypothetical (ASP 029-28-10, pp. 1-2).[2] Consequently, in an ensuing note "On 'Affirmations,' " he refined his conception to reflect Carnap's suggestions. According to Schlick's later treatment, the expression "Here now white," construed as an affirmation, is to be understood as the response of an experimental subject to questioning about his immediate perceptions (Schlick 1979b [1935b], 409). And this new formulation, Schlick felt, in no way compromised his earlier contention that affirmations were indubitable. His argument depends critically on the demonstrative character of affirmations: "What is common to all these statements is that they contain *demonstrative* terms having the meaning of a present gesture, i.e., their rules of use stipulate that in making the statement in which they occur, an experience occurs, attention is directed to something observed" (Schlick 1979b [1935a], 385). Because of their demonstrative character, affirmations are not, strictly speaking, expressions of the physical language. Nonetheless, they are translatable into proper physicalistic expressions, though the resulting translations could never possess the key epistemic features — incorrigibility and indubitability — that characterize affirmations. For the grammar of affirmations is such that it makes no sense to attach any note of doubt or uncertainty to them: "[W]e notice a peculiar feature of the grammar of affirmations, in that by addition of such words as 'perhaps,' 'probably,' 'seemingly,' 'possibly,' and the like they are turned into meaningless statements, whereas with hypotheses the use of these words is always permitted, and indeed called for" (Schlick 1979b [1935b], 409). Schlick thought this grammatical fact of the matter demonstrated that the usage he characterized as "affirmations" was neither hypothetical nor artificial, but a usage occurrent in common discourse — at least in experimental situations.[3] He writes: "There is thus a use of such sentences as 'Yellow here,' 'There are two lines in the visual field', etc., in which it would be *absurd* (that is, contrary to the accepted rules), to speak of error or deception: and where *this* usage prevails, I call the statement an 'affirmation' " (ibid., 409–10). And though Carnap had momentarily endorsed a similar idea in his earliest protocol essays, he had his doubts (Carnap 1932b, 176).[4] In particular, he thought the demonstrative character of affirmations was "peculiar," and he worried about the possibility of formulating a logically intelligible — or syntactically correct — account of how demonstrative statements are connected with hypothetical ones (ASP 029-28-14, p. 1).

Indeed, by the lights of Carnap's *Logical Syntax of Language,* Schlick's analysis of affirmations was strictly incoherent. Although it is not feasible, in the present context, to fully describe the close relationship between Carnap's evolving conception of protocols and his emerging analysis of language, a few points must be mentioned (Oberdan 1990, 26–30). The first is that the whole idea of treating observation in terms of protocol *sentences* was founded on what Carnap called the "thesis of metalogic," roughly speaking, the claim that all philosophical contentions (that are not nonsense) are metalinguistic claims about expressions of language and their syntactically specifiable features (Carnap 1932a, 435n).[5] Further, the metalogic thesis served to isolate philosophical pseudotheses — that is, statements that *seem* to concern substantive issues determined by matters of fact but are actually about language or its internal structure. Carnap called these metalinguistic statements masquerading as factual claims "pseudo-object sentences," and his detailed analysis of their disguised character became a centerpiece of his philosophy during the era of his *Logical Syntax* (Carnap 1937, §74; Oberdan 1992, §3). The problem with such sentences is that they implicitly suggest that the source of linguistic meaning lies outside language, leading to mistaken hypostatizations and implying the wrong-headed notion that the structure of language is somehow derived from the nature of the given. Moreover, any attempt to justify the choice of a language by reference to anything outside it is misguided. This further claim, that the choice of a specific language is a conventional decision, that it cannot be justified or determined by anything outside language, is the thesis of Carnap's Principle of Tolerance (Carnap 1937, 51–52). Thus, the two chief principles of Carnap's early 1930s philosophy functioned together in the diagnosis and dissolution of philosophical pseudotheses. The metalogic thesis isolated statements of the material mode that purport to concern substantive matters but really pertain to language and its logical features, while the Tolerance precept explicitly disclaimed any relation between logical matters and substantive, extralinguistic ones.

Carnap thought "sentences about meaning"—that is, assertions about the relations between language and the nonlinguistic—were prime examples of such pseudoclaims (Carnap 1937, §75; Oberdan 1992, §3). This category of philosophical pseudotheses included correspondence accounts of truth like the one Schlick hoped to explicate by means of his notion of affirmations. Cast against the background of Carnap's analysis of pseudo-object sentences, it is no wonder the physicalists suspected that Schlick's epistemological foundationalism, his belief that all epistemic warrant is derived from certain privileged statements (affirmations), was founded on a parallel semantic doctrine that was both logically and metaphysically reprehensible. This "semantic foundationalism" would be characterized by the idea that the meanings of all scientific statements are derived from the meanings of the epistemologically foundational ones. The significance of nonfoundational statements would be derived, via inferential relations, from affirmations. Affirmations, in turn, would directly acquire their own significance from the experiences that confirm them and that they represent. Ultimately, this line of thought would lead to a hypostatization of the referents of the foundational statements, entailing that only those languages representing the experiences described in affirmations are significant. The resulting view would be just the sort of metaphysical hypostatization, based on a mistaken view of meaning, that Carnap identified in his analysis of pseudo-object sentences; as such, it would fly in the face of the fundamental principles of the physicalists' philosophy of language.

Neurath suggested as much in "Radical Physicalism and the 'Real World,'" when he pointed out that Schlick's defense of the correspondence view of truth depends on a conception of "the one, true reality," or "the real world" (Neurath 1983 [1934], 106–8). Neurath had long been critical of doctrines of truth, which he considered attempts to set off "language as a whole" from "experience," "the world," or "the given" (Carnap 1983 [1934], 61). And he apparently thought Schlick regarded the chunks of the world to which bits of language supposedly correspond as elements of a determinately structured reality given independently of the language used for its description. Carl Hempel advanced this allegation when he echoed Neurath's remark that the "facts" to which true statements are supposed to correspond, and with which they are supposedly "compared," are "meaningless duplications" — that is, "nothing but the result of a re-doubling metaphysics" (Hempel 1935a, 55; 1935b, 95). Hempel further argued that treating "facts" as the measure of the truth of propositions suggests that language must somehow represent the intrinsic structure of the "facts," raising "pseudoquestions" that seem to concern genuine, verifiable issues but are really a matter of syntactic conventions (Hempel 1935b, 95). Accordingly, the physicalists' complaint about Schlick's defense of correspondence boils down to the charge that it implicitly involves a pernicious metaphysics in a way that violates the spirit of Carnap's philosophy of language. When added to Carnap's doubts about the syntactic intelligibility of affirmations, it appears that Schlick's epistemology offended against both the metalogic thesis and the Principle of Tolerance, the leading precepts on which the physicalists' conception of language, and in turn their epistemology, were founded.

2. Form and Content

The charge that Schlick's semantics was foundationalist is vitiated by his well-known doctrine of implicit definition, which he used to account for the formation of scientific concepts in the first (1918) edition of his *General Theory of Knowledge.* Instead of the Machian empiricist explanation of scientific concept-formation as an abstraction from experience, or the Kantian view of concepts derived from the faculties of intuition and understanding, Schlick borrowed the notion of implicit definition from David Hilbert's work in geometry to argue that the system of scientific concepts is an autonomous network, consisting of elements that are simultaneously defined by their interrelations (Schlick 1975 [1925], 33). Strictly speaking, a concept is said to be implicitly defined if it is a primitive introduced in a set of axioms. Thus introduced, scientific terms "acquire meaning only by virtue of the axiom-system, and possess only the content it bestows on them. They stand for entities whose whole being is to be bearers of the relations laid down by the system" (ibid., 34). Of course, the reason Hilbert introduced implicit definitions in the first place was the same reason that Schlick used them in his discussion of concepts: to display a method of concept-formation that guarantees concepts are completely empty of intuitive content. But in the context of geometry, Hilbert's goal was to assure a ground of absolute certainty for geometric proofs while, in epistemology, Schlick's aim was to show how science develops abstract tools of sufficient generality and adequate precision to account for scientific knowledge. One immediate consequence was that intuition—either in the pure form that founded Kant's doctrine of the synthetic a priori or as the experiential basis of all knowledge in Mach's philosophy — was diminished in epistemological import.

A more significant result is that Schlick was faced with a stark contrast between two distinct levels of cognitive activity. This dichotomy, expressed in Schlick's *General Theory of Knowledge* as the difference between concepts and intuition, would, in one guise or another, remain a central feature of his thinking at least until the time of the protocol-sentence controversy.[6] It soon became prominent in Schlick's thinking as the distinction between "form" and "content," shortly after the second edition of *General Theory* appeared in 1925. It was at this time that Schlick read Carnap's latest manuscript, currently called "Konstitutionstheorie" but eventually to be immortalized as the *Aufbau* (Carnap 1967 [1928]). Schlick even mentioned to his long-time confidante, Albert Einstein, that it stands "on a very high plateau" (EC 21.591). A few months later, Schlick reported to Einstein that he was deeply immersed in the logical writings of Frege, Russell, and Wittgenstein, expressing his hope that these works would lead to a thorough reform of philosophy or even a complete overthrow of it as superfluous (EC 21.596, 2). Schlick was particularly awed by the *Tractatus,* which he thought "the deepest and truest" work of the new philosophy (EC 21.599, 1). These studies convinced Schlick that he had much to learn, for now his own earlier epistemology appeared to him as "primitive" and "immature" (ibid.).

The immediate effect of these influences was that Schlick developed, in the late 1920s and early 1930s, a conception of language based on his earlier doctrine of implicit definitions, expanded and modified with notions borrowed from other thinkers. Beginning with the essay "Experience, Cognition, and Metaphysics," Schlick broadened his account of scientific concepts to include ones expressed in everyday language (Schlick 1979b [1926], 100–3; 1979b [1938], 291–95). First he adapted Wittgenstein's doctrine of internal relations to show that the concepts of everyday discourse are formally related in much the same way as concepts introduced by implicit definitions. Then, the Tractarian account of tautologies and contradictions was appropriated to establish that meaning relations are purely formal and devoid of material content. The immediate result was a "structural" or "formal" conception of language, according to which the significance of any expression is founded on its logical relations to others. This implies that linguistic meaning cannot be explicated in terms of any external determinants, accessed via some privileged epistemic mode, like “acquaintance” in Russell’s philosophy or “intuition” in Bergson’s (Schlick 1979b [1938], 318–20). The flip-side of this coin was Schlick’s Thesis of the Incommunicability of Contents, the claim that only the structural relations obtaining among experiences may be successfully communicated, while the content of experience, the “greenness” of the color green or the distinctive odor of the smell of wood smoke, must forever remain ineffable (Schlick 1979b [1926], 102–3; 1979b [1938], 295–6).

Nor was Schlick the only positivist seduced by the idea that experience provides the content and only its form is represented by scientific knowledge. Carnap himself had placed such a distinction at the center of his *Aufbau* and even endorsed the idea that the content of experience was itself incommunicable (Carnap 1967 [1928], §67). But Carnap’s attempt to separate the formal and material components of scientific knowledge was not wholly successful, for the basic structural relation among experiences, “recollected similarity” (*Aehnlichkeitserinnerung*), was itself material or empirical (Friedman 1987, 530–33). In the immediately following years, Carnap focused on the analytic-synthetic distinction as a means for explicating the insight that lay behind the form-content distinction. Of course, this later project was doomed if, as the *Tractatus* dictated, the formal element cannot itself be independently represented or described. In fact, one of Carnap’s earliest syntactical insights was that the method of arithmetization developed by Gödel provided just the means needed for expressing the formal: by arithmetizing syntax one could describe the structure of a language in that language, thus isolating — and, despite the *Tractatus,* expressing — its formal component (Carnap 1937, §18; ASP 081-07-17; 081-07-18; 081-07-19; Oberdan 1992, §2).

Ultimately, Carnap repudiated all renderings of the form-content distinction that implied an Incommunicability Thesis, which he considered a vestigial form of psychologism. So when Edgar Zilsel, representing the Schlick wing of the Vienna Circle, criticized the early discussions of protocols among the physicalists, Carnap responded by condemning, in no uncertain terms, the very idea of ineffability on which Zilsel’s critique was founded (Carnap 1932c, 181). Carnap’s barbs were undoubtedly directed at Schlick and would have pierced the “form and content” philosophy Schlick was currently promoting. But by the time Schlick entered the fracas over protocols, he had thought through the details of a conception of language that incorporated a conventionalist component with a plausible account of empirical content and the determination of truth. Gone from Schlick’s thought are the disastrous notions of form and content, relics of an earlier stage in the evolution of his thought. Likewise reduced in philosophical import was the implausible, counterintuitive, and highly dubious Thesis of the Incommunicability of Contents.[7] In their place Schlick substituted a thoroughly holistic conception of language that incorporated an endorsement of conventionalist themes, a critique of pseudotheses much like Carnap's, and an explicit disavowal of the elements of his earlier philosophy that had provoked the physicalists' charges of "metaphysics."

3. Grammar and Conventions

In Schlick's later conception, propositions serve the theoretical function of truth-bearers, in explicit contrast with the notion of a "sentence" or "propositional sign" as the "outer structure," "word-complex," or "sequence of linguistic signs" that may, under certain conditions, be used to make a statement or express a proposition (Schlick 1934–35, 63–64; 1979b [1935c], 408). Roughly, a proposition or statement is a sentence used in accordance with determinate grammatical or logical rules, for only when its use is governed by rules that regulate its employment does a sentence *express* (or *become*) a proposition. A meaningful proposition is identified with the linguistic function it serves in the language, and the sentence to which the meaning is attached may then be regarded as a place-holder for that particular linguistic function. The function itself is constituted by the rules of the language that pertain to that sign. Then it follows that, in general, it is only in virtue of its grammar that an expression is significant, that it can be used, perhaps in conjunction with other signs, to say something. Accordingly, it would be misleading to speak of the significance or meaning of any expression, on its own, without reference to the language whose grammatical rules govern its use (Schlick 1934–35, 67). For obviously two different sign-sequences, in distinct languages, may mean the same thing, and a given syntactic structure could easily have different meanings relative to different sets of grammatical rules, like the term "simultaneous" in classical and relativistic mechanics (Schlick 1979b, 444).

Moreover, Schlick maintained that the choice among alternative grammars is arbitrary or conventional (Schlick 1934–35, 67). Thus, in his note "On 'Affirmations,'" Schlick insisted that the rules of grammar were all "attached by convention" to the expressions they govern (Schlick 1979b [1935c], 408). Again, in "Meaning and Verification," he pointed out that the rules of language "are not found anywhere in nature, but are made by men, and are, in principle, arbitrary" (Schlick 1979b [1936a], 466). That is, the rules of logical grammar "are not facts of nature which could be 'discovered' but they are prescriptions stipulated by acts of definition" (ibid., 464). Similarly, he emphasized in "Are Natural Laws Conventions?" that the grammatical rules "in their totality form the grammar of the scientific language. . . . They are the only conventions" (Schlick 1979b [1936b], 443). Schlick thought only grammatical or logical rules are conventional, and not scientific theories, laws, hypotheses, or singular statements. This distinguishes Schlick's conventionalism from views that emphasize the conventional element in all scientific claims on the grounds that no contention can be definitively verified by experience. Indeed, Schlick thought it was inappropriate to even call this view 'conventionalism,' since it is a trivial variety, of little philosophical interest (ibid., 438).[8]

Schlick further emphasized that conventionalism characterizes not only the special, technical languages employed in particular scientific disciplines but everyday language as well. While he thought the clearest cases of linguistic conventions could be drawn from scientific contexts, conventions are not necessarily the product of due deliberation but might just as well be conceived on the model of the diachronic development of natural, "home-grown" languages as well (Schlick 1934–35, 98–99). Grammatical rules are conventional, whether they arise through express stipulations or are taken up through use (Schlick 1934–35, 86). Nor is this to deny that biological, sociological, or psychological factors — to say nothing of all the obvious physical ones — exert tremendous pressure on the forms languages assume at various points in their evolution. For even though natural languages develop in certain ways because of determinate external influences, any of the features they thereby acquire could, in principle, be changed, if only we chose to do so. Since "the validity of a convention is of our own making," a particular usage or practice will be accepted and established only so long as it suits our needs and purposes (Schlick 1979b [1936b], 438). Schlick's recognition of the conventionality of grammar suggests that his philosophy at the time of the protocol-sentence controversy was founded on themes similar to the one expressed in Carnap's Principle of Tolerance. Yet it is natural to wonder whether Schlick's conventionalism was anything other than a trivial semantic variety, admitting conventional choices only between languages that do not differ in any substantive logical respects. Then Schlick's conventionalism would be a far weaker species than Carnap's, which countenanced conventional decisions between languages differing as radically as a constructivist one (like Language I in *The Logical Syntax*) diverges from the language of classical mathematics (Language II). However, it is evident from Schlick's treatment of philosophical pseudoproblems that he thought the decision to adopt a language was conventional, even when its logical structure differs significantly from the alternatives.

4. Pseudotheses

Perhaps the reason that the conventionalist strain in Schlick's later thought has been neglected is because its chief function in his philosophy was in the analysis and dissolution of pseudotheses, an application he misleadingly advertised as a defense of the ill-starred slogan "Meaning is the method of verification."[9] And while his critique of philosophical pseudostatements was framed with sufficient generality to apply to a variety of typically "metaphysical" theses, like Platonism, psychologism, and phenomenalism, it is perhaps best known from his essay "Meaning and Verification," where it was applied to the thesis of solipsism (Schlick 1934–35, 24–25, 96–100; 1936a/1979b, 471–79). The solipsist's claim,

(Q) I can only feel my pain,

contrasts with the patently contingent assertion,

(P) I feel pain only when the body *M* is hurt.

In common usage, (P) has a clear-cut sense: the expression "I feel pain" is a meaningful description or report of immediate experience independent of the empirical procedures for finding out when a particular body — called *M* in this case — is hurt. That is, the sense of (P) is the same as the contingent assertion that my feelings of pain are always accompanied by physical hurt to the body *M*. Interpreted this way, (P) is similar to a whole series of trivial truths relating individual experiences and physical events in a particular body, like the fact that I do not have any visual impressions when *M*'s eyes are closed, I do not hear when *M*'s ears are plugged, and so on. Of course, these psychophysical correlations are purely contingent, and (Q) might be similarly interpreted as a contingent statement concerning two independently determinable things: what I feel and my pain. Then it is still logically conceivable for beings just like us in every other respect to feel pain when some other body is hurt. Suppose, for instance, that the nerve endings in the foot of a person *A* were wired, via some imagined electronic gadgetry, to the cerebral cortex of a person *B*. When *A*'s foot is hurt, *B* could meaningfully (and even truly) say that he feels pain in *A*'s foot. Yet this happenstance is exactly what is denied by (P) and (Q), interpreted as contingent assertions.

If, however, these counterfactual circumstances obtained — if, that is, one occasionally felt pain in another's body — (P) would certainly be false and (presumably) (Q) would too; opposed to them would be the true claim:

(R) I can feel somebody else's pain as well as my own.

Here, (R) contradicts (P) and (Q), thus allowing that I can feel pain not only when the body *M* is hurt but when some other body is injured as well. No longer would the body *M* be the only one whose states affect my perceptions. Moreover, the solipsist's thesis, asserting the inviolable privacy of sensations, would obviously be a contingent claim that could, in principle, be falsified by experience.

Note, first of all, how far Schlick's thought has progressed from his "form and content" philosophy, especially its nearly mystical incommunicability thesis. His earliest arguments for this claim, in his essay "Experience, Cognition, and Metaphysics," based it on "the logical doctrine of implicit definitions," insinuating that the thesis is a logical claim and that it is logically necessary that contents are incommunicable (Schlick 1979b [1926], 100–101). Later, in "Form and Content," he argued that contents are *essentially* incommunicable, founding his contention on the inconceivability of the idea that, somehow, the privacy of sensation might be abrogated and thus intimating that it is logically impossible to separate the subject from his experience (Schlick 1979b [1938], 319). Now, however, in his treatment of solipsism, he grants that the privacy of experience is no more than a contingent happenstance, so that it is logically possible that a given experience might have occurred to someone else (Kraft 1953, 44). This radical departure from his earlier views is of fundamental significance, for if the nature of an experience is independent of its contingent owner, individual experience may now be mobilized to provide warrant for scientific assertions, without compromising the public character of the scientific claims so warranted.

The crucial point is that, even if (R) were true, even if it were possible to feel pain in another's body, the solipsist would not admit that it is correct to say,

I feel someone else's pain,

but only that,

My pain is in someone else's body. (Schlick 1979b [1936a], 474–75)

But by denying that (Q) is falsified, the solipsist implicitly admits its meaninglessness.[10] Thus (Q) is meaningless not because, as things stand, it is unfalsified, but rather because the solipsist refuses to admit that his thesis could be falsified in any logically possible circumstance. By isolating it from possible circumstances of falsification, the solipsist admits he regards (Q) as a definitional stipulation; as he uses the term, every pain is to be regarded as belonging to the person who experiences it, regardless of whose body is affected. In short, the solipsist has denied the word "my," as it occurs in (Q), any significance at all, for there would be no sense in saying "I have pain" or calling a sensation "my pain" unless we could meaningfully replace "I" and "my" in these contexts with "he" and "his" (ibid., 477–48). Consequently, the thesis is, for the solipsist at any rate, an empty tautology, an analytic proposition (Schlick 1934–35, 97).

Of course, analytic statements, for Schlick as for the other positivists, are just object-language effects of linguistic rules. So even though Schlick cited the lack of verifiability as a mark of philosophical pseudotheses, the reason he thought they were meaningless, that (in the present case) solipsism is not a genuine thesis, is that they are treated, by their proponents, as unfalsifiable on logical grounds, in every logically conceivable circumstance (Schlick 1934–35, 97). For Schlick, statements isolated from the possibility of falsification are object-language "misexpressions" of grammatical rules rather than genuine assertions, just as, in Carnap's philosophy, pseudo-object sentences are really metalinguistic sentences and not authentic "real-object" ones.[11] Eschewing Carnap's strict dichotomy of sentences into syntactical and material, Schlick instead argued from the unverifiability (specifically, the unfalsifiability) of alleged pseudotheses to their grammatical (and, hence, analytic) character. And just as Carnap argued that the formal mode translations of philosophical theses should be regarded as proposals to adopt a certain language form, so Schlick contended the solipsist's thesis was not a bona fide contingent claim but rather an attempt to introduce a particular mode of speech (ibid., 97).

Moreover, Schlick granted, like Carnap, that the choice of a mode of speech, though strictly arbitrary or conventional, may nonetheless be guided by pragmatic considerations. The choice of a particular form of language is never compelled by extralinguistic affairs but is arbitrarily decided and may be subsequently revised. Nor is this to deny that the texture of experience renders some choices more felicitous than others and so fosters the adoption of certain language forms rather than others. And the features of experience that render one language form more advantageous than others are of a very simple kind, like the ones represented by commonsense correlations linking perceptual states with bodily ones (Schlick 1934–35, 97–98). Of course, these traits of experience are merely contingent, and if experience were other than it is, we would undoubtedly adopt other ways of speaking. Recognizing all this is, for Schlick, the key to understanding philosophical error, which arises when someone mistakes a philosophical thesis, a convention induced by a fact, for the linguistic representation of that fact (Schlick 1934–35, 102). And even if the solipsist's thesis is construed as a recommendation to adopt a mode of speech in which (Q) is a recognized convention, there are no pragmatic reasons to recommend the solipsist's proposal. Not only does it engender pseudotheses like the one just discussed; it generally gives the mistaken impression that experience is subjective, and necessarily so.

5. The Application of Language

The conclusion of Schlick's critique of solipsism vitiates the physicalists' complaints about the phenomenalism in Schlick's epistemology and his insensitivity to conventionalist themes. This in turn colors the other differences between Schlick and the physicalists in previously undetected shades. If the physicalists' diagnosis of philosophical error underlying Schlick's empiricism is wrongheaded, his departure from physicalist orthodoxy over the issue of truth might likewise be founded on substantial philosophical reasons. Yet these reasons are scarcely evident in some of Schlick's remarks, like his flippant response to Hempel when he affirmed that propositions could indeed be compared with facts:

> I have often compared propositions with facts; so I had no reason to say it couldn't be done. I found, for instance, in my Baedeker the statement: "This cathedral has two spires," I was able to compare it with 'reality' by looking at the cathedral, and this comparison convinced me that Baedeker's assertion was true. (Schlick 1979b [1935b], 400)

Of course, the level of Schlick's defense in this passage is just that at which Samuel Johnson "refuted" Bishop Berkeley — appropriate, perhaps, but nonetheless unedifying.[12] Yet beyond Schlick's taunting sarcasm lies at least one salient remark illuminating the insight he was trying to capture: "By the way, it is easy to express in a purely formal way my opinion that facts and propositions can be compared: words denoting symbols and words denoting other things may occur in the same sentence" (ibid., 402). By affirming the need for statements that are neither syntactical (referring exclusively to linguistic expressions and their syntactical features) nor real-object sentences (couched entirely in terms denoting extralinguistic things), Schlick explicitly confronted the strict dichotomy of the formal and material modes of speech underlying the physicalists' epistemology. To oppose the physicalists' syntacticism, Schlick promoted "mixed-mode" sentences that would be metalinguistic, for they refer to linguistic expressions, but are not strictly formal, since they also refer to nonlinguistic things that cannot be specified in strictly syntactical terms.

At the time of the fracas over protocols, Schlick insisted that, in addition to the rules of logical syntax, there must also be "application rules," distinguished by the fact that they connect linguistic expressions with definite conditions for their use, "real situations" or "experiences" in which word-combinations are applied (Schlick 1934–35, 57–58, 64–65, 67, 69–70; 1979b [1935c], 408; 1979b [1936b], 442). Schlick's point is simply that an adequate account of nonformal languages, like those used in science and everyday life, requires not only logical rules like the ones that prohibit the application of two different color-terms to the same object at the same time (like "Nothing is both red and blue") but also ones to indicate when the color of a given object is to be called "red" rather than, say, "blue." Since application rules are part of grammar, they are conventional and the relations they govern are necessary, for it is an essential feature of an expression's meaning that it is to be applied thus and so. Schlick's account of application rules was offered in support of his claim that his own conception of a grammar is more comprehensive than Carnap's notion of a logical syntax (Schlick 1934–35, 57–58). In particular, application rules provide an account of indexical or demonstrative expressions, like "I," "you," "here," "now," and so forth (ibid., 63). The propositions expressed by sentences containing these (and similar) expressions depend on varying factors of the context of use, like the time of the utterance, the identities of the speaker and audience, indicated items of the immediate environment, and so forth (Schlick 1934–35, 65–66, 69–70; 1979b [1935c], 408). Although Carnap himself acknowledged that his *Logical Syntax* approach was restricted to "languages which contain no expressions dependent upon extralinguistic factors," he thought his method was sufficient to deal with such expressions, by replacing them with person-, place-, and time-designations (Carnap 1937, 168). And it would, of course, be plainly incorrect to deny that indexicals and demonstratives can be force-fit into the Procrustean mold of Carnap's thoroughgoing syntacticism. But it is equally clear that an account developed along the lines sketched by Schlick in his treatment of application rules would be far more natural and satisfactory.[13]

Moreover, this last contrast between Carnap's conception of logical syntax and Schlick's notion of a grammar is clearly the reason Schlick thought good sense could be made of demonstrative utterances in the reporting of experience. But the sweeping epistemological conclusions Schlick drew from that usage are just as patently incorrect. Any attempt to make sense of affirmations as Schlick characterized them is bound to meet with failure, for it is impossible to combine the nonhypothetical character of affirmations — their incorrigibility and indubitability — with their use to communicate observational results.[14] Although Schlick's account of the epistemic character of affirmations — their incorrigibility and indubitability — comes to nought, apart from this contentious element of his epistemology, his treatment otherwise provides a plausible account of the empirical foundations of scientific knowledge. By grounding his theory of knowledge on a conception of language incorporating conventionalist themes, his views were invulnerable to the strongest criticisms leveled by the physicalists. By insisting that grammar must include rules for the application of language, his conception of language contains the means for the empirical confirmation of scientific claims.

6. Empiricism

The protocol-sentence controversy is a complex topic, and it would be presumptuous to suppose it could be treated with full justice in a single essay. Nor, for that matter, has the attempt been undertaken here. Rather, the aim has been to isolate and — to the extent possible — eliminate some of the misunderstandings that have impeded an accurate perception of Schlick's views. Many of the objections raised by the physicalists — especially Neurath and Hempel — were based simply on their lack of familiarity with Schlick's earlier writings, especially his classic *General Theory of Knowledge.* Carnap, on the other hand, understood Schlick — and his work — much more deeply, and it is largely for this reason that he appears the most sympathetic of Schlick's antagonists. Nor is Schlick entirely free of responsibility for the misunderstandings of his views. Not only are his notions about the certainty of the empirical basis pointless; using affirmations as a device to explain them made matters worse. If Schlick's motivations for requiring certainty of the empirical basis were murky in the first place, introducing affirmations to show how it might be achieved rendered the issues hopelessly obscure. Viewed in perspective, affirmations themselves may be the principal obstacle to the accurate perception of Schlick's views. For it appears that affirmations had little function in Schlick's philosophy other than to guarantee the certainty of the foundations of empirical knowledge. Without understanding why the basis is, or ought to be, certain, affirmations seem rather gratuitous. But there was another reason for introducing affirmations, one that was obscured by Schlick's discussion of certainty and consequently neglected by subsequent students of the protocol-sentence controversy.

Schlick introduced affirmations not only to guarantee the certainty of the basis but to ensure that it was empirical, thus securing a role for experience in scientific knowledge. As far as he was concerned, the physicalists, for all their table-banging endorsements of empiricism, had failed to address the question of the role played by experience in the adoption of protocols. This lacuna might be covered up by insisting it constitutes no failure on the part of the physicalists but represents, rather, their naturalism. Certainly, both Neurath and Carnap thought that the question of how protocols come to be accepted is a matter best left to empirical science, especially behaviorist psychology. But they never required naturalistic explanations to assign experience an essential role. Rather, the door was left open to philosophically disastrous explanations like "Protocols are accepted because the social status of the endorsing scientist is enhanced," or "Such and such a protocol was accepted because a law was passed enforcing its endorsement under penalty of a stiff jail sentence." No wonder Russell quipped that, on the physicalists' account, truth may be determined by the police (Russell 1940, 148). Thus Carnap gave a purely sociological account of the acceptance of beliefs by the scientists of a particular "culture-circle" in his reply to Zilsel (Carnap 1932c, 179–80). And one can only be amused when Neurath complained that Popper's account failed to explain how protocols are themselves confirmed, for Neurath himself never devoted any attention to this issue in his own writings on protocols (ASP 029-09-87, 1-2).

Although naturalism may have provided Neurath with the motivation for neglecting the role of experience in the acceptance of protocols, it seems doubtful that it was the only — or even the principal — reason Carnap never discussed the matter. It would appear, rather, that Carnap's failure to allot experience any role in his account of protocols depends on his strict syntacticism, which expressly denied the very possibility of a role for experience. And the best evidence for this is his willingness, even eagerness, to explain how statements are accepted on the basis of "confrontation with observation," once he endorsed Tarski's semantics and thereby abandoned his syntacticism (Carnap 1935, 124–25).

One might think that — apart from the nonissue of the doctrine of certainty — Carnap's desertion of syntacticism left him quite close to complete agreement with Schlick. By the time of Carnap's public endorsement of Tarski's work, he agreed with Schlick that language is conventional, that philosophical pseudotheses arose from a failure to appreciate the conventionality of language, and that experience played a role in the acceptance of scientific beliefs. Then it would appear that, apart from a few remaining differences over relatively minor matters, Carnap and Schlick had reached, by the end of 1935, a consensus on essential precepts. But this conclusion, however neatly it resolves the protocol-sentence controversy, misses the most important single issue separating the two sides in the dispute.

For even when all the misunderstandings are cleared away, when Schlick's silliness about certainty is dismissed, and Carnap abandoned his syntacticism, one key difference still divided the philosophies of the two leading protagonists in the protocol-sentence controversy. This difference concerns the relative status, within Schlick's and Carnap's respective philosophies, of conventionalist and empiricist precepts. In both their philosophies, the thesis that language is conventional is a predominant principle, so that it is a fundamental tenet for both that the choice of a language is arbitrary. For Schlick, it was equally fundamental that languages must be applicable in experience, a point that is clear from his discussion in "Meaning and Verification," as well as his insistence that application rules are a necessary part of grammar. Thus even though the choice among alternative languages is conventional, the choice is limited to languages that are empirically applicable. One can choose to employ any one of a number of alternative languages, but all the alternatives must be constituted in such a way that they can be used to report experience. In other words, since application rules constitute an essential part of the grammar of any language, the conventional choices range only over empirically applicable languages.

In contrast to Schlick, Carnap did not allot empiricism the same status as conventionalist tenets, but an inferior, subordinate one. That is to say, Carnap recognized that one may choose to adopt any language whatsoever, regardless of whether it was applicable to experience or not. If Coffa is right in assessing Carnap's attitude toward conventionalism as a higher order semantic fact of the matter, then empiricism — in Carnap's scheme — is a lower order conventional choice (Coffa 1991, 317ff.). And this seems to be Carnap's express point in his discussion of empiricism in "Testability and Meaning," where he discusses alternative choices of an empiricist criterion of significance, characterizing the different alternatives available to the empirically minded (Carnap 1936–37,

§27). Thus, Carnap's empiricism is itself a conventional choice, unlike his conventionalism, which was a higher-order semantic fact of the matter. For Schlick, however, both conventionalism and empiricism were equally fundamental tenets, on a par philosophically. And even though Carnap never expressly contrasted empiricism and conventionalism until months after Schlick had been assassinated, Schlick seems to have perceived this difference in their respective philosophies all along. If the issue in the protocol-sentence controversy concerns the question of how language relates to experience, then this issue must be understood in terms of the respective roles of conventionalist and empiricist precepts in a genuinely scientific philosophy.

Notes

Earlier versions of this essay have been circulating for more years than I can remember, much less care to admit. As a consequence, I have benefited from helpful suggestions and criticisms from a number of people, too numerous to recall or enumerate, whom I would like to thank for their assistance. The errors that remain are, of course, mine. Research was funded by the National Science Foundation and a Clemson University Research Grant.

1. Uebel treats only Neurath's rejections of correspondence and renders his grounds deeply philosophical. It should be noted, however, that Neurath's criticism, as recounted by Uebel, expressly depends on formulations of the correspondence account in vogue circa 1930 (Uebel 1992, 86–88). I regard Neurath's naturalism as more generally opposed to any logical doctrine of truth as a prioristic.

2. After seeing the German manuscript of "On 'Affirmations,'" Carnap noted, in a letter to Neurath, Schlick's claim to have defended physicalism in his *General Theory of Knowledge* and wondered why none of the physicalists had noticed this before (ASP 029-09-66, p. 1). In reply, Neurath scoffed, granting only that Schlick might have vaguely anticipated physicalist sentiments but had opposed his own and Carnap's detailed and precise formulations of the position (ASP 029-09-65, p. 2). Carnap, however, leapt to Schlick's defense, first by giving Neurath (in a prompt reply) the page numbers of Schlick's *General Theory* where physicalism was defended and then, in a subsequent letter, by explaining why Schlick's early discussion was not merely a "vague anticipation" of current physicalism but the thesis of physicalism itself (ASP 029-09-61, p. 1, and 029-09-54).

3. Of course, it is unlikely that it would be *grammatically* incorrect to doubt an experimental response. But cf. n. 14 below.

4. Carnap unambiguously endorsed the apparently contradictory ideas that (1) protocols are indubitable; (2) protocols are translatable into physicalistic language; and (3) no physicalistic statement is indubitable or infallible. Coffa regards this as evidence that Carnap's views on protocols were, by 1932, "already unstable" (Coffa 1991, 358). But Coffa overlooks the fact that Carnap clearly thought that the translation of a protocol need not share its "logical form," including whatever epistemological features it may possess. Although a protocol may be incorrigible and indubitable, its proper translation into physicalistic language need not be. Once this feature of Carnap's understanding of the

translation of protocols is recognized, the apparent inconsistency in Carnap's early views evaporates. Cf. also Uebel 1992, chap. 7.

5. Carnap's "pragmatic" interpretation of protocols and his rejection of their semantic analysis are consequences of his *syntacticism,* the idea (expressed in the metalogic thesis) that the syntactic analysis of language is wholly sufficient for logical and philosophical purposes (Oberdan 1990, §2).

6. One criticism of Schlick's early dichotomy of concepts and intuitions addresses the question of how scientific theories, couched in terms of formal concepts, are to be related to intuitions for the purposes of testing and confirmation. Although this is not the place to assess the merits of this criticism, it is obvious that Schlick expected the notion of a convention to provide the solution to the problem of the link between concepts and intuitions (Friedman 1983, 507–12; Coffa 1991, 172–79).

7. The thesis of the incommunicability of contents lingered in Schlick's philosophy until 1933 (see Schlick 1987, 143). But the repudiation of the thesis is an important consequence of the analysis of solipsism Schlick developed in "Meaning and Verification," and its import in Schlick's later philosophy is nothing like its earlier significance.

8. Although the conventionalism of the Quine-Duhem thesis is reminiscent of the viewpoint Schlick outlined in his early exposition of relativity theory, the differences should be carefully noted (see Howard 1984, 617–18).

9. "Meaning and Verification" was composed during the summer and fall of 1934; Schlick first wrote Carnap about it in June and reported, on the first of November, that it was completed (though needed to be shortened) (ASP 029-28-10, p. 2 and 029-28-04, p. 1). Both regarded it as an essential amplification of the essay "On the Foundations of Knowledge." The same may well be said of two other essays Schlick wrote at this time, especially "On the Relation between Psychological and Physical Concepts" and "Are Natural Laws Conventions?" (Schlick 1979b [1935a]; Schlick 1979b [1936b]).

10. As Schlick readily acknowledged, much of the inspiration for his argument against solipsism was due to conversations with Wittgenstein and reading the *Blue Book,* as recounted by Peter Hacker (Hacker 1986, 218ff.). But what Hacker largely overlooks is that, unlike Schlick, who sought to generalize the analysis of pseudotheses exemplified in the dissolution of solipsism, Wittgenstein was only interested in questions about the grammar of phenomenal discourse, or "phenomenology" as he called it in *The Big Typescript,* where he devoted §§94–100 to the subject. Perhaps because of this, Hacker also misses the unity and coherence that Schlick's argument possesses and that Wittgenstein's — if it can even be called an "argument" — lacks (Wittgenstein 1933, 88–96; 1958, 63–70; cf. also Moore 1954–55).

11. This suggests two important points. The first is the radical difference between Carnap's conception of philosophy ("the logic of science") as an a priori discipline consisting of specific doctrines and Schlick's conception of philosophy as the activity of explaining and elucidating meanings. About this matter, much needs to be said, especially concerning the history of the philosophical function of "elucidations" in the thought of Frege, the early Wittgenstein, and the Vienna Circle (cf. Oberdan 1992, §1). Second, Carnap's analysis of pseudo-object sentences could not survive the demise of the *Syntax* program, and, indeed, he did not formulate a suitable replacement until the 1950 essay "Empiricism, Semantics, and Ontology" (Carnap 1937, 233–40; and 1956 [1950], 206–13). Schlick's conception, on the other hand, was fully integrated with an understanding of language that was independent of syntacticism.

12. Nor was Schlick's the first "stone-kicking" rejoinder to the physicalists' claim that propositions cannot be compared with the facts. When Wittgenstein first read "Die physikalische Sprache . . . ," he nearly choked. "Of course there is a confrontation of propositions

with reality." If someone says, "There are six chairs in this room," his statement can be compared with reality by looking in the direction of the chairs and counting them (Waismann 1967, 209).

13. Y. Bar-Hillel has especially emphasized the difficulties concerning the immediate applicability of Carnap's *General Syntax* to natural languages (Bar-Hillel 1963, esp. 123). What Bar-Hillel calls "pragmatics" is just that domain of the analysis of language in which Schlick's application rules come into play.

14. The idea behind Schlick's construction of affirmations is that the possibility of error may be reduced by the substitution of demonstrative expressions for nonindexical ones. Thus the affirmation "Here now white" is less susceptible to error than the corresponding physicalistic protocol, "A white spot appeared on the screen at such and such a time and place," simply because there are fewer points at which failure of reference or description can occur. The meaningful use of a demonstrative depends on successful reference, and it then follows that the nonhypothetical character of affirmations can only be guaranteed by the elimination of all nondemonstrative expressions. A fortiori, the form "Here now this" is even less vulnerable to error than "Here now white"; indeed, any possibility of error arising through misdescription of the observed color would be excluded. Schlick actually argued that the possibility of this type of error does *not* show that affirmations are hypothetical. Schlick's idea was that, since the error is one of linguistic usage — the speaker deviated from normal usage — it does not follow that affirmations are therefore dubitable. There is, however, little reason to characterize such error as "linguistic," rather than an empirical misidentification, apart from Schlick's insistence to the contrary. Consequently, any attempt to make sense of Schlick's account of affirmations must disregard it and consider the results of construing affirmations as statements containing *only* demonstrative expressions. Then the proper form of an affirmation would be "Here now this," where the constituent expressions respectively indicate the place and time of the utterance and some item of the immediate environment (presumably a material thing or property thereof). This form would guarantee the nonhypothetical character of affirmations, for no doubt could ever be attached to such an assertion: it is indubitable that the identified item is present, or else it could not have been indicated.

But the incorrigibility and indubitability thus established are hardly enough to establish Schlick's claims about the nonhypothetical character of affirmations. For the incorrigibility and indubitability of affirmations — construed as constructed only from demonstratives — obtain only for the speaker at the time of the utterance. For that matter, a statement constructed exclusively from demonstratives could never be used to communicate anything. The element of certainty thus isolated at the foundations of knowledge is purely subjective and does not attach to any statements that could be used for intersubjective communication.

Joia Lewis Turner

Conceptual Knowledge and Intuitive Experience: Schlick's Dilemma

> Nothing harms inquiry so much as the pronouncing of an *ignorabimus.*
>
> Moritz Schlick 1925

A number of scholars have recently found it increasingly difficult to locate the so-called received view in the original work of the Vienna Circle philosophers. Honest examinations of the written work of Schlick, Carnap, and Neurath, for example, fail to show unbending allegiance to the picture of triumphantly verified scientific claims standing on unchallenged bits of sense-information, towering over the rubble of discarded metaphysical speculations. I have argued elsewhere (1988, 1989, 1990) that Schlick in particular was not only clearly opposed to what he called the "strict positivism" of Ernst Mach, with its rigid, sensation-driven constraints on scientific theories, but remained throughout his life devoted to a type of realism that acknowledged the continual tinkering and adjusting of scientific theories in the light of novel "affirmations" of nature's ways. The fact that we are able to do this at all was an *empirical* reality for Schlick, a glorious and contingent fact that made the human quest for understanding as rewarding in its possibilities as it was humbling in its limitations. While it is indisputable that Schlick harbored a positivist's impatience with "intuition-based" philosophical claims, his unabashed delight in working through the philosophical implications of new scientific theories, such as relativity theory, makes it difficult to avoid the conclusion that he was dedicated to finding *sounder* solutions to the very problems he thought were being so poorly dealt with by other philosophers, rather than rejecting the business of answering philosophical questions out of hand. There is, however, a profound tension running through Schlick's philosophical work, which reveals both the extent to which he took his philosophical mission seriously and the difficulty of integrating his optimism about science with the age-old frustrations of establishing an epistemological foundation for that scientific knowledge. His insistence on the possibility of a purely conceptual, nonintuition-based bank of knowledge claims was clearly in conflict with the empiricist agenda that required the source of such claims to be in the raw data of sense-experience. Rather than assume that this could and did occur in some facile and uninteresting manner, Schlick tried on several occasions to demonstrate the type of relation that must be assumed for conceptual knowledge to be based on intuitive experience. Yet one of the ideas most strongly defended by Schlick throughout his life was a complete separation between conceptual knowledge and intuition. This was also an important part of the realism he supported, that only an implicitly defined conceptual system could represent objective reality, fully free of the subjective and inexpressible elements of sense-experience.

The separation of the intellect from the senses has been a pivotal issue throughout the history of Western philosophy. Generally the separation has been between two *types* of knowledge, one considered the better and the other the lesser sort. Philosophers have usually leaned toward one of the following claims: either our vague, sensory impressions compare unfavorably to abstract, precisely definable ideas, or the simplicity and immediacy of our sense-experiences show up the rest of our knowledge claims as mere hypotheses. Schlick's distinction between concepts and intuition differs, however, in that he refused to accept the notion of intuitive knowledge at all. It was, for him, a contradiction in terms. His distinction was therefore between knowledge and not-knowledge, rather than between two grades of knowledge.

This kind of distinction is not incompatible with certain forms of rationalism, in which genuine knowledge comprises *only* abstract concepts or ideal forms. How then does Schlick effect this division as an empiricist who believes in the sensory foundation of conceptual knowledge? It is evident from his earliest writings that he was concerned with the fact that knowledge was somehow "ultimately grounded" on intuition, though he was at least initially much more interested in separating the two than in relating them to each other. It becomes clear, however, that he spent a great deal of his philosophical time and energy working out consequences of this extreme position that would fit his beliefs about the nature of empirical knowledge.

I have isolated three of Schlick's attempts to deal with the relationship between conceptual knowledge and intuition. The first comes from his early writings on relativity theory (see Schlick 1915 and 1917 in vol. 1 of his *Philosophical Papers* [1979a]). Schlick here described how our scientific and objective conceptions of space and time result from our subjective experiences. The second account comes from his later writings on form and content. Following Wittgenstein's emphasis on the *communicable* versus the *incommunicable,* Schlick described how our publicly verifiable knowledge encompasses, while not strictly including, our individual and private sense-experiences. The third attempt of Schlick's to deal with the relationship between knowledge and intuition that I will look at comes from his theory of *Konstatierungen* (affirmations) as the foundation of empirical knowledge.

It is important first to look at the historical context of Schlick's epistemological work and particularly at the views of Hermann Helmholtz and Max Planck, both of whom influenced Schlick a great deal.

1. Historical Antecedents

Alberto Coffa noted that "Schlick was the first to attempt a systematic formulation of the picture of knowledge implicit in Helmholtz's writings" (Coffa 1991, 171–72).[1] Helmholtz had demonstrated that mid–nineteenth-century science supported Kant's fundamental assumptions. Research in physiology, particularly by Johannes Müller, showed that stimulation of a nerve resulted in a sensation specific to that nerve, regardless of the type of external cause involved in the stimulation. Kant's distinction between the phenomenal and the noumenal was vindicated here: there was no way to infer the exact nature of the external causes of our internal experiences. Similarly, research in physics, including Helmholtz's own work on the conservation of force, validated the existence of a mind-independent, mechanically regulated reality, shedding light on Kant's noumenal realm as physiological research shed light on the phenomenal realm.

According to Helmholtz, our sense-perceptions bring us information of the lawful connections in nature but do not perfectly reconstruct the external world. Rather, as signs of changes in the external world, they are images only in the sense that they represent succession in time (Helmholtz 1869, 243). We get at the formal properties of the otherwise unknowable, noumenal realm through the "code language" of our perceptions. Here is Helmholtz's link between knowledge and reality: "If the same kinds of things in the world of experience are indicated by the same signs, then the lawful succession of equal effects from equal causes will be related to a similar regular succession in the realm of our sensations" (Helmholtz 1878b, 385).

Schlick also accepted the main features of Planck's physical realism.[2] These features were: (1) that mature science should lose its originally anthropocentric character, and (2) that a unified world-picture, in purely quantitative terms, was the goal of science. Planck began with Helmholtz's picture but then insisted that we give up our original perceptual formulations of the formal element in favor of purely quantitative, nonanthropocentric terms.

Schlick's earliest epistemological writings show his interest in this progression from the qualitative to the quantitative.[3] For example, as science progresses we learn "to speak quantitatively of frequency instead of alluding to qualitative differences of pitch" (1979a, 31). This progression from the qualitative nature of information grounded in sense-experience to the quantitative concepts of science was not just an exercise in economical descriptions of nature, however.[4] Rather, in Schlick's words, "our knowledge seems thereby to have penetrated more deeply into the areas of reality in question, to have detached them from sensory intuition and its contingencies, and to have presented the world of objects in greater independence of an apprehending consciousness" (ibid., 26).

Using scientific concepts, rather than inner experience, to get to reality was in no way universally accepted among philosophers, then as now. Helmholtz had written of the unhappy relationship between philosophy and the natural sciences during the first half of the nineteenth century, due to the influence of the philosophies of Schelling and Hegel: "Basically the cause of the trouble was the profound antithesis between the methods used to justify [each of the] positions.... The discord, however, did not continue long with its initial bitterness. Through a rapid series of discoveries the natural sciences proved to everyone that they contained a healthy seed of exceptional fertility" (Helmholtz 1894, 513). By 1911 Schlick was able to write that the current hostile attitude toward philosophy was "psychologically explicable as a residue from the period during the last century when the special sciences... had to defend themselves against the pretensions of philosophers" (1979a, 104). Schlick devoted many passages in his *General Theory of Knowledge* (1918; 2d ed. 1925; trans. 1975) and early articles to criticism of what he called the "intuitionist philosophies" of Bergson and Husserl, among others. It is not surprising that Schlick would fight antiscience sentiments, given his background in physics and his interest in developing a new epistemology consistent with contemporary science; what is surprising is that Schlick would also find fault with scientifically oriented philosophies. In particular, Schlick directed a great deal of criticism toward the positivism of Mach. According to Schlick, neither reality nor knowledge should be restricted to the given, as Mach proposed.

Both Helmholtz and Planck had criticized their positivist colleagues as well. According to Helmholtz, the positivist request for *description* was not sufficient. Thus, Kirchhoff had delineated the task of mechanics as describing completely and in the simplest possible way the movements occurring in nature. Helmholtz felt that a mere description of the relations among the phenomena ignored the genuinely occurring lawfulness in nature: "I would add to the position expressed by [Gustav] Kirchhoff that the most complete and simplest description can be given only when we formulate the laws which lie at the foundation of phenomena" (Helmholtz 1894, 523).

Further investigation can help us see that "the occurrence of such a phenomenon is dependent only upon certain objective conditions and is completely independent of occurrences in our mind" (ibid.). Likewise, Planck had also written that Mach's restriction of reality to sense-impressions was unwarranted: more than *empirical adequacy* is required if science is to reflect "real natural events which are quite independent of us" (Planck 1909, 23). Planck also had even remarked that Mach's philosophy is alien to scientific research, which demands a *"constant* world-picture, independent of changing times and people." Coffa sums up the difference between these two views as follows:

> Planck's picture of science was the mirror image of that of Mach and other positivists. For Mach, truth in science is entirely a matter of sense and sensibility, and the farther science goes in the direction of insensible inhuman postulation, the farther it departs from truth and reality. For Planck, on the other hand, the more closely attached we remain to human experience, the more likely it is that the picture science is offering is a distorted guide to reality as it is when not experienced. (Coffa 1991, 182)

Schlick followed Planck in criticizing any kind of theory that stopped with relations found among sense-experiences. In fact, very often Schlick gave his reasons for making the distinction between knowledge and intuition as being important primarily by *"removing all prejudices from philosophy against science"* (1979b, 324). This is a criticism of any philosophy that elevated intuition above the conceptual nature of scientific knowledge.

Schlick described the difference between knowledge and intuition in his *General Theory of Knowledge* as follows: only the former involved the process of coordinating concepts into systems of interlocking judgments. If *x* is *known,* in other words, we must be able to place it within a conceptual system. If we "merely experience" *x,* on the other hand, we confront an object without comparing or relating it to anything else.

In making this distinction, Schlick was most concerned with identifying science with the precise nature of genuine, conceptual knowledge and divorcing it from the vague and ephemeral nature of momentary sense-experience. The problem is, of course, that while science does indeed rely on the most sophisticated conceptual schemes that we invent, it also requires the constant input of our sensory apparatus in order to develop, test, and compare empirical theories. How could Schlick so devalue the sense-experience that provides us with the ability to verify scientific theories?

One way to make sense of this is to say that Schlick simultaneously held two views of sense-experience: a "good" sense, which he identified with *verification* (whatever caused us to make perceptual judgments in verifying a hypothesis), and a "bad" sense, which he identified with *intuition* (whatever occurred when we take in our surroundings in an apparently unconscious manner). Since he never admitted to holding two distinct definitions of sense-experience, it is fairer to say that Schlick felt the tension in his position and was concerned on several occasions to attempt to bridge the gap that he had wedged between concepts and intuition, as we will see.

Coffa considered Schlick's joint acceptance of these two views to be the major problem in his early philosophy: Schlick mistakenly equated Helmholtz's formal element in intuition with Planck's elimination of subjective content. Both Helmholtz and Planck were part of the Back to Kant movement among scientists in the late nineteenth century, but they differed in their interpretations and their development of the Kantian picture more than Schlick realized: "The divergence between these two branches of scientific neo-Kantianism is clear. Whereas, under Helmholtz's program, one aims to remove content from the given in order to derive a purely structural picture of things in themselves, under Planck's project one aims to remove the human sensibility" (Coffa 1991, 182). By equating these two processes, Schlick could conclude that removal of the subjective, phenomenal elements of sensation from our accounts of reality gives us the formal characteristics of the world. In Coffa's words, "Only if one confuses, as so many did and still do, the psychological or the phenomenal sensible content of a representation with the semantic content of a statement of fact could one draw the absurd conclusion that a picture without subjective, phenomenal elements — such as Planck called for — must also be a purely formal, contentless, structural characterization" (ibid.).

2. The Objective/Subjective Account

An early attempt to explain how abstract and precise concepts are produced from the vague, fleeting experiences of individuals can be found in Schlick's writing on relativity theory. In *Space and Time in Contemporary Physics* (1917), Schlick distinguished the various psychological experiences of extension and duration, which each individual has, from the objective conceptions of space and time used in the natural sciences. According to Schlick, we must always distinguish between the *"direct experience of duration* [which is] an everchanging intuitional factor" and *"physical time, which only signifies an arrangement having the properties of a one-dimensional continuum"* (1979a, 261). The same distinction must always hold for our notions of space as well; Schlick similarly found that we must separate "the spatial as *intuitively representable extension* and the spatial as a *system of ordering natural objects,* achieved with the aid of pure concepts" (1975, 251). The resulting picture, then, is as follows: "In conjunction with objective time, physical space is designated by the four-dimensional scheme ... which in mathematical language can simply be treated as the manifold of all number quadruples x_1, x_2, x_3, x_4" (1979a, 261). As in the case of knowledge and intuition in general, it is easier to state the differences between objective and subjective conceptions of space and time than their relationship to each other. In this case, however, Schlick stated clearly that our objective conceptions of space and time are *grounded* on our subjective intuition and explained this relationship in terms of the "uniform functional relations" found among our differing subjective conceptions of space; these relationships then give us the structure of objective space. According to Schlick,

> These spaces are essentially dissimilar and incapable of comparison with one another; but they have, as our experiences teach, a perfectly definite uniform functional relation to one another. Tactual perceptions, e.g., correlated themselves with visual perceptions. A certain correspondence exists between the two spheres; and through this correspondence it is possible to arrange all spatial perceptions into one scheme, this being just what we call objective space. (ibid., 262)

It is interesting to compare Schlick's account with Helmholtz's claim that our senses are signs of reality. While Helmholtz's picture required a phenomenal realm that provided us with a minimal but nonetheless direct indication of the structure of the noumenal realm, Schlick's account has a proliferation of untrustworthy subjective views of objective reality. We must then work to find the structural relations among these differing views in order to find the abstract features of the real world. Here again we see a result of applying Planck's devaluation of the senses to Helmholtz's picture. However, Schlick's account was also meant to reinforce the existence of a mind-independent reality, and he emphasized that even though there are many subjective experiences of space-time, still there is only one objective world: "The identically same objects of this world are found again in the most varied relations to elements of the world of consciousness. For the concepts that are applied to transcendent objects are defined by means of relations to or correlations with the given" (1975, 276). How exactly do we correlate our various perceptual spaces in such a way that we obtain the structure of objective space? Schlick was eager to show that his account of the relationship between the subjective and objective was corroborated by Einstein's work: we "see that we encounter just that significance of space and time which Einstein has recognized to be essential and unique for physics" (1979a, 263). It was Einstein's "coincidences of events" that Schlick was most interested in here; they provided a method of constructing objective space out of subjective experiences:

> In order to fix a point in space, we must in some way or other, directly or indirectly, *point* to it: we must make the point of a pair of compasses, or a finger, or the intersection of cross-wires, coincide with it. . . . Now these coincidences always occur consistently for all the intuitional spaces of the various senses and for various individuals. It is just on account of this that a "point" is defined which is objective, i.e., independent of individual experiences and valid for all. (ibid., 262–63)

Schlick focused here on the particular issue of the spatial coincidence of points, but this is at best an incomplete answer to the question of how we are able to correlate all of our subjective experiences with one another and arrive at objective conceptions of space-time that are accessible and confirmable by all observers. This account relies both on the existence of 'uniform functional relations' among our subjective experiences and on the possibility of determining coincidences of events. It is interesting to note here that much later, Schlick published a paper (Schlick 1935a, in 1979b) in which he drew again on this notion of coincidences. In this case he expressly broadened the project to relating the coincidences to *all* other events ("pains, colors, pleasures, memories," and so on [1979b, 424]). Since physical properties are *measurable,* it followed for Schlick that "every measurement springs from a counting, and can in the last analysis always be traced to a numbering of 'coincidences,' where by a coincidence is to be understood the spatial coming together of two previously separated singularities of the visual or tactual fields" (ibid., 424). Schlick went on here to invoke the by now familiar *thesis of physicalism* but was careful to point out that the fact that psychological propositions could be translated into expressions containing solely physical concepts was an empirical, factual claim rather than a philosophical discovery. For Schlick, the translatability of psychological propositions followed from the fact that the objectivity and universality of the physical language rested on the "data of experience" (ibid., 427). In other words, the systematic relationships among our experienced 'coincidences' and all other events *determined* the world of experience for us. For example, relating the physical and psychological concepts of color, Schlick stated that "every difference in colour quality goes hand in hand with a difference in the optical segment of the nervous system" (ibid., 428). We group resembling properties by "intensive similarity" and attach a common name, such as a color, to these experiences. Incorporating visual and tactual coincidences into intersubjective space then produces the physical, and the objective ordering of these coincidences *is* simply the physical space-time order (ibid., 425). Schlick even makes the "coincidences" the sine qua non of physical concepts: "[T]he essential feature of physical concepts is that they are arrived at by selecting out of the infinite variety of events a special class, namely these 'coincidences,' and describing their inter-relationships with the help of numbers. Physical magnitudes are identical with the number-combinations which are thus arrived at" (ibid., 425–26).

3. The Form/Content Account

After Schlick's move to Vienna in 1922, he began to focus on two new distinctions, both of which he identified with his distinction between genuine knowledge and mere intuition. The first distinction was between form and content; the second, between what is communicable and what is incommunicable.

Schlick first discussed these two new distinctions in an article entitled "Experience, Cognition, and Metaphysics" (1926, in 1979b). Schlick stated here that what makes knowledge communicable at all is that it is "by nature addressed to pure form" (1979b, 103). What we are able to communicate to each other therefore are only the formal aspects of our system of knowledge. What is not communicable, on the other hand, are the nonformal aspects of knowledge, or what is privately "known only through immediate experience" (ibid., 99). Since experience is content, according to Schlick (ibid., 103), this new distinction of form versus content coincided with what is communicable and what is not:

> The differences between structure and material, between form and content, is, roughly speaking, the difference between that which can be expressed and that which cannot be expressed. The fundamental importance for philosophy of that which is vaguely indicated by this distinction can hardly be exaggerated. We shall avoid all the typical mistakes of traditional philosophy if we bear in mind that the inexpressible cannot be expressed, not even by the philosopher. (ibid., 291)

Once again, it is important to keep in mind that Schlick was always eager to find new ways to separate an acceptable scientifically based philosophy from a speculative metaphysics.

In the objective/subjective account, Schlick focused on *how* concepts are produced from intuition; in the form/content account, he focused on a different aspect of the relationship between the two: "[T]he fact that intuition, immediate awareness, or, as we should rather say, the mere presence of content, is indispensable for all knowledge . . . has no significance whatsoever, for it is indispensable for everything; it is the ineffable ever present fundament of all else, also of knowledge, but this does not mean that it is itself knowledge" (ibid., 322). Schlick now identified intuition with a much broader notion, a generalized "content" that is indispensable for all knowledge but only by virtue of the fact that it is indispensable for everything. In the preceding quote, he also identified structure versus material with form versus content. Is he really still dealing with one and the same distinction in all of these variations?

Schlick constructed the list shown in figure 1 to help us recognize the various terms he used to distinguish between what does and does not constitute genuine knowledge (ibid., 324). He emphasized immediately after this that "the main result of this discussion is that it clears the air of all prejudices against scientific knowledge and its method" (ibid., 324).

Schlick gave several examples to illuminate his account of form versus content. One concerns the communication of a claim about the color of an object, such as "This leaf is green." Since what can be communicated involved form only, and not content, we can only express the "logical structure" of the green color. Specifically, the "logical structure" of a claim about an object's color, Schlick says, is given by the internal relations "which determine [a color's] place in the system of qualities" (ibid., 294). The system of qualities in this case is the color spectrum of white light: "Evidently it belongs to the intrinsic nature of our green that it occupies a definite position in a range of colours and in a scale

Figure 1

Intuition	*Knowledge*
only one term	two terms
enjoyable	useful
living	thinking
presentation	explanation
acquaintance	description
inexpressible	expression
that which is ordered	order
content	form

of brightness, and this position is determined by relations of similarity and dissimilarity to the other element (shades) of the whole system" (ibid., 293). Similarly, all of our sense-experiences belong to systems of qualities: "In this way every quality (for instance, the qualities of sensation: sound, smell, heat, etc., as well as colour) is interconnected with all others by internal relations which determine its place in the system of qualities. It is nothing but this circumstance which I mean to indicate by saying that the quality has a certain definite logical structure" (ibid., 294). Physical structure here would appear to be like Helmholtz's noumenal structure.

What cannot be communicated, on the other hand, is what Schlick called the "ineffable quality" of the greenness itself. This comes under the heading of content and can be supplied only by the listener. Each perceiver is responsible for "filling in" the communicated logical structure with privately experienced sensations. Schlick goes on to say that even if the communication is between a seeing and a blind person, the situation is the same: the blind person will fill in the inconceivable reference to a color with a memory of some sort of content from his own experience, possibly drawing on one of the other senses. In other words, a congenitally blind person could fill in the communicated logical structure with a remembered sound rather than a color. (We are all, then, in the position of a blind person when spoken to.)

It is difficult to know exactly how to understand Schlick's form, or "logical structure." In this last example, he is not referring to Helmholtz's noumenal structure, which may be revealed to us in the equations of theoretical physics; the color spectrum of white light, which provides the color green with its niche in the "system of qualities," is not going to account for much in a blind person's experience. On the other hand, Schlick cannot make logical structure dependent on an individual's perceptual abilities, since no communication would then be possible at all. Schlick's structure has to do instead with the agreed upon symbols and rules by which we communicate with each other. It is "located" in the public domain, not in objective reality. It is an intersubjective notion.

For our present purpose, however, we may note that Schlick is not trying to *ground* concepts on intuition in this account. Rather, he has described their very separate roles in the process of communication in general. The major problem with this account is that Schlick makes it very hard to distinguish between purely mathematical theories and empirically interpreted theories (see Friedman 1983). Schlick asks, "What is the stuff which must be added to the empty frame of a deductive system in order to make a *science* of it?" and he answers, "There seems to be but one possible answer to this question, namely, 'the purely formal structure must be filled with *content*—it could not be anything else, because there is nothing else' " (1979b, 331). But if the content with which we fill in the communicated structure is private, we cannot then communicate the difference between an uninterpreted formal system and an empirically interpreted one. Schlick was ambivalent about the linguistic status of private content and continued to be in the next account as well, as we shall see.

4. The Foundational Account

Schlick's last attempt to bridge the gap between conceptual knowledge and intuitive experience comes from his most original work, his theory of *Konstatierungen.* These "affirmations" of reality were to function as a novel kind of foundation of empirical knowledge and play a role as well in verification. We have seen Schlick's concern with the grounding of concepts on intuition in his objective/subjective account and also his attempts to describe the separate roles of concepts and intuition in the communication process. It was in the last few years of his life that he attempted a more detailed explanation of what actually occurs at the moment in which sense-experience elicits a judgment from us. As is well known, Schlick's colleagues, Carnap and Neurath, had retreated from a position asserting the incorrigibility of basic empirical statements to a position asserting the conventional choice of basic statements. Schlick agreed with his colleagues that all statements of science were hypotheses and therefore refutable. But he did not agree that there was no way to isolate a particular set of fundamental propositions. The only way he could hold these apparently contradictory views was to suggest that these fundamental propositions were not part of science. Schlick's affirmations of reality are the *"only synthetic propositions which are not hypotheses"* (ibid., 387). They have in common with analytic propositions the property of being understood and verified simultaneously. They are epistemically privileged by virtue of the fact that they are asserted by *oneself* and *in the present.* Further, they are never selected by an individual. They are, rather, *the* set of propositions that each individual naturally counts on as incorrigible. They have the "positive value of absolute validity, and the negative value of being useless as an enduring foundation" (ibid., 386). And, according to Schlick, it is *Konstatierungen* that connect science with reality: all scientific statements are hypotheses, but *Konstatierungen* are the "unshakable points of contact between knowledge and reality. . . . In no sense do they lie at the basis of science, but knowledge, as it were, flickers out to them, reaching each one for a moment only, and at once consuming it. And newly fed and strengthened, it then flares on toward the next" (ibid., 387).

We may recall here the "good"-verification and "bad"-intuition aspects of sense-experience: here Schlick seems to allude to these in giving the positive and negative values of *Konstatierungen.* It is also important once again to remember the context of Schlick's writing. He was accusing his colleagues of relativism, of rationalism, and most of all of promoting a coherence theory of truth, all of which were extremely negative positions, for Schlick.

It is clear that, as statements, Schlick's *Konstatierungen* fall prey to all of the problems inherent in his colleagues' original definition of protocol propositions. In particular, they will be translatable into the corrigible statements of science and will thus lose their incorrigibility. Although Schlick referred to his *Konstatierungen* as statements, however, he also gave them nonlinguistic properties: "It is clear what role is played in [the actual procedure of science] by assertions about the 'immediately perceived.' They are not identical with statements written or remembered, i.e., with what could properly be called 'protocol propositions,' but are the *occasion* for framing them" (ibid., 381). If Schlick's affirmations are understood as nonlinguistic entities, purely as parts of the perceptual process, then at the very least his position is consistent. They are not part of science by virtue of the fact that they are not part of language at all. In fact, the separation of concepts from intuition precludes the possibility of *Konstatierungen* being part of language. There remains the ambiguity about whether Schlick's foundational units were meant to have strictly linguistic or nonlinguistic properties. But if we assume that Schlick's focus was still on attempting to explicate the relationship between what is *experienced* and what is *known,* rather than on whether his foundational units had linguistic properties, then his lack of discussion on this point is understandable. There is a passage in the *General Theory of Knowledge* in which he criticizes contemporary thinkers like H. Münsterberg and A. Messer for confusing what can be *known through experience* with what *can be experienced* (see Münsterberg 1900, 72; Messer 1901, 121). Münsterberg had defined the mental as that which can be experienced only by a single subject and the physical as that which can be experienced in common by several subjects (see Schlick 1975, 297). What is wrong with this, according to Schlick, is an equivocation on the expression "can be experienced": "experience" with respect to the private domain cannot be equated with intersubjective agreement in the public domain. Schlick noted also that Mach had attempted a similar distinction between what is immediately present for everyone (the physical) and what is immediately given to just one person (the mental) (see Mach 1905, 3d ed., 6), but, Schlick noted, "on this definition, there is absolutely nothing that corresponds to the physical" (1975, 298). Schlick was here promoting a view he called "epistemological monism," which was meant to counter what he referred to as the various "interactionist positions" on the psychophysical problem.[5] There was no reason, according to Schlick, to suppose that the system of quantitative concepts must fail in regard to the given world of qualities known by direct acquaintance. To tie this into the present discussion, it may be that Schlick's later foundational units were *intended* simply to be the interface between what was commonly considered to be the psychological and the physical, the subjective and the objective, the private and the public, and even the analytic and the synthetic. If Schlick was genuinely "monistic" about all of these types of distinctions, then his *Konstatierungen* would therefore be expected to appear ambiguous with respect to these familiar dichotomies.

It may be instructive to note here that in his 1935 paper on the relation between psychological and physical concepts (written the year after his article on the foundations of knowledge), Schlick wrote that we are often confused by what we mean by the "spatial"; while ideas and perceptions may have to do with what is spatial, they are themselves nonspatial.[6] But the reason we make assertions like this is, according to Schlick, because the words "perception" and "idea" are themselves ambiguous: "By them one can refer either to the content, that which is given [*une donnee actuelle*], or to the event, the act of perception, which is characterized as a 'mental process' and concerning which there is indeed no question of 'extension' " (1979b, 422). Schlick was clearly still frustrated with this apparent ambiguity between the content of the idea and the act of perception. His goal in this later paper was, however, to emphasize the *empirical* rather than the logical character of the intersubjective agreement we share with respect to physical concepts.[7]

•

We have seen in each of Schlick's accounts of the relationship between the conceptual and the intuitive that while his determination to make the distinction may have outweighed his success at clarification, in each case he was very concerned to show the inadequacy of philosophies that ignored the objective conceptions of science. However, perhaps his inability to clearly explicate the distinction stems from deeper reasons.

First, what is the distinction between concepts and intuition a distinction *of*? When philosophers downgrade the kind of knowledge found in one in favor of the other, they make a distinction about types of knowledge. But Schlick's distinction, as we have seen, is a distinction between knowledge and not-knowledge. We must then look to another domain to find the category for this division. If concepts and intuition do not share the property of knowledge, what do they share? We need to look for some more general property of human information processing that would include nonlinguistic as well as linguistic aspects of cognition.

Second, perhaps Schlick's ambivalence about whether or not the subjective and intuitive should be granted linguistic status is due to the fact that there simply is no good way to determine at what point linguistic activity begins or ends in the cognitive process. Perhaps research will bring us new categories, and old distinctions will drop out of the picture. If this turns out to be the case, then Schlick's ambivalence about language and his unsuccessful attempts to clarify this distinction would be, in a certain sense, vindicated. The view that cognition must be seen in a context wider than language use is forcefully promoted today by both philosophers and scientists, particularly those interested in a long-term evolutionary story. Patricia Churchland has noted that "sentence-crunching"

> is certain to have been a cognitive latecomer in the evolutionary scheme of things, and it must have knit itself into the preexisting nonsentential cognitive organization, or . . . evolved out of preadaptive nonsentential structures. To be sentence-crunching "all the way down" implies either that cognition must have been sentence-crunching "all the way back," which is implausible, or that sentence-crunchers have no cognitive heritage from earlier species, which is also implausible given the evolution of the brain. (Churchland 1986, 388)

Paul MacLean, who has published extensively on the "triune brain," has found that the neocortex apparently relies on the earlier evolved neomammalian or limbic system of the brain for not only a

> sense of self, of reality, and the memory of ongoing experience, but also a feeling of conviction as to what is true or false. This presents a problem of crucial epistemological significance because there is no evidence that the limbic structures of the temporal lobe are capable of comprehending speech, nor is there any basis for inferring a capacity to communicate in verbal terms. Hence, it would appear that the manufacture of belief in the reality, importance, and truth or falsity of what is conceived depends on a mentality incapable of verbal comprehension and communication. (MacLean 1990, 578–79)

This type of research would support Schlick's efforts insofar as he meant to *describe* the process of "affirmation" as accurately as possible and would undoubtedly suggest new conceptual frameworks for discussion as well. We should not forget that Schlick and his Vienna Circle colleagues knew their science. They came to questions of logic and language with a thorough knowledge of the scientific theories of their day, and if they were here with us today we can be sure that they would be extremely excited about current physical theories as well as current research in the neurosciences. They would be looking for ways to inject new theories into old epistemological problems, and they would not be afraid to let go of distinctions that no longer served any purpose. Schlick would be the first to applaud our efforts to do the same. He cared deeply about both philosophy *and* science, and in this we may find in him the quintessential philosopher *of* science: doing justice to our most sophisticated physical theories *meant,* for Schlick, that they would figure prominently in the answers to our philosophical questions. For Schlick, an *ignoramus* (we do not know) in the scientific realm need not entail an *ignorabimus* (we shall not know) in the philosophical. And it would be very much in the spirit of his work to allow a new scientific theory to lead us to a sounder epistemology.

Notes

1. Schlick gave a lecture entitled "Helmholtz the Epistemologist" in 1921 (in Schlick 1979a), and he also coedited, along with Paul Hertz, an edition of several of Helmholtz's most important works.

2. Schlick completed his dissertation on the reflection of light in a nonhomogeneous medium under the direction of Max Planck at the University of Berlin in 1904.

3. See his 1910 essay "The Boundaries of Scientific and Philosophic Concept Formation" in Schlick 1979a.

4. See my 1988 essay on Schlick's criticisms of Mach's principle of economy.

5. "Whatever is real is open to designation by quantitative concepts" (Schlick 1975, 331).

6. "The idea of a red triangle is itself neither red nor triangular" (Schlick 1979b, 421–22).

7. He asked his readers here to imagine a world where intersubjective agreement occurred in the psychological rather than the physical realm. Inhabitants of this world would never find law-governed regularities among external events. Rather, they would achieve universal agreement in the intersensual and intersubjective formation of psychological concepts (see 1979b, 432–33). The purpose of this science-fiction exercise was, once again, to emphasize that we are only contingently and fortunately *able* to find agreement among our knowledge claims about the external world.

Alan W. Richardson

From Epistemology to the Logic of Science: Carnap's Philosophy of Empirical Knowledge in the 1930s

This essay takes its title from a paper that Rudolf Carnap gave at the 1935 Paris Congress for the Unity of Science. In that paper Carnap (1936a) invited the participants in the congress to view scientific philosophy as having entered a third stage. The first stage of scientific philosophy was the rejection of metaphysics, "the transition from speculative philosophy to epistemology" (ibid., 36). The second stage was the rejection of the synthetic a priori and the consequent adoption of empiricist epistemology.[1] Of the third stage he had this to say: "The task of our current work appears to me to consist in the transition from epistemology to the logic of science. In this, epistemology is not, as were metaphysics and *a priorism* before, completely repudiated, but rather purified and decomposed into its constituent parts" (ibid.).[2] The constituent parts were psychological and logical, which according to Carnap had been previously confusedly mixed together. The logic of science or the logical syntax of scientific language — the nascent program of logical empiricist philosophy of science — was to resolve the confusions inherent in epistemology by taking up the logical questions while assigning the psychological questions to the empirical work of psychologists. Philosophy of science was, then, epistemology purified into the wholly analytic study of the logical relations of scientific language systems.

This is a striking vision of the relation of philosophy of science to traditional epistemology, for it suggests some important properties of Carnap's philosophy of science. First, philosophy of science is not just epistemology applied to *scientific* knowledge. Rather philosophy of science is the logically acceptable replacement project for epistemology. Second, there is no room alongside philosophy of science for an analytic metaepistemology that, say, analyzes judgments of the form "*S* knows that *p*" to uncover the conditions of knowledge. The judgments that are the concern of philosophy of science aren't judgments like (1) but rather like (2):

(1) Otto knows that there is a red cube in front of him.

(2) The confirmation of theory, *T*, on evidence, *e*, is *r* ($0 \leqslant r \leqslant 1$).

Agents with propositional attitudes like belief, presupposition, or knowledge are notable by their absence from Carnapian reconstructions of the verification of scientific theories, which is the hallmark issue of empirical knowledge for the logical empiricists. Third, Carnap's call to purify epistemology in this way is his reaction to his understanding of the epistemological project of *Der logische Aufbau der Welt* (Carnap 1928a [1967]; henceforth, the *Aufbau*). Thus, his account of the lessons to be taken from the failure of this work are very different from the ones that Quine (1953, 1969) has urged. This suggests that Carnap's objections to and understanding of the project of the *Aufbau* are rather different from Quine's as well. Indeed, I think this is a key to understanding why the famous Carnap-Quine debate on analyticity is so frustrating—their conceptions of the business of philosophy are, in the end and not withstanding the notable convergence of views on many points, fundamentally at odds, and neither can understand the force of the other's claims.[3]

My purpose in this essay shall be to present in broad outline some important features of this movement from the epistemology of the *Aufbau* to the logic of science in Carnap's works from 1928 to 1937. I shall concentrate mainly on three salient documents. The first is the *Aufbau* itself, as this is the document that most perspicuously presents Carnap's vision of the task of epistemology in the first and second stages of scientific philosophizing. The second is the 1932 *Erkenntnis* piece translated as the "Unity of Science" (Carnap 1932a [1934]; henceforth, "Unity"). This essay strikes me as occupying a halfway house between the *Aufbau* and the logic of science—it is, if you will, the *Aufbau* in the formal mode of speech. Finally, I shall discuss Carnap's main early document within the logic of science, the 1936–37 *Philosophy of Science* article, "Testability and Meaning."[4]

1. The Central Tension of the *Aufbau*

My interpretative perspective on the *Aufbau* takes off from the revisionary work on it by Susan Haack (1977), Alberto Coffa (1985, 1991), Werner Sauer (1985, 1989), and especially Michael Friedman (1983a, 1983b, 1991, 1992a, 1995).[5] That is to say, I align myself with the new school of thought that finds significant Kantian elements in the constitutional project that Carnap undertakes in the *Aufbau*.[6] How can this be when the project clearly is to define all of the concepts of science on the basis of sensation — a project that seems to be the high-water mark of strict empiricism?

Interestingly, Carnap's own view of the project of the *Aufbau* is not that it is strict empiricism. Indeed, as an epistemological school of thought, the *Aufbau* scarcely mentions empiricism.[7] The empiricism that seems so evident to contemporary readers of the *Aufbau* can be subsumed under Carnap's notion of subjective idealism, which stresses the primacy of private sensation as the foundation for knowledge. Carnap's official view is, however, that the *Aufbau* is not an endorsement of subjective idealism. Rather the constitution-theoretic view of epistemology is neutral with respect to all the issues dividing this project from all other traditional epistemological schools.[8] Unlike each of these traditional schools, the constitutional project is, according to Carnap (§178), uninfected with metaphysics, and these schools only disagree with one another on these metaphysical issues. Thus, Carnap is not endorsing traditional subjectivist empiricism any more than he is any other traditional epistemological project.

This may loosen the grip of the strict empiricist interpretation of the *Aufbau,* but at the cost of making the neo-Kantian interpretation suspect as well. For transcendental idealism is among the projects that Carnap claims to be infected with metaphysics. Nevertheless, I take it that the epistemological project is more continuous with neo-Kantianism than with strict empiricism. This can be seen in the vocabulary Carnap uses to motivate the project. Moreover, Carnap's logic-based rejection of the metaphysics of neo-Kantianism can itself be seen as a development of the general line of scientific neo-Kantianism that views its project in the words of Ernst Cassirer (1923, 261) as providing "a logic of objective knowledge."[9]

Thus, at the most basic level the new interpretative school asks us to consider what questions about knowledge Carnap is asking and how the constitutional system on autopsychological basis is meant to answer them. The stress on Kantian interpretation, then, receives its force from the way Carnap motivates the project in quotations like the following:

> Since the stream of experience is different for each person, how can there be even one statement of science which is objective in this sense (i.e., which holds for every individual, even though he starts from his own individual stream of experience)? The solution to this problem lies in the fact that, even though the *material* of the individual streams of experience is completely different, or rather altogether incomparable, since a comparison of two sensations or two feelings of

> different subjects, as far as their immediately given qualities are concerned is absurd, certain *structural properties* are analogous for all streams of experience. Now, if science is to be objective, then it must restrict itself to statements about such structural properties, and, as we have seen earlier, it can restrict itself to statements about structures, since all objects of knowledge are not content, but form, and since they can be represented as structural entities. (§66)

The epistemological problem of the *Aufbau* is, then, to show how objective knowledge is possible given the subjective origin of knowledge in individual experience. This Kantian-sounding problem is given a Kantian-sounding answer in the stress Carnap puts on the notion of the form of experience. The form of experience is identical for us all and is sufficiently rich to allow the definition of the concepts of science as themselves purely structural entities. It is the logical form of experience that allows us to ascend from a purely subjective world of primitive experience to an objective world of science.[10]

The difference from Russell's empiricism in the External World Project is clear. The point is not to reduce everything back to acquaintance but precisely to show how we can get beyond mere acquaintance to objective knowledge appropriately so-called. For Carnap, sensation is primary in the causal order but not appealed to as privileged in the epistemic order. In a slogan we can say: for Russell, all knowledge is knowledge by acquaintance; for Carnap, all knowledge is knowledge by description. This is so because to fulfill the demands of quotations such as the one from §66 above, Carnap must provide what he calls a "structural definite description" for an object if it is to be an object of knowledge. Such a structural definition description uniquely picks out an object solely by virtue of its logical relations to the other objects of science. If it is possible, this avoids any and all appeal to a private but epistemically privileged relation of acquaintance or ostension.

My mission here is not to marshall evidence for this interpretation of the *Aufbau* directly but to indicate how attention to Carnap's concern with objectivity can lead to a rethinking of his development of philosophy of science. This is because if we take the problem of objectivity and its relation to logic seriously, something other than the failure of the definitional procedure outlined by Quine turns out to be the crucial problem of the work. Carnap's overarching difficulties can be seen in the following circumstance: having set up his task as showing how objective knowledge is possible starting from subjective experience, Carnap ultimately gives us two solutions to this problem. The notion of objectivity itself finds no univocal place within Carnap's constitutional epistemology.

Carnap's first attempted account of the objectivity of scientific knowledge invokes a notion of objectivity that is constructed in the system itself. This is the notion of the intersubjective world, which is, Carnap (§149) tells us, "the object domain of science." Having constructed the intersubjective world and shown how to intersubjectivize at all levels of the system, Carnap then goes back to his announced project, intimated in the quote from §66 above, of objectivity as purely structural definite description. Thus, he engages in the attempt to define away the basic relation, *Rs,* as the last nonlogical primitive in the system (§§153–55). When this is done, the system finally provides rules for translating every scientific claim into a sentence in which only logical notions appear. This translation is then the role of the system in the rational reconstruction of knowledge — with the system in hand one can exhibit the propositions of science as objective because purely structural. On this account, no objective/subjective distinction is drawn internal to the system; rather the system as a whole provides the conceptual framework that first allows the explicit display of the objective nature of scientific judgment.

There are many interesting aspects of the construction of the intersubjective world, especially in its connections to Carnap's pre-*Aufbau* writings, that we must skip over here. Some salient features of the process of intersubjectification should be noted, though. Most importantly, the construction of the intersubjective world ineliminably goes through the construction of the world of physics as a pure world of numbers — that is, as values of quantitative physical concepts at space-time points governed by mathematically expressed physical laws:

> The perceptual world is constructed through the assignment of sense qualities; from it we must distinguish the *world of physics,* where physical state-magnitudes are assigned to the points of the four-dimensional number space. This construction has the purpose of formulating a domain which is determined through *mathematically expressible laws.*... [T]he necessity of constructing the world of physics rests on the circumstance that only this world, but not the perceptual world, can be made intersubjective in an unequivocal, consistent manner. (§136)

This difference in intersubjectivizability, and hence the ineliminability of the world of physics, is due to two principal reasons. First, other epistemic agents are first introduced into the system as physical objects in the world of physics. As the whole question of objectivity and subjectivity only makes sense when such a plurality of agents is at hand, the epistemological problem only can be raised within the system at this point. Second, intersubjectivizing is based on coordinate transforms and the ability to locate intersubjective objects in the worlds of physics of various agents that correspond to the objects in the world of physics of oneself. Roughly, I can locate a physical object, *a,* via its spatio-temporal relations to me as given in my world of physics; I can do the same to locate an agent, *M.* I can, therefore, determine *a*'s spatio-temporal relations to *M* and then determine the object in *M*'s world of physics that has that relation to *M,* call this object, "a^M." Thus, *a* and a^M intersubjectively correspond. A similar procedure works for the rest of the agents, and then the intersubjective object for *a* is just the abstraction class of the intersubjectively corresponding objects *a,* a^M, a^N, and so on. The precise mathematical relations and the shared laws of physics are required for this process to get started.

Once started, though, intersubjectivizing goes through for all other scientific but nonphysical object domains:

> We said earlier that this intersubjective correspondence does not hold for the lower constructional levels, but only for the levels beginning with the construction of the space-time world, while for the lower levels we could only show constructional analogy. However, after the intersubjective correspondence, which was first introduced for the world of physics, has now been accomplished for the psychological world, it gives us a thoroughgoing correspondence of all objects of S and S_M. (§147)

Thus, intersubjective objects are constructible for all the scientific domains, and, hence, the intersubjective world is the world of science.

Although I have no time to argue it in detail here, it seems to me that this construction of the objectivity of science via the purely mathematically expressible relations of physics that takes us beyond the merely qualitative and private relations of sense-experience is just the story about objectivity toward which Carnap's pre-*Aufbau* works point. Interestingly, those works all deny the explicit definability of the concepts of physics on the experiential basis. And, indeed, Carnap rehearses these antireductionist arguments in the crucial section in which he constructs the world of physics:

> [T]he assignment of a quality to a world-point in the perceptual world does not determine which structure of state-magnitudes is to be assigned to the neighborhood of the corresponding physical world-point of the world of physics; the assignment of this quality merely determines a class to which this structure must belong. It is clear that the

physico-qualitative correlation cannot be free from the imprecision which attaches to the perceptual world generally. (§136)

So, Quine (1969) is right in finding that the constructional procedure changes dramatically just at the crucial stages as we move outside the autopsychological realm. Rather than draw the Quinean conclusions about reductionist empiricism, I think the more interpretatively appropriate questions are these: Given that Carnap was not a reductionist before or after the *Aufbau,* why does he saddle himself with the project of giving explicit definitions in the constitutional system? and What problems are there for the construction of the intersubjective world that lead to his second notion of objectivity as pure structure? I shall argue that the answer to the first question is not that Carnap moved to radical empiricism. Rather the answer lies in considering the logical resources Carnap has at his disposal and the role they are meant to play in epistemology. We can see this by first considering the second question.

Carnap's motivating epistemological distinction is between the objective and the subjective. The subjective is the realm of the essentially private, the phenomenal, the experiential "raw feels" that stand outside but are the starting point of knowledge. The objective, by contrast, is the realm of the communicable, the conceptual, the knowable. If the problem of the subjective is its ultimate privacy, then there is an obvious sense in which intersubjectivity, the possibility of genuine agreement and disagreement, should be the appropriate notion for objectivity. Yet I think that Carnap's own construction of the intersubjective world misfires as an attempt to capture his motivating distinction. Certain broad features of that distinction simply are not adequately captured.

The first problem is that experiential relations are ultimately intersubjectivized — among the objects in the intersubjective world is the intersubjective object corresponding to the elementary experiences of each agent. This is, moreover, as it must be — for psychology is not only a science of experience (among other things) but the science to which we look to recover the primitive structural features of experience. That is, experience is the scientific object, the structure of which is the epistemologist's professional concern. So experience — the quintessentially subjective — is an intersubjective object of science: "[T]he constructed objects are objects of conceptual knowledge only *qua* logical forms which are generated in a certain way. Ultimately, this holds also for the basic elements of the constructional system.... It is only... as constructed objects that they become objects of knowledge in the proper sense of the word, in particular, objects of psychology" (§178).

Moreover, objectivity as intersubjectivity leads to *formal* aspects of individual constitutional systems counting as subjective: "[M]ost statements about the form of the construction of an object in S or S_M [the system for M] must be described as subjective in S or S_M, respectively" (§148). For example, the form of the construction of my body in my system is vastly different from the form of the construction of the intersubjectively corresponding object (my body from M's point of view) in M's system. Thus, formal aspects of the constitution of objects, as given in the definition of the constitutional system, are the ones that turn out to be subjective. This stands rather at odds with Carnap's claim that the objective is the structural while the subjective is the contentful in such quotes as the one from §66.

A bit of expansion of this point may be in order.[11] It could be claimed that the construction of the intersubjective world and the distinction Carnap makes between intersubjective correspondence and constructional analogy actually capture the import of our quote from §66: each stream of experience stands in a relation of constructional analogy, and this plays the role of the logical form in common to them that underwrites the possibility of objectivity, now via intersubjectivizing.[12] Even leaving aside worries that *Rs* does not stand in constructional analogy to other streams of experience as constructed in the system, there is a problem. The relation of constructional analogy plays only a minimal role in the construction of the intersubjective world. The structure that is actually exploited in intersubjectivizing is not the structure common to the streams of experience due to their constructional analogy but rather the structure imposed upon the private perceptual worlds by the space-time manifold, the methodological posits of the constructional principles mapping the sensations onto the points of that manifold, and the mathematical relations of the laws of physics. Those structural features common to the constructionally analogous streams of experience are necessary but insufficient for the intersubjectivizing. This indicates that there is a general mismatch between the idea expressed in §66 and elsewhere and the procedure of intersubjectivizing.

There is an oddity involved in constructing the objective/subjective distinction inside the system in any case. For, as a distinction within the system, it becomes a scientific distinction. This does not square well with the task of epistemology as a rational reconstruction of science that exhibits its objectivity, in accord with the other vocabulary Carnap employs in the *Aufbau* to express his epistemological ambit. There is a task to be performed on or for science by the epistemologist in giving the constitutional system. This task cannot be understood as the drawing of a distinction within science or the constitutional system itself.

If we are suspect about the scientific status of our epistemological distinction between objectivity and subjectivity, where are we to go? Well, of course, Carnap argues for the dispensability of metaphysical notions in the *Aufbau,* so it is best to avoid locating the distinction there. This leaves only one remaining possibility: the distinction is a logical one. This idea is the one that motivates the project of purely structural definite descriptions. Ultimately, it leads to Carnap's curious attempt to eliminate the basic relation, *Rs,* itself:

> A purely structural statement must contain only logical symbols; in it must occur no undefined basic concepts from any empirical domain. Thus, after the constructional system has carried the formalization of scientific statements to the point where they are merely statements about a few (perhaps only one) basic relations, the problem arises whether it is possible to complete this formalization by *eliminating from the statements of science these basic relations* as the last, nonlogical objects. (§153)

There are two crucial problems to this project also, however. First, as Carnap notes (§§153–55), the answer to the question of whether the basic relation can be eliminated seems to be no. For it must be eliminated via a definite description, which carries a uniqueness requirement. The idea is to define the basic relation as the one and only relation that allows a rather complex construction higher up in the system. But, working in type-theory, Carnap can prove the existence of too many relations for uniqueness to hold. Basically you just define a permutation mapping, *f,* of the elex onto themselves and then define *Rs′* such that:

$$Rs'(e_i,e_j) \text{ iff } e_i = f(e_k);\ e_j = f(e_l) \text{ and } Rs\ (e_k,e_l)$$

This relation is obviously isomorphic to *Rs,* and hence anything constructible from *Rs* has a structural analogue definable in the same way from *Rs′*. So the uniqueness condition fails. Carnap attempts to finesse this by smuggling in a new logical notion—foundedness—which basically makes the notion of an experienceable or ostendable relation into a primitive of logic.[13]

Second, even supposing that this had been successful, it is a very strange thing at which to want to succeed anyway. For the project clearly threatens to erase the distinction between empirical and logical truth—all objective judgments become logical truths. No one after Leibniz, not even Carnap, really wants that. My diagnosis is that there are two good ideas combined here that cannot be disentangled from this bad one because Carnap is thinking about logic as Russellian type-theory and in a rather naive way.

The first good idea is this: epistemological notions like "construction" can be captured as metalogical concepts like "explicit definition in the language of *Principia Mathematica*." This is the crucial move in the elimination of metaphysics, for it isn't science, but rather epistemology, that is typically infected with metaphysical pseudoproblems. If we can reconstrue the point of epistemology without recourse to old, metaphysically loaded terminology of traditional epistemology, we will have deprived the metaphysicians of their entering wedge. Carnap states this view clearly in *Scheinprobleme*:

> It has frequently been emphasized that the epistemological quest for the justification or reduction of a cognition to others must be differentiated from the psychological question concerning the origin of a cognition. But this is only a negative determination. For those who are not satisfied with the expressions "given," "reducible," "fundamental," or those who want to eschew using these concepts in their philosophy, the aim of epistemology has not been formulated at all. In the following investigations we propose to give a precise formulation of this aim. It will turn out that we can formulate the purpose of epistemological analysis without having to use these expressions of traditional philosophy. We only have to go back to the concept of implication (as it is expressed in if-then sentences). This is a fundamental concept of logic which cannot be criticized or even avoided by anyone: it is indispensable in any philosophy, nay, in any branch of science. (Carnap 1928b, §1)

Thus, it is, I would claim, this reinterpretation of epistemological terminology as logical, and not a conversion to radical empiricism, that guides the movement to explicit definability. This, not the conventionalist accounts favored by Carnap in his early work on physics, simply is the only notion of concept introduction found in the logic with which he is working. Quine, of course, is correct in noting that this induces a mismatch between the official account of concept introduction as explicit definition and the actual constitutional procedures of the construction of the physical world. But Carnap's endorsement of explicit definitions is new in the *Aufbau* and actually is a milestone on his road to empiricism. Having adopted this purely logical vocabulary for epistemology, Carnap no longer can express the epistemological point of the conventionalism of his pre-*Aufbau* work, although he continues to employ its methodological tenets in the construction of the physical world. His early disagreement with radical empiricism can no longer find expression in his logic-based epistemological vocabulary. Since the disagreement between strict empiricism and neo-Kantianism can no longer be expressed, Carnap takes the issue to be merely apparent and remains neutral.

The second good idea is that logical truth is, so to speak, a framework constitutive notion. Any meaningful question presupposes an antecedent logical framework that provides the meanings of terms. The logic of *Principia* plays this role generally, but we want to extend this idea so that we can give principles that constitute the meaning of empirical concepts. That is, to get objectivity in empirical science, we must extend the notion of logical truth beyond type-theory to analytic judgments that constitute the meanings of empirical terms. Only on the antecedent availability of such framework-constituting principles can any empirical judgment be understood and can its epistemic (confirmational) status be assessed.

Unfortunately, the view of logic that derives from Russell asks one not to think in terms of object language/metalanguage distinctions. The logic of *Principia* is the universal logical language in which everything can be said. The logician, moreover, is in the business of providing definitions and proofs in this logic rather than about it. Thus, these two good ideas, which require basically metalinguistic notions — definability in a system or analyticity for an empirical language — are not adequately disentangled from the idea that these meaning-constitutive principles must come out as logical truths as logical truth is understood in type-theory. Hence, Carnap embarks on the odd quest for structuralization of all of science.

So Carnap has one outstanding problem in moving beyond the *Aufbau* — his motivating epistemological vocabulary has found no univocal place in the philosophy of the *Aufbau*. His solution to this difficulty has two steps. First, he adopts a metalogical perspective, which clarifies the role of logic in epistemology. Second, he ultimately drops the motivational distinction between objective knowledge and subjective experience from his philosophy — this gloss on the problems of knowledge is seen as an expression of the inherent confusion of the project of traditional epistemology.

2. The *Aufbau* in the Formal Mode of Speech

Carnap's movement to the metalogical perspective is well known. This was the key idea of his famous sleepless night of January 1931 and was influenced by his close association with Gödel and Tarski as well as his increasing concerns in philosophy of mathematics.[14] I shall pass over the difficult interpretative issues in philosophy of mathematics and instead look at how the metalogical perspective informs Carnap's work in epistemology of empirical knowledge.[15] The most interesting piece for this investigation is his 1932 *Erkenntnis* article "Die physikalische Sprache als Universalsprache der Wissenschaft" (revised and translated as "The Unity of Science" in 1934).

The main differences between the *Aufbau* and "Unity" are two. First, Carnap is far less concerned about giving definitions of physical concepts on experiential basis or vice versa in "Unity" than in the *Aufbau.* Indeed, although constitution theory is mentioned in "Unity" (Carnap 1932a [1934], 33), nothing like the definitional project of the *Aufbau* is undertaken at all. Second, "Unity" systematically employs the formal mode of speech. Thus, it would be more in the spirit of "Unity" to have expressed the first point as: in "Unity," unlike in the *Aufbau,* Carnap shows no particular interest in providing definitions of predicates of the physicalist language on the basis of those of the protocol language. The chief similarity is that Carnap's main task in "Unity" is to defend the idea that the physicalist language, conceived as a pure coordinate language, is both intersubjective and universal, as is the language of unified science in the *Aufbau.* This claim requires investigations into the relation between the physicalist and protocol languages. But these investigations are not directed at the role of protocols in verification. This is mentioned only very much in passing, and a Duhemian line is taken:

> [P]rotocol statements can be deduced by applying the rules of inference to sufficiently extensive sets of singular statements (in the language of the scientific system) taken in conjunction with laws of nature.... Scientific statements are not, in the strict sense, "verified" by this process. In establishing the scientific system there is therefore an element of convention, i.e. the form of the system is never completely settled by experience and is always partially determined by conventions. (ibid., 49)

Rather, the main focus is the translatability of the protocol language into the physicalist language and the question of what role subjective protocols have in intersubjective science.

The switch to the formal mode of speech has many advantages. First, locating language as the domain of logic simplifies the antimetaphysical points of the *Aufbau.* In the *Aufbau,* there were detailed investigations into whether notions such as "essence" or "reality" could be constructed. After the move to the formal mode of speech, Carnap can simply seek to dissuade nascent metaphysicians by indicating how questions about the "essence" or "reality" of, for example, numbers cannot be expressed in the formal mode. In this way, metaphysics can be shown to lie outside of logic and, hence, philosophy. More than this, the formal mode of speech allows us to see more easily that questions of logical deducibility, definability, and so on are the central considerations of a philosophy of empirical knowledge. Theses such as the unity of science cease to look like old-fashioned metaphysics with a bad conscience. The unity of science is now understood as a thesis about the representational capacity of the physicalist language.

Despite the clarity that is achieved by the formal mode of speech, Carnap's discussion of the protocol language in "Unity" still shows a concern with the given in experience and, hence, psychology. According to "Unity," the protocol language still exhibits features that reflect the close connection between it and the elementary experiences of the *Aufbau*. First, there is a fact of the matter of how the protocol language is formed, although the precise nature of it is not yet settled: "In the present state of research it is not possible to characterize this [protocol] language with greater precision, i.e., specify its vocabulary, syntactical forms and rules" (ibid., 45). Presumably this is because the protocol language exactly captures the given in its given form.[16] In the *Aufbau*, Carnap undertook material investigations into the findings of Gestalt psychology to inform him of the originary structure of experience. While he does not undertake these investigations in "Unity," it seems that psychology still provides the starting point for epistemology.

The protocol language is, thus, subjective. This is true in the following senses. First, each of us has our own protocol language. Second, each protocol language is a distinct language from the physicalist language. Third, after physicalizing the protocol languages, they can be shown to be disjoint sublanguages of the physicalist language: "[T]he protocol language is a sub-language of the physical language. The statement previously made..., that the protocol languages of various persons are mutually exclusive, is still true in a certain definite sense: they are, respectively, nonoverlapping sub-sections of the physical language" (ibid., 88).

Moreover, an agent can discover the inferential connections between the physicalist language and his own protocol language. This relies on the "contingent fact" that our protocols have certain ordinal properties that allow their qualitative determinations to be single-valued functions of physical-quantity determinations:[17]

> *A* can discover which physical determination (or class of physical determinations) corresponds to a definite qualitative determination in his protocol language.... That determinations of this kind are theoretically always possible is due to the fortunate circumstance (an empirical fact, not at all necessary in the logical sense) that the pro-

tocol has certain ordinal properties. This emerges in the fact of the successful construction of the physical language in such a fashion that qualitative determinations in protocol language are single-valued functions of the numerical distribution of coefficients of physical states. (ibid., 60–61)

Finally, this type of mapping is independent of an agent. That is, the ordinal properties of the protocol language necessary for the map from the physicalist to the protocol language are the same for all of us. This too is a happy accident:

> The determined value of a physical magnitude in any concrete case is independent not only of the particular sensory field used but also the choice of the experimenter. In this we have again a fortunate but contingent fact, viz. the existence of certain structural correspondence between the protocols of the various experimenters. . . . Physical determinations are valid inter-subjectively. (ibid., 64–65)

In "Unity," Carnap explicitly draws the obvious consequence of these considerations: there are two contingent psychological facts about human experience that ground the very possibility of intersubjective science. He states: "It may be noticed however that these facts, though of empirical nature, are of far wider range than single empirical facts or even specific natural laws. We are concerned here with a perfectly general structural property of experience which is the basis of the possibility of intersensory Physics and intersubjective Physics" (ibid., 65). The possibility of objective scientific laws is contingent upon the truth of two particular psychological facts. Or to put the point more along the official Carnapian lines: the possibility of science considered as the system of intersubjectively valid sentences of the physicalist language depends upon the contingent truth of two sentences stated in that language.

Clearly, Carnap has gotten himself further embroiled in the confusions we saw in the lack of a univocal place for epistemology in the *Aufbau.* Carnap continues to hold out for an epistemological perspective outside of science from which to answer general, still rather Kantian questions of the possibility of intersubjective and universal science. Yet the answer to such questions is now firmly and explicitly in terms of the a posteriori forms of our psychological experience. But Carnap cannot have it both ways. If his questions are sensible questions and lead to logical investigations of empirical knowledge in accord with his official view of epistemology, then the answers in terms of empirical psychology are insufficient. Transcendental questions do not allow of empirical answers. If the answers suffice, then the questions themselves are empirical questions and the motivations for the logical view of epistemology are utterly lacking. Carnap seems caught between transcendental epistemology and empirical psychology with no room again for his logical reorientation of epistemology.

"Unity" may well be the apex of Carnap's confusions about his epistemological project. It also provides us with an object-lesson in how philosophically misleading Carnap's talk of "the logical structure of experience" (or of the protocol language) has been throughout the works from the *Aufbau* to "Unity." This phrase is trying to do too much work and ambiguously covers the following divergent ideas. First, we are dealing with logical structure and, hence, as philosophers we are simply engaged in the type of noncontroversial, analytic, logical work we see in mathematical logic. Second, this structure is yet meant, following the neo-Kantians, to play the philosophical role of the Kantian synthetic a priori — it is the guarantor of objectivity in empirical science. But, finally, as it is simple psychological experience with which we are dealing, this structure is simply known empirically. It is a simple matter of fact that we have structurally similar streams of experience. This is a morass of confusion with the structure of experience simultaneously trying to fit into logical, transcendental, and empirical roles. As an epistemology, this seems hopeless.

3. Beyond Epistemology

As we have seen, the distinction between the formal and material modes of speech was meant to carry with it the rejection of metaphysics at the time of "Unity." That is, the formal mode of speech was endorsed as the only appropriate phraseology for the philosopher, and metaphysics disappeared by being inexpressible in that mode. Let us consider a bit more carefully, however, what cannot be formulated in the formal mode of speech in "Unity." Consider the following two quotations, the second of which is the formal "translation" of the first:

> The simplest statements in the protocol-language refer to the given, and describe directly given experience or phenomena, i.e. the simplest states of which knowledge can be had.
>
> The simplest statements in the *protocol-language* are the protocol-statements, i.e. statements needing no justification and serving as the foundation for all the remaining statements of science. (ibid., 45)

It is clear that reference to the "given" or "experience" or "the phenomena" only occurs within the material mode of speech. These locutions disappear without remainder in the formal mode.

Consider, however, the properties claimed for the protocol sentences in "Unity" once again. They are the private end points of the verification of scientific claims. They do, however, have content, and this content can, as a matter of course, be expressed in the physicalist language. But under this translation, the content is a public state of affairs and cannot play the role of verification anymore. In short, the protocol sentences have inherited many of the puzzling features of the autopsychological realm of the *Aufbau.* Our epistemological story requires that they be subjective, private sentences in contrast to the intersubjective physicalist language. But it also requires that the physicalist language exhaust the sayable and, hence, what is said in the protocol languages. The problematic of the *Aufbau* persists — the investigation of the connections between the protocol and physicalist languages captures the story of the relation of objective knowledge to private experience.

But Carnap's account of the epistemological role of the protocol sentences is guided by considerations that can only occur in the material mode of speech. Now, however, such an account is not only confused but also officially unsayable. The formal concerns of the Carnapian philosopher can no longer make any reference to the "given" or, indeed, the objective/subjective distinction at all. Although it is not taken in "Unity," the way out of this morass is, however, clear. Carnap must simply let the other shoe drop. Having officially adopted a formal account of philosophy that cannot express his motivating epistemological distinction, he can simply drop the distinction. This is precisely what he does in his response to Neurath, "On Protocol Sentences" (Carnap 1932d [1987a]).[18]

In the first striking use of the linguistic pluralism attainable in the metalogical or syntactic view — soon to be given its famous formulation in the Principle of Tolerance in syntax — Carnap severs any and all connection between protocol sentences and the "form of experience." The protocol language is no longer bound by the need to capture the structure of the given. The protocol-sentence debate ceases to be one over the facts about the protocol language and its relation to the physicalist language — it comes to be a series of proposals for the construction of languages for science:

> Neurath opposes certain features of the view about protocol sentences I advocated in my article on the physicalistic language. He wants to contrast it with another view according to which protocol sentences

> are in a different form and are manipulated according to other procedures. My opinion here is that this is a question, not of two mutually inconsistent views, but rather of *two different methods for structuring the language of science both of which are possible and legitimate.* (Carnap 1987a, 457)

The investigations of the structure of these languages proceed unconstrained by any psychological investigations meant to determine the epistemological starting point of knowledge in the experientially primitive. Indeed, Carnap is most attracted to a generally Popperian view that merely selects for any given scientific inquiry a class of sentences of the physicalist language that will serve as the protocol sentences. Of this option he writes:

> Every concrete sentence of the physicalistic language can serve under certain circumstances as a protocol sentence. (ibid., 465)

> [W]ith this procedure no sentence is an absolute endpoint for reduction. Sentences of all kinds can if necessary be reduced to others. Reduction proceeds at any given time until one arrives at sentences that one acknowledges by decision. Thereby everything takes place in the intersubjective, physicalistic language. (ibid., 467)

Two things go by the wayside with this change in view. First, we lose the need for thinking of the protocol language as outside the language of science because it is to play the role of the subjective and private. Carnap envisions the possibility of the protocol language simply forming part of the system language. All that is required is the connection between the term "protocol language" and those sentences of the system language chosen to be the end points of verification. Second, the very notion of an antecedent, linguistic-form-independent "form of experience" itself is recognized as a pseudoconcept:

> [I]t is not a matter of indifference how we formulate the question [of the theory of knowledge]. For the formulation in the formal mode of speech speaks of sentences and makes us thereby attentive to the circumstance that the question is still not complete, that is, that a statement of which language the question relates to is necessary.... In contrast to this the contentful formulation, which speaks of "the form of the phenomena," can easily lead to the dangerous error that there is such a thing as an absolute, linguistic-form-independent, final given structure of the phenomena, which one need only simply intuit and take up. (Carnap 1936a, 39)

In the formal mode of speech, there is simply no way to designate such a notion of the form of the given as such. Nor can we express how such an alleged structure to the given could provide material constraints on the range of available languages for science. As such, Carnap comes to recognize all epistemological questions as questions from within specified linguistic systems. Indeed, epistemological questions are interpreted as logical questions about the language so specified—questions of consequence relations and other notions definable from consequence such as analyticity or synonymy.

Famously, Schlick and his fellow right-wingers in the Vienna Circle thought Carnap had given up his empiricism in his flight from the absolutism of the protocol sentence in tight connection to the form of experience.[19] Indeed, one might think worse thoughts—given the seeming lack of constraints on languages, not only must rationalists and their fellow epistemological travelers come back into the fold, but room must be made for Hegel, Heidegger, and all their metaphysical friends. This latter view is certainly wrong. For the Principle of Tolerance does not lead one to tolerate Heidegger (much less Hitler, *pace* Quine).[20] This is because the Principle of Tolerance is constrained not by extralinguistic facts but by the principle of syntax, which presents the task of the philosopher:

> In the foregoing we have discussed several examples of negative requirements (especially those of Brouwer, Kaufmann, and Wittgenstein) by which certain common forms of language — methods of expression and inference — would be excluded. Our attitude to requirements of this kind is given a general formulation in the *Principle of Tolerance: It is not our business to set up prohibitions, but to arrive at conventions.... In logic, there are no morals.* Everyone is at liberty to build up his own logic, i.e. his own form of language, as he wishes. All that is required of him is that, if he wishes to discuss it, he must state his methods clearly, and give syntactical rules instead of philosophical arguments. (Carnap 1937, §17)

Heidegger does not recognize his duty to give us precise syntactic rules rather than philosophical arguments, and his language is not reconstructible as following precise syntactic rules. This is the import of the famous passage in "The Elimination of Metaphysics" (Carnap 1932e [1959]) in which Carnap seeks to understand Heidegger's principle that "Das Nichts selbst nichtet."

But what of empiricism? Certainly the syntactic perspective is wider than strict empiricism. That is, syntactic considerations of language structure can go forward independently of concern about the observability or verifiability of any predicate or sentence in the language. Indeed, Carnap does just that for Languages I and II of *Syntax* itself, since he is there concerned with the mathematical richness of the languages — largely because he is concerned with their suitability as syntax languages. But has Carnap's Principle of Tolerance simply severed all connection between the languages of science, including the protocol languages, and the experiential beginning of empirical knowledge upon which the empiricist, as well as the neo-Kantian, insists?

In a certain sense, yes. All questions of epistemology do become questions of logical relations within particular languages. No external perspective from which to inquire about which languages get the protocol sentences "right" or in the "appropriate relation to experience" is countenanced. The type of empiricism that Schlick wants to urge is, in Carnap's view, simply a metaphysical dogma.

But empiricism does not simply drop out for Carnap. In "Testability and Meaning" (Carnap 1936–37) empiricism is recognized not as a thesis stated in some language or other. Nor is it a thesis constraining the syntactic well-formedness of all languages. Rather empiricism becomes a proposal for the use of certain languages as languages for empirical science or the reconstruction of empirical science:

> It seems to me that it is preferable to formulate the principle of empiricism not in the form of an assertion — "all knowledge is empirical" or "all synthetic sentences that we can know are based on (or connected with) experience" or the like — but rather in the form of a proposal or requirement. As empiricists, we require that descriptive predicates and hence synthetic sentences are not to be admitted unless they have some connection with possible observations, a connection which has to be characterized in a suitable way. (ibid., §27)

In empiricist languages the logical notions such as consequence, analyticity, reducibility, reducibility of confirmation, introductive chains, and so on, map onto the traditional empiricist concerns with the confirmational basis of science via this "suitable" connection to experience. But what can such a suitable connection to experience look like for Carnap?

Carnap's formulation of the proposal or requirement for empiricist languages in the quote above is, as is usual in the contexts in which he discusses rather than implements his ideas, a pseudo-object sentence. Hence, it is strictly speaking meaningless. In practice, the requirement takes the form that the primitive predicates of a language for science be observable. Observability is, however, not a syntactic or logical notion; it is a notion from "psychology, ... more precisely, the behavioristic theory of language" (ibid., §11). Carnap's rough formulation of the notion of observability is as follows:

> A predicate 'P' of a language L is called *observable* for an organism (e.g. a person) N, if, for suitable arguments, e.g. 'b', N is able under suitable circumstances to come to a decision with the help of few observations about a full sentence, say 'P(b),' i.e. to a confirmation of either 'P(b)' or '–P(b)' of such a high degree that he will either accept or reject 'P(b)'.... This [psychological] explanation is necessarily vague. There is no sharp line between observable and non-observable predicates. (ibid., §11)

The notion of observable predicates gives the starting point for confirmation — the basic sentences of these predicates are accepted independently of the acceptance of any other sentences. The concept of confirmation in empiricist languages is then defined on the basis of observability. A sentence is confirmable if it stands in the relation of reducibility of confirmation to the class of observable predicates (ibid., §11). Thus, the confirmation of an observation sentence is defined as zero or one depending on whether that sentence is accepted or not. In this way, we are provided with the stock of basic predicates and the base clause of the definition of degree of confirmation, and these then feed into the general logical machinery of the language that allows the introduction of new predicates on the basis of certain rules from the primitive ones, the assignment of confirmation values to the higher level sentences given the acceptance of certain protocol sentences, and so on.

Now, it may seem that there is still an admixture of the psychological in the logic of science in this use of the psychological notion of observability. Indeed, the ability to recommend empiricist languages does require the availability of observable predicates and, thus, the idea that the definition of an observable predicate in the behavioristic theory of language picks out a nonempty set. Also, epistemological vocabulary is defined on the basis of this psychological notion in empiricist languages.[21] But, first, from the point of view of epistemology, "observability" is simply a primitive notion for predicates. The observability of any given predicate is not a matter for epistemological concern. As philosophers of science we look to our psychologist friends to design the experiments that will reveal the predicates that are observable. Second, from the epistemological perspective, the degree of confirmation of a sentence and not the definition of "confirmation" is most important. And the degree of confirmation is not defined on the basis of observability. Roughly, we are simply given a stock of predicates and the base clause of the degree of confirmation such that the primitive sentences involving those predicates get values zero or one depending on whether such sentences are accepted or rejected.

Does this admixture of psychology in the philosophy of empirical knowledge not confute Carnap's alleged sharp separation between the two? Moreover, does it not rely on a perspective on issues of knowledge that goes beyond Carnap's austere formal mode of speech? I think we can answer each of these questions in the negative. First, Carnap can give logically precise accounts of epistemological vocabulary such as confirmation; this is done for individual languages and not from some global point of view. Moreover, the reliance on psychology here is not in the form of some translinguistic matter of fact about confirmation of empirical knowledge. Rather it is in the form of a process of regimentation of knowledge claims into a particular type of language. Thus, the reliance on psychology in helping to delimit the useful class of languages that are recommended as languages for the reconstruction of knowledge does not amount to an antecedent philosophical perspective from which to issue judgments about metaphysical matters of fact.[22]

Moreover, the morass in which Carnap found himself in "Unity" has disappeared. The move to observable predicates indicates that there is no *logical structure* of experience that falls within the purview of epistemology and stands in a problematic relation to psychology. Psychology provides the stock of observation predicates in empiricist languages, but the epistemological work for science is done in the logical connections of those predicate (and sentences containing them) to scientific predicates and sentences generally. As there is no logical structure to experience in this view, the whole Kantian cast of the project has been completely erased.

At the same time, the reliance on psychology for reconstruction of science in empiricist languages has become much less problematic than the talk of "subjective experience" previously. The earlier talk had the oddity of being epistemological talk about that which stood outside of objective knowledge. Thus, it has the feel of an extrascientific, purely philosophical domain of inquiry. With the move to the logic of science, this has simply changed into the reliance of empiricist epistemology on the science of psychology. This does not, moreover, obviate the need to reconstruct psychology itself so that its theoretical terms and claims are shown to be in logically precise relations to observable predicates. Even the notion of "observation predicate" itself, as this is a theoretical notion, can be clarified in this way. It is a characteristic feature of Carnapian philosophy of science that it is a gradual process that uses the science at hand and the resources of logic to perform a clarificatory task that moves science forward.[23]

Thus, the whole epistemological question motivating the *Aufbau* — How do we achieve objective knowledge despite the subjective beginning of knowledge in private sensation? — has completely collapsed. For now the idea that knowledge rests on observation is understood quite differently. The way that Carnap had understood this previously was that the epistemically primitive vocabulary is autopsychological: namely, "Otto perceives a red dot in the middle of the visual field," or, indeed, "Recollection of similarity holds between e_4 and e_{23}." That is, the observational base of knowledge was interpreted as requiring that the epistemically primary language was a private language of personal experience. Thus, the question of how to bootstrap into the intersubjective language of science becomes preeminent. The movement to observation predicates indicates that this is misconceived. Basic sentences involving observation predicates are not subjective or private — observation predicates simply form a subclass of the class of physicalist predicates of the common physicalist language.[24]

The logic of science shows, then, how to investigate the epistemologically interesting questions of the confirmation of scientific theories, the introduction of scientific concepts, the observational base of science, and so on, without interpreting these questions as showing how to go from subjective experience to objective knowledge. Thus, the objective/subjective distinction, which was the principal distinction motivating the epistemology in the *Aufbau,* is unnecessary. Carnap's central embarrassment in the *Aufbau* is that the objective/subjective distinction finds no univocal place either inside or outside the system. The change in his view of logic and the movement to the logic of science show how we can reconceive the business of philosophy so that this distinction can happily be done without.

Notes

The research leading to this essay was supported by a National Science Foundation Postdoctoral Fellowship (Number DIR-9105217) at the University of Pennsylvania, a Postdoctoral Fellowship supported by the Leverhulme Foundation at the University of Keele, and a Postdoctoral Fellowship at the University of California, San Diego. Prior versions of this essay were presented at a one-day workshop on Carnap's philosophy in the 1930s at the University of Sheffield and at an HPS seminar at King's College London. I thank both audiences for helpful discussion and, respectively, David Bell and David Papineau for the invitations. Special thanks to all my fellow workshop participants in Minnesota, especially Michael Friedman and Thomas Ricketts, for stimulation and inspiration both at the workshop and generally.

1. It is not unimportant that the rejection of metaphysics precedes the adoption of empiricism in Carnap's view. This provides indirect support for the interpretation of the *Aufbau* presented below.

2. All translations from previously untranslated work are my own. For work that has appeared in translation, I use the published translation.

3. A defense of this claim lies beyond the scope of this essay. For some suggestions along this line see the final section of Ricketts 1982.

4. Thus, I shall be presenting Carnap's development largely in isolation from the work of his colleagues. Work on this period in Vienna Circle philosophy is proliferating, and, with particular reference to the protocol-sentence debate, connections between Carnap and others are made in Uebel 1992b and Oberdan 1993. I do think the pressures internal to his own project were the principal motivators to Carnap's change in view, however, and those are the ones I seek to illuminate here.

5. My views on the connections between Carnap and neo-Kantianism are found in Richardson 1990, 1992a, 1992b. Some further connections between the *Aufbau* and Kantian philosophy can be found in Webb 1992. Compare also remarks on various neo-Kantian perspectives on the *Aufbau* in Uebel 1992, chap. 2.

6. Throughout this essay references to the *Aufbau* and other works in which Carnap uses section numbers will be by those section numbers only.

7. Where it is mentioned and endorsed, in §181, it is endorsed only as against rationalism and with an immediate and important disclaimer that the *Aufbau* is not "raw empiricism." Rather, the project stresses the importance of "form elements of cognition." These formal elements point toward the neo-Kantian interpretation given in the text below.

8. Throughout this essay I render Carnap's *Konstitution* and related words indifferently as "constitution" or "construction." The subtle shades of philosophical meaning between these words shall not concern us here. See Coffa 1985 and Friedman 1992b for more discussion of these terms.

9. For more on the epistemological project of Carnap and its connections to the neo-Kantians, see especially Friedman 1992b; Sauer 1985; and Richardson 1992a, 1992b.

10. This is not strict Kantianism since this form is now logical form alone. The emphasis on logical form is due to the collapse of the intuition/concept distinction as it is found in Kant. Without intuition, formal notions are tied to concepts alone and thus to logic. This path was well worn by the neo-Kantians in their flight from intuition. See literature cited in n. 9 above.

11. I was pressed on this point by Michael Friedman and Thomas Ricketts at the workshop. I hope these remarks go some way to assuaging their doubts.

12. For the difference between constructional analogy (as in, for example, the analogy between my construction of my body and *M*'s construction of *M*'s body) and intersubjective correspondence (as in, for example, the relation between my construction of my body and *M*'s construction of my body), see §146.

13. See Michael Friedman 1987 for more details of this problem situation.

14. On the sleepless night, see Carnap 1963a, 53–54.

15. I actually don't think these two projects can be separated in this way, but I shall remain silent on the connections. For a preliminary account of the situation as I see it, see Richardson 1994. Other important recent articles on Carnap's philosophy of logic and mathematics in the *Syntax* period include Friedman 1988, 1992a; Goldfarb and Ricketts 1992; Sarkar 1992; Ricketts 1994; and Creath 1995.

16. This claim and the whole interpretation of "Unity" given below may be considered problematic due to Carnap's consideration of a materialist protocol language (1932a

[1934], 47). "Unity" was written at a time of great ferment of interaction with Neurath, who was a devoted physicalist from the start. Uebel (1992b, 99) dates a first draft of "Unity" to September 1930, that is, even before the move to metalogic. Thus, I think the possibility of a materialist protocol language is a very late addition to the text, probably dating to well after Carnap had seen Neurath's objections to the original draft. I do not believe that the drafts of "Unity" exist, however. More importantly, regardless of these conjectures, I would simply claim that the materialist protocol language does not support the type of epistemological talk that Carnap engages in in the text. Hence, I take it that it does not represent the view of the Carnap of the body of the text. I thank Michael Friedman for pressing me on this point also.

17. Note that this is still a unidirectional claim. Carnap does not claim that the values of the physical state-magnitudes are single-valued functions of the qualities of the protocol language for any agent.

18. For a very helpful chart that gives the order in which various documents were written during this period, see Uebel 1992, 98.

19. See Schlick 1934; Oberdan 1993; and Uebel 1992b, chap. 8.

20. See Creath 1991, 241, for Quine's letter to Carnap expressing his concern that the Principle of Tolerance will lead to tolerating Hitler.

21. Thanks to Barry Gower for insisting that I get these claims right.

22. Of course, this problem is further alleviated with the movement to semantics, already half taken in "Testability and Meaning." Once semantics is in place, the philosopher may with impunity speak of both logical forms and empirical matters. See Ricketts 1996 for worries that the movement to semantics makes Carnap's rejection of metaphysics untenable.

23. See, for example, Carnap's remarks in the last section of *Logical Syntax.*

24. Carnap (1936–37, §16) argues for the methodological importance of this distinction between perception terms and observable predicates.

PART V

AFTERWORD

Ronald N. Giere

From Wissenschaftliche Philosophie *to Philosophy of Science*

Most current research on the origins of logical empiricism deals with developments before 1938. This is appropriate because that year marks the bitter end of scientific philosophy in Europe. With the *Anschluss* in March 1938, Austria ceased to exist as a separate nation and Czechoslovakia was threatened. There was no place left in the German-speaking world for the scientific philosophers. By the end of 1938, almost everyone who was going to leave had done so.[1] Feigl, who had received his Ph.D. under Schlick in 1927, had already been at the University of Iowa since 1931. Carnap had joined the faculty at the University of Chicago in 1936, where he was soon joined by Carl Hempel. Reichenbach was on his way from Istanbul, where he had found refuge in 1933, to his new position at the University of California in Los Angeles. Philipp Frank was a guest of Percy Bridgman at Harvard.

It is not the mere fact of a historical break that has led those working on the history of logical empiricism to concentrate on the European phase of its development. Many of those engaged in this research are themselves products of the logical empiricist movement in its post–World War II forms. It is difficult for anyone to regard their own familiar professional development as history. The prewar period, by contrast, has until very recently remained largely unknown to contemporary philosophers of science, the subject of autobiographical remarks and disciplinary founder myths rather than of genuine historical scholarship.

The technical nature of scientific philosophy makes it difficult for anyone not trained in philosophy of science to investigate its development. Yet in large part because their initial training has been as philosophers rather than historians, most of those now investigating the early history of logical empiricism tend to approach their subject as intellectual historians practicing what historians of science used to call "internal" history. The personal fortunes of the historical participants, as well as the larger social and cultural context of the time, remain in the background. Even these internal histories, however, contain indications of relevant background conditions. So there are fuller histories of this earlier period yet to be written. But one must begin somewhere, and the intellectual history is surely a good place to start.

If we now contemplate future studies focusing on the development of logical empiricism in North America, it is already clear that a rather different balance must be struck right from the start. The European origins of logical empiricism are not intellectually continuous with its later development in North America. But the cause of this discontinuity was clearly not primarily intellectual. It was the forcible dislocation of many of the major participants from the culture of German-speaking Europe during the interwar years to the English-speaking world of North America beginning around 1933. It is with this fact that any future history of logical empiricism in North America must begin.

The social facts are dramatic. In 1930, following publication of the Vienna Circle's manifesto "Wissenschaftliche Weltauffassung: Der Wiener Kreis" (Carnap, Hahn, and Neurath 1929), these scientific philosophers represented just one among many modernist intellectual movements in the German-speaking world and operated largely outside the German philosophical establishment. In 1960, just prior to publication of a volume entitled *Current Issues in the Philosophy of Science* (Feigl and Maxwell 1961), logical empiricism was clearly the dominant philosophy of science in North America.[2] So the overriding "external" question is this: How, between 1930 and 1960, did a dissident European movement advocating the replacement of much established German philosophy by *Wissenschaftliche Philosophie* transform itself into the dominant tradition for philosophy of science in North America?

As one seriously interested in the history of logical empiricism in North America, but not personally engaged in uncovering that history, I would like here to raise some general issues, pose some specific questions, and suggest some hypotheses that might be examined by future historians of this period. My motives are more than historical. Being able to see logical empiricism as the contingent historical development it surely must have been may allow future philosophers of science to put aside some old arguments and devote more of their energies to developing new approaches to understanding modern science. In addition, recovering the history may establish connections with earlier traditions containing forgotten resources useful to the contemporary enterprise.

1. Why Did the European Origins of Logical Empiricism Remain So Long in Relative Obscurity?

I begin with an issue already noted above. The European origins of logical empiricism remained in relative obscurity until well after it had lost its unchallenged position as *the* North American philosophy of science. Why was this so? The study of key texts has typically been the primary mechanism by which intellectual movements are transmitted. This was clearly not the case with logical empiricism. Carnap's *Aufbau* (1928a), the major work to come out of the Vienna Circle, did not appear in English translation until 1967 (Carnap 1967), just three years before his death. Reichenbach's reputation in Germany rested primarily on his analyses of relativity theory as found in his *Relativitätstheorie und Erkenntnis Apriori* (1920), *Axiomatik der Relativistischen Raum-Zeit-Lehre* (1924), and *Philosophie der Raum-Zeit-Lehre* (1928). It was this work that persuaded physicists such as Einstein, Laue, and Planck to sponsor him for a position at the University of Berlin.[3] Yet none of these works appeared in English during Reichenbach's lifetime, with the latter appearing finally in 1958, five years after his death. Again, Schlick's major work, *Allgemeine Erkenntnislehre* (1925), did not appear in English for fifty years (Schlick 1975). Wittgenstein's *Tractatus* (1922), by contrast, appeared in English translation already in 1922. Why was the case so different for the founding works of logical empiricism?[4]

My hypothesis is that the scientific philosophers, such as Carnap and Reichenbach, realized that their future, if they were to have a future, lay in North America. And they realized, quite rightly, that works like the *Aufbau* and *Relativitätstheorie,* which were written in the context of a cultural, scientific, and philosophical tradition that did not then exist in North America, would not be much appreciated in the North American context. So they put their efforts into other projects, ones better suited to their new intellectual and cultural environment. Carnap focused first on formal semantics (1942) and then on logical probability (1950b), which drew on his work in semantics. Reichenbach devoted his Istanbul years to a general epistemological work, *Experience and Prediction* (1938), which he wrote in English to practice for his hoped-for emigration to the United States. Then he revised his book on probability for an English edition (1949b), also writing a logic text (1947) and technical treatises on quantum mechanics (1944), causality (1954), and the direction of time (1956)—the latter two published only after his death.

Even if I am mistaken about the motivations of leading logical empiricists during the 1930s and 1940s, the fact remains that the development of logical empiricism in North America proceeded in virtual ignorance of the major early writings that defined the European movement for scientific philosophy. Logical empiricism in North America was to a considerable extent a new creation — built on the old foundations, to be sure, but styled for a new audience so that what appeared in public view in North America was something noticeably different from what had existed in Europe.

So what were the texts that first defined logical empiricism in North America? For the general philosopher and countless students, Ayer's *Language, Truth and Logic* (1936) provided an Anglicized version of the movement. Carnap's "Testability and Meaning" was originally published in the United States in 1936–37 and Reichenbach's *Experience and Prediction* in 1938. Then there were the early monographs in Neurath's *International Encyclopedia of Unified Science* (Neurath, Carnap, and Morris 1955), such as Carnap's *Foundations of Logic and Mathematics* (1939) and Ernst Nagel's *Principles of the Theory of Probability* (1939). And of course there were the major works on probability by the original leaders of the movement, Reichenbach's *The Theory of Probability* (1949b) and Carnap's *Logical Foundations of Probability* (1950b). By the time the first volume of Minnesota Studies in the Philosophy of Science (Feigl and Scriven) appeared in 1956, logical empiricism had pretty well become the established philosophy of science in North America.

I suspect a large role was also played by the early *Readings:* Feigl and Sellars's *Readings in Philosophical Analysis* (1949) and Feigl and Brodbeck's *Readings in the Philosophy of Science* (1953). For the most part the papers reprinted in these "readers" dated from the mid-1930s through the 1940s, after the migration to North America was well under way. Moreover, many of the articles by the scientific philosophers themselves were originally published in English, such as Carnap's "Testability and Meaning" (1936–37), Schlick's "Meaning and Verification" (1936a), and Reichenbach's "The Philosophical Significance of the Theory of Relativity" (1949a). The paucity of writings by the original scientific philosophers before 1936 might be explained simply by the editors wanting to provide their readers with the most up-to-date materials. But how then do we account for Frege's "On Sense and Nominatum" (1892), Russell's "On Denoting" (1905), G. E. Moore's "Hume's Philosophy" (1922), C. I. Lewis's "The Pragmatic Conception of the *a Priori*" (1923), C. J. Ducasse's "Explanation, Mechanism, and Teleology" (1926), or P. W. Bridgman's "The Logic of Modern Physics" (1928)?[5]

It is a striking feature of these "readers" that they included papers by a number of Americans regarded as sympathetic to the movement: Lewis White Beck, May Brodbeck, R. M. Chisholm, Morris Cohen, C. J. Ducasse, Adolf Grünbaum, Sidney Hook, C. I. Lewis, Edward Madden, Paul Meehl, Ernest Nagel, W. V. O. Quine, Wilfrid Sellars, B. F. Skinner, K. W. Spence, W. T. Stace, and C. L. Stevenson. Also well represented are British philosophers such as C. D. Broad, W. C. Kneale, G. E. Moore, and of course Bertrand Russell. No one can deny that the scientific philosophers were genuinely internationalist, outward looking, and inclusive whenever possible. Moreover, they were explicit in advocating philosophy as a collective enterprise with many individuals contributing to its overall achievements. It is equally undeniable that this attitude was very beneficial to the institutional success of the movement in its new environment.

Another even more "external" explanation for why much of the spirit of the "Wissenschaftliche Weltauffassung" was left behind in the transition to North America is the simple but absolutely essential matter of institutional affiliation. The scientific philosophers had few institutional ties to philosophy in Europe. Reichenbach's position in Berlin, as noted, was in physics. When Carnap left Europe, he held a chair for natural philosophy in the division of natural sciences at the German University in Prague. Feigl and Hempel never held any academic appointments in Europe. It was clear, however, that if they were to have academic positions in the United States, it would have to be as members of philosophy departments. To achieve this status they had to soften the rhetoric of *Wissenschaftliche Philosophie* and become philosophers of science.[6]

2. Hempel's Explications

For most academics, even most philosophers, the individual who best personified logical empiricism in North America was neither Carnap nor Reichenbach, but Carl Hempel. Hempel was originally Reichenbach's student in Berlin, but he also spent time in Vienna. He was caught with his dissertation not yet completed when Reichenbach was dismissed in 1933. He nevertheless completed his degree in Berlin the following year with Wolfgang Köhler, the Gestalt psychologist, serving as his official supervisor.[7] After spending some time in Belgium, he became Carnap's assistant at the University of Chicago in 1937 before moving to his first teaching position at Queens College in New York. Virtually all of his professional life was spent in the United States. Hempel's early papers, "Studies in the Logic of Confirmation" (1945) and "Studies in the Logic of Explanation" (1948, with Paul Oppenheim), effectively defined what by 1960 were arguably the two most active areas of research in North American philosophy of science.[8] Neither topic had clear antecedents in the European writings of the scientific philosophers. Nor did the method employed, *explication.* That was a recent creation of Carnap's, first mentioned in the opening sections of *Meaning and Necessity* (1947, but written in 1942–44, 7–8) and discussed at length in the opening chapter ("On Explication") of *Logical Foundations of Probability* (1950b).[9] Of course the idea of philosophy as the logical analysis of scientific concepts had long been central to scientific philosophy. But part of that idea, particularly for Schlick and Reichenbach, was that the concepts were connected with particular *scientific theories,* as the concepts of space and time are central to relativity theory. Explication, as practiced by Carnap and Hempel from the 1940s onward, had no such connection to any scientific theory. The concepts to be analyzed were general, methodological concepts supposedly common to all the sciences. Even in the case of confirmation, the fact that Carnap's final quantitative confirmation relation had the structure of mathematical probabilities was a *conclusion* of the analysis, not its starting point. For Hempel's purely qualitative confirmation relation, and for explanation, there is no associated scientific theory at all. The constraints on the analyses were provided by first-order logic and one's methodological intuitions expressed as initial conditions of adequacy for the resulting explicatum.

My questions here are: How does Carnap's 1940s notion of explication connect with his earlier views on the nature of philosophical analysis? What motivated him to develop this method in the way he did, when he did? These are questions of antecedents about which I know little. About the *consequences* of adopting the method of explication, however, I have some definite hypotheses. One is that philosophical analysis as practiced by logicians and philosophers of science becomes comparable to philosophical analysis practiced by British "ordinary language" philosophers following the spirit of G. E. Moore and the later Wittgenstein.[10] One ends up comparing the analyses produced with prior intuitions through the construction of examples and counterexamples. The difference was that the result of explication is a new, more precise concept, a *replacement* for the original vague concept, more suitable for philosophical purposes. Wittgensteinian analysis, by contrast, supposedly leaves everything as it was, except that one should no longer be tempted to ask senseless questions. By the time of the Schilpp volume on Carnap's philosophy (1963), Carnap, ever tolerant, tried to see formal and informal analysis as complementary. All philosophy consists of conceptual analysis; only the method and tools employed may differ.[11]

Another consequence of regarding the explication of general methodological concepts as the primary task of the philosopher of science was an increasing separation between philosophy of science and the content of the sciences. People trained in philosophy, but with little knowledge of any science, could write article after article with titles like "The Paradoxes of Confirmation" or "The Symmetry between Explanation and Prediction." Carnap and Hempel cannot, of course, be blamed for the ensuing trivialization of much of the philosophy of science. No one can be expected to be that prescient. But when Kuhn came on the scene with a wealth of examples from the history of real science, the impact was amplified by comparison with the then existing discussions in the philosophy of science concerned with black ravens and the shadows of flagpoles.[12] As an indirect consequence of the challenge Kuhn posed for the philosophy of science, many philosophers of science in the 1970s attempted to reconnect the philosophy of science with real science. This effort was particularly noticeable in the philosophy of physics, in philosophical inquiries into the nature of probability and statistical inference, and a little later in the then newly reemerging philosophy of biology. To their eventual detriment, however, these efforts proceeded in virtual ignorance of debates over the proper relationship between science and philosophy that had taken place during the 1920s and 1930s.

3. Probability and Induction

Following the publication of Reichenbach's *The Theory of Probability* (1949b) and Carnap's *Logical Foundations of Probability* (1950b), probability and induction became major topics in the philosophy of science. The problem of induction had not played a significant role in the development of scientific philosophy before 1933. The scientific philosophers were of course much concerned with how experience and language connect with the world and how the structure of experience could possibly reflect the structure of the world. But these questions were understood in a Kantian framework: How is objective scientific knowledge possible? They were not questions about the inductive warrant for claims about the real world. Even Carnap's *Aufbau,* we now realize, was primarily concerned with questions about the structure of objective knowledge, not about the empirical warrant for knowledge claims.[13] What accounts for the change in perspective?

My hypothesis is that a major factor in the change of perspective was the change in location and philosophical climate. Induction had been a major problem in the empiricist tradition since Hume and, in spite of Kant, continued to be a major problem for Mill and then for Russell. It had ceased to be a problem in German philosophy after Kant. The Hegelian influence on both British philosophy and on the American pragmatists had pretty much been overcome by Moore and Russell, and by Dewey, by the 1930s. Traditional British empiricism had reasserted itself. It was this tradition, I think, into which both Carnap and Reichenbach inserted themselves. Whether this was in any sense part of a conscious effort at assimilation I do not know. It was nevertheless a highly adaptive action.

Carnap recounts that he began to think more systematically "about the problems of probability and induction" in the spring of 1941 while on leave from the University of Chicago at Harvard University. He recalls lectures by Richard von Mises and Feigl as being especially influential (Schilpp 1963, 36). It was in the spring of that year that he "began to reconsider the whole problem of probability" (ibid., 72), a reconsideration that led him to reread John Maynard Keynes's *Treatise on Probability* (1921). The result was first his paper "On Inductive Logic" (1945) and then *Logical Foundations of Probability* (1950b). I do not know of any further insights into his motivations in the published literature.

Reichenbach's case is more complex. The following story is told about Reichenbach. When the Nazis marched in and shut down the University of Berlin, Reichenbach is said to have exclaimed (a German equivalent of): "Now I understand the problem of induction!"[14] One thinks immediately of Russell's story of the man and the chicken in chapter 6 ("On Induction") of *The Problems of Philosophy* (1912). This story about Reichenbach may well not be true. But it is true that Reichenbach's published work on induction dates from his 1933 paper, "Die logischen Grundlagen des Wahrscheinlichkeitsbegriffs," which contains an outline of his famous "pragmatic" justification of induction.[15] It also seems to be true that Reichenbach was very proud of this argument and wanted to include it in a monograph on probability in the *International Encyclopedia of Unified Science*. But the editors, including Carnap and Neurath, had already engaged Ernest Nagel to do the monograph on probability (Nagel 1939) and wanted Reichenbach to do one on space, time, and relativity. He refused. As a result there was no contribution by Reichenbach on any subject.[16] I suspect that Reichenbach was consciously weighing what sort of publication would best further his prospects for a professorship in the United States.[17]

Probability is another matter. Reichenbach's dissertation, completed in 1915, was on the application of probability to the physical world. His *Wahrscheinlichkeitslehre* was published in the Netherlands in 1935.

The problem of induction in its Humean form appears only in the final section, in which he outlines his pragmatic justification. This last section could well have been added after Reichenbach left Berlin.[18] In any case, this final section (§80, comprising ten pages) grew into a completely new chapter 11 ("Induction," comprising fifty-three pages) in the 1949 English edition. In the preface, dated May 1948, Reichenbach refers to his proof of the justification of induction as originating "some fifteen years ago," which would put it in 1933. Reichenbach and Carnap engaged in both cooperation and competition all their professional lives. I do not know the extent to which either operated in the years immediately preceding publication of their books on probability. The controversy generated after publication in fact contributed positively to both their reputations, and to the vitality of logical empiricism, in North America.

4. Discovery Versus Justification

At the end of the opening chapter ("A Role for History") of *The Structure of Scientific Revolutions* (1962), Kuhn notes that his remarks "may even seem to have violated the very influential contemporary distinction between 'the context of discovery' and 'the context of justification.'" These are about the only explicit references to logical empiricism that appear in his book. Referring to this and other unnamed distinctions, he goes on to say: "Rather than being elementary logical or methodological distinctions, which would thus be prior to the analysis of scientific knowledge, they now seem integral parts of a traditional set of substantial answers to the very questions upon which they have been deployed." I think Kuhn had it exactly right. The discovery-justification distinction was a substantial presupposition of the logical empiricist understanding of science as it developed in North America. He was wrong, however, in thinking it contemporary, and he underestimated the depth of commitment the distinction commanded.

Ironically, most of the critical literature on this distinction implicitly honors it by considering only its legitimacy and not inquiring into its origins. The source of the distinction "context of discovery"/"context of justification," formulated in just these words, is, as is widely recognized, Hans Reichenbach's *Experience and Prediction,* published by the University of Chicago Press in 1938, almost certainly through the good offices of Charles Morris. As noted above, it was written in English during the years 1934–37 at the University of Istanbul where Reichenbach, along with fifty or so other former German professors (including

Richard von Mises), found refuge in Mustafa Kemal's new republic after being dismissed from his post in Berlin in early 1933. But versions of the distinction had been common in German philosophy for at least fifty years. For example, Rudolf Hermann Lotze, a neo-Kantian, used a version in the 1870s to distinguish questions regarding the psychological genesis of spatial representations from questions regarding the validity of geometrical knowledge of such representations (Hatfield 1990, 163–64).[19]

Reichenbach introduces his version of the distinction in the very first section of *Experience and Prediction,* where he sets out the task of epistemology as he conceives it. Here he is primarily concerned to distinguish epistemology from psychology. This he does by associating the concerns of epistemology with those of logic and then drawing on the long tradition, emphasized by Frege, of distinguishing logic from psychology. It will, he writes, "never be a permissible objection to an epistemological construction that actual thinking does not conform to it" (1938, 6). This presentation proceeds in a manner suggesting Reichenbach does not think his distinction needs much defense, and he does not seriously attempt to provide one.

Interestingly, the distinction reappears only once in *Experience and Prediction,* and then only briefly, near the end of the final chapter on probability and induction. Here he writes:

> What we wish to point out with our theory of induction is the logical relation of the new theory to the known facts. We do not insist that the discovery of the new theory is performed by a reflection of a kind similar to our expositions; we do not maintain anything about the question of how it is performed — what we maintain is nothing but a relation of a theory to facts, *independent of the man who found the theory.* (ibid., 382; emphasis added)

He then cites the example of Einstein and the general theory of relativity. The contrast between "the man who found the theory" and a logical "relation of theory to facts" exactly parallels that between the contexts of discovery and justification. My conjecture is that part of the significance of the distinction for Reichenbach at this time was its implicit denial that the character of a person proposing a scientific hypotheses has anything to do with the scientific validity of the hypothesis proposed. This applies, in particular, to being a Jew, from which it follows that his dismissal from his position in Berlin, as well as the persecution leading to Einstein's resignation, had been in principle unwarranted.[20]

My only supporting textual evidence for this conjecture appears in the final paragraph of a paper on the state of logistic empiricism in

Germany, which Reichenbach published in 1936 in *The Journal of Philosophy.* This paper is as much an advertisement for his own views, and a criticism of Carnap, as it is a survey of the movement he champions. There he writes: "Science, surely, is not limited to national or racial boundaries; we prefer to stand for this historical truth, in spite of all the pretensions of a certain modern nationalism" (Reichenbach 1936, 160). The "racial boundaries" include those between Jews and others, and "a certain modern nationalism" can only be National Socialism. In philosophical terms, Reichenbach seems to have made it a precondition for any scientific epistemology that it rule out the possibility of Jewish — or any culturally identifiable — science. But more than this, it seems that separating questions of the origins of ideas from questions of their validity was, for Reichenbach at that time, a matter as deeply personal as it was philosophical. And this sentiment was surely shared by everyone in the movement.

If one is going to insist on so strong a distinction between discovery and justification, one is obliged to produce a theory of justification to back it up. That Reichenbach did. His own theory of induction does satisfy the precondition that the justification of a hypothesis be independent of its origin. His rule of induction operates as a relationship between purely formal aspects of a fixed set of data and a single hypothesis — a relative frequency in a finite sequence of occurrences and a postulated limiting relative frequency, respectively. There is simply no place in such a formal relationship for any aspects of the wider context to enter into the calculation. The very different accounts of scientific inference developed by Carnap, and by Popper, also satisfy Reichenbach's requirement, and for similar reasons.

It is interesting to speculate on what would have been Reichenbach's reaction to N. R. Hanson's writings on the "logic of discovery" around 1960 (Hanson 1958, 1961). Hanson took Reichenbach's distinction to imply, for example, that Kepler was not thinking "logically" or "rationally" as he worked his way toward the hypotheses we now know as Kepler's laws. In retrospect this seems a gross misunderstanding of Reichenbach's intent. Reichenbach was concerned to distinguish, as was common among German philosophers since Kant, between psychology and logic. And he wanted to assert that the personal characteristics of scientists are irrelevant to the validity of their ideas. Claiming that there was no rationale in Kepler's attempts to formulate the laws of planetary motion seems to have been far from his intent.

There may well exist additional documentary evidence regarding Reichenbach's personal motivations for insisting on a distinction between discovery and justification around 1935. I doubt that he explicitly foresaw, and maybe he was never consciously aware of, the usefulness of such a distinction for a German immigrant in the United States in the subsequent years. For the distinction says: Don't think about the fact that I am a German immigrant or speak with an accent; just consider the validity of my ideas. That had to be a very useful stance for anyone in his position.

5. What Happened to Pragmatism?

The story of the success of logical empiricism in North America coincides with the story of the decline of American pragmatism. In 1930 pragmatism was the leading, though not the only, philosophical program in the United States.[21] By 1960 it was much diminished as an identifiable philosophical school and was no longer well represented in leading departments of philosophy.[22] The historical issue here is the extent to which these two phenomena were causally related. In particular, how much did the success of logical empiricism contribute to the decline of pragmatism? Could these have been cotemporal but causally disconnected phenomena? My hypotheses is that they were somewhat connected in the way suggested but that there were also independent factors at work.

It is a matter of historical record that members of both movements initially viewed each other as philosophical allies. Philosophers identified with pragmatism, including Charles Morris, Ernest Nagel, and W. V. O. Quine, visited their European counterparts in the 1930s. These same people were soon thereafter instrumental in securing academic positions in America for their erstwhile hosts, including both Carnap and Reichenbach. Russell (1939) and Reichenbach (1939) both contributed to the first Schilpp volume, *The Philosophy of John Dewey.* And selections from the writings of Dewey and several other pragmatists were included in Feigl's *Readings.* Morris, in particular, sought to unify the two movements both through his writings (Morris 1937) and by becoming an editor and sponsor of Neurath's *International Encyclopedia of Unified Science.* So what happened in the following decades to change the philosophical climate so dramatically?

In spite of similarities, there were also deep differences between pragmatism and scientific philosophy, including differences concerning the nature of philosophy itself. By 1929, when he turned seventy, Dewey was a philosophical naturalist and to some extent even an evolutionary naturalist (Dewey 1910). There was, for Dewey, no special sort of philosophical knowledge, particularly none that could provide any sort of foundation or ultimate legitimization for the sciences. Rather, our understanding of evolutionary biology and psychology provided a basis for an understanding of scientific inquiry itself. What was special to philosophy, for Dewey, was the task of bringing to moral and political inquiries the conclusions and methods of the sciences. Moreover, he was far less concerned with the truth of scientific conclusions than with their usefulness for solving current societal problems.

In retrospect, one could describe the program of scientific philosophy as one of naturalizing that part of philosophy consisting of Kantian, or neo-Kantian, metaphysics. More specifically, in a scientific philosophy, an autonomous philosophical understanding of arithmetic and geometry, of space, time, and causality, is abandoned in favor of a *scientific* understanding of these concepts. But the rest of philosophy itself is not naturalized; it is transformed. Scientific philosophy becomes the logical analysis of the language, concepts, and theories of the sciences, an enterprise that, like modern mathematical logic itself, takes place in the philosophically autonomous realm of the analytic a priori. And this, for the logical empiricists, came to include scientific epistemology. That conception is clearest in the case of Carnap's inductive logic but holds also for important aspects of Reichenbach's more pragmatic, decision-oriented account of primary induction.[23]

Part of our question, then, is: How did a naturalistic pragmatism incorporating an empirical theory of inquiry get replaced by a philosophy that regarded induction as a formal relationship between evidence and hypothesis? In the 1950s, many philosophers of science would have given such a question very short shrift. Many would simply have asserted that pragmatism was mistaken and logical empiricism correct. Induction just *is* a formal relationship between evidence and hypotheses. That is why logical empiricism took over. But such a response will not suffice in the 1990s. Ever since Quine advocated naturalizing epistemology (Quine 1969), philosophical sentiment has been moving back in Dewey's direction. So the question now is why philosophers then *believed* that pragmatism was so obviously wrong and logical empiricism so obviously right.

Answering this question will require a close historical look at philosophical debates in the 1940s and 50s. What, for example, was the reaction of various philosophers to the opposition between the pragmatic conception of truth and the combination of Tarski's definition of "True in L" together with Carnap's distinction between truth and confirmation (Carnap 1949)? Reichenbach (1939) accused Dewey, as well as Peirce, of failing to distinguish between the contexts of discovery and justification. Was this judgment widely shared and taken as a ground for dismissing Dewey's *Logic: The Theory of Inquiry* (1938) — published the same year as Reichenbach's own *Experience and Prediction* (1938)?

More general explanations are also possible. One possibility is that pragmatism had played itself out. There may simply have been a lack of interesting new problems to engage the attention of new graduate students. The logical empiricists, by contrast, soon developed a large budget of new problems and new logical techniques with which to attack them. So we may have here a case of a stagnant research program overwhelmed by a vigorous new program. Again, following a decade of economic depression and four years of war, the postwar period was one in which there was a strong desire in all areas of life to put the past behind and begin anew. Pragmatism was past; logical empiricism was new.[24] Moreover, logical empiricism identified itself with the new physics that, by making possible a nuclear bomb, had ended the war in the Pacific. It fit in well with the generally positive image of science following the war.[25]

Other less innocent cultural factors may also be relevant. Pragmatism, particularly as exemplified by Dewey, was a social philosophy. As such, it was bound up with various American social movements in the 1930s. Dewey himself visited Japan and China in 1919. He also visited the Soviet Union in 1928 and later published a series of articles that earned him labels like "Bolshevik" and "Red" in conservative segments of the American press. Then, in the mid-1930s, he went to Mexico as head of the Commission of Inquiry into the Charges against Leon Trotsky. The verdict of "not guilty" led others to label him a "Trotskyite."[26] Later, in 1939, Dewey argued against American involvement in the war (Dewey 1939a). Could it be that pragmatism exhausted itself in the social and political battles of this period?[27]

More ominously, could pragmatism to some extent have been a victim of anticommunism and McCarthyism in the decade following the war? Those are just the years when the balance between pragmatism and logical empiricism seems to have shifted strongly toward logical empiricism.[28] If this is at all correct, there is considerable irony in the situation. In the European context, the scientific philosophers were socially every bit as radical as the American pragmatists. Many had ties to European socialist and communist parties. That might initially have been part of the bond among members of the two movements. But these facts were left buried in the past. Once in America, the logical empiricist philosophers of science pretty much stuck to their p's and q's.[29] Their recent experience, after all, had surely convinced them of the destructive power of nationalistic political movements.[30]

6. The Next Generation

The primary means by which intellectual movements grow in the academic world is through the recruitment of graduate students who will constitute the next generation of professors. For whatever reason, following the war the pragmatists were not very successful in recruiting or placing new students. By contrast, even an impressionistic survey reveals the logical empiricists to have been extraordinarily successful.[31] Among Reichenbach's students at UCLA were Hilary Putnam, later longtime professor of philosophy at Harvard, and Wesley Salmon, who himself had many students both at Indiana University and at the University of Pittsburgh. Carnap had few students at Chicago, but among them was Richard Jeffrey, who eventually went to Princeton. While still at Queens College, Hempel taught both Adolf Grünbaum and Nicholas Rescher, both of whom had emigrated from Germany with their parents in the 1930s. Hempel later moved to Yale, where he became Grünbaum's dissertation director, and then to Princeton, where he had many more students. Grünbaum and Rescher both had many students at the University of Pittsburgh. Feigl, of course, had many students at Minnesota. Among American followers, Quine at Harvard and Nagel at Columbia both had many students. Among Nagel's students were Patrick Suppes, who himself trained many students at Stanford, Henry Kyburg at the University of Rochester, and Isaac Levi, who ended up back at Columbia.[32] The picture would be even richer if one considered less central figures such as Gustav Bergman, Max Black, and Arthur Pap.

That a number of students of the early logical empiricists ended up in positions where they could also train graduate students is itself of considerable significance. When the postwar baby boom generation reached college age in the 1960s, universities expanded, and the new positions in philosophy thus created were filled by analytic philosophers and philosophers of science trained in the tradition of logical empiricism. The continued vitality of logical empiricism into the 1980s owed much to this further consequence of World War II.

In this regard it is enlightening to contrast the fortunes of logical empiricism as an academic philosophy of science with the scientific philosophy of "the Vienna Circle in exile" led by Philipp Frank at Harvard beginning in 1939 (Holton 1992, forthcoming).[33] Frank first formed a series of meetings including scholars from many fields, but mainly the sciences, under the heading "Inter-scientific Discussion Group." In 1947 he created the Institute for the Unity of Science under the auspices of the American Academy of Arts and Sciences in Boston. Meetings and conferences were held, many attended by eminent scholars such as George Birkhoff, Wassily Leontief, Harlow Shapley, B. F. Skinner, and Norbert Wiener. The only possibility for training graduate students in the philosophy of science, however, was through Quine in the department of philosophy. So Frank's direct influence on subsequent developments in the philosophy of science was minimized. But his indirect influence was great. When Frank's institute was dissolved in 1958, it was succeeded by what became the Boston Colloquium for Philosophy of Science, organized by Robert Cohen and Marx Wartofsky at Boston University.[34]

There is a common mythology, more common outside philosophy than within, that following publication of Kuhn's *Structure of Scientific Revolutions,* there was a dramatic shift toward a more historical and less logical approach to the philosophy of science. The above analysis shows why this has to be a myth. By the 1960s, the majority of philosophers of science had been trained in the tradition of logical empiricism with its emphasis on logical analysis. One cannot quickly move from a logical to a historical mode of doing philosophy of science. A few philosophers of science shifted their focus to questions concerning conceptual change and the development of science, issues that had not been on the agenda of logical empiricism. Many more moved slowly away from the particular doctrines and projects of logical empiricism, often by moving closer to work in the sciences themselves. Thus, although in the 1990s there are very few philosophers of science who would identify themselves as logical empiricists, the majority are still pursuing topics and employing means of analysis that are historically continuous with those of logical empiricism. That may be the best measure of the success of logical empiricism as a philosophy of science.[35]

7. Conclusion

In this look at the development of logical empiricism in North America I have emphasized context over content. This is justifiable because many readers will already be familiar with the content and have thought little about the context. A satisfactory history of these developments will, of course, have to put the content into the context. But so as not to lose sight of the importance of the context, I would recommend one keep in mind the following related historical counterfactuals: imagine that the Social Democrats rather than the National Socialists had come to power in Germany in 1933 (and thus that World War II never happened). What would have been the fate of *Wissenschaftliche Philosophie* in Germany, Austria, and throughout the world? What would have been the fate of

American pragmatism? And what would now be the complexion of the philosophy of science in North America?

Notes

1. Thomas Uebel reminds me that Kurt Gödel did not leave for the last time until 1940.

2. Herbert Feigl was both one of those listed as members of the Vienna Circle in the 1929 manifesto and one of the editors of the 1961 volume that represents a high-water mark for logical empiricism in North America. Carnap and Reichenbach, followed by Hempel, were undoubtedly the intellectual leaders of logical empiricism in North America, but Feigl, more than anyone else, created the institutional basis for the movement.

3. The nature and circumstances of Reichenbach's appointment in Berlin were complicated. Apparently the original plan was to create for Reichenbach a position as *ausserordentlicher Professor für Naturphilosophie.* This appointment was rejected by a faculty vote of twenty-six to eight, with five abstentions (Hecht and Hoffmann 1982, 659). This vote seems to have included members of the philosophy faculty. The reasons for rejecting Reichenbach included his liberal political views, as evidenced by his involvement with socialist student groups after World War I, but his outspoken antimetaphysical philosophical positions, which privileged scientific theory over philosophical metaphysics, also played a role. In the end, Laue and other physicists secured for him an appointment in the physics faculty as *nichtbeamteter ausserordentlicher Professor* assigned to teach "die erkenntnistheoretischen Grundlagen der Physik." This appears to have been an untenured (*nichtbeamtet*) position. I thank Don Howard for helping me sort out this issue.

4. One possible but mistaken answer is that these works did not need to be translated because everyone who needed to read them could do so in the original German. World War II accelerated a disengagement with German culture that had begun with World War I. Before 1914, of course, the large numbers of German immigrants, particularly from around the years 1848 and 1871, ensured that German language and culture were well represented in North America. By the end of World War II, the numbers of students in North America learning the German language had declined dramatically. Thus few students in the postwar period could read the founding documents of scientific philosophy in the original even if they had been so inclined, which most apparently were not.

5. In the case of Carnap, there is another possible explanation urged by Alan Richardson. It is simply that by 1936 Carnap regarded his earlier work as mistaken and thus not worth translating. Richardson cites an unpublished letter written by Carnap in 1938 expressing very negative views of the *Aufbau* to an inquiring Nelson Goodman. And it is, of course, very plausible that Feigl would have consulted with Carnap about what works should appear in his *Readings.* On the other hand, that a philosopher later regards an important earlier work as mistaken is not typically taken as grounds for later philosophers not to read it. The *Tractatus* again provides a clear contrast.

6. This applies doubly to the socialist rhetoric of Neurath's "Wissenschaftliche Weltauffassung" (Carnap, Hahn, and Neurath 1929), which would have been damaging to the movement in the America of the 1930s, 1940s and 1950s.

7. This information comes from a conversation with Hempel in Princeton in the spring of 1992.

8. The truth of this claim depends, of course, on how one defines "most active areas of research." I am quite convinced that a count of articles or citations would support this claim but do not have actual numbers to back up that conviction. For a relatively recent overview of the extensive literature on explanation, which demonstrates the enduring interest in the topic, see Salmon 1989. No similar review exists for confirmation.

9. I owe the realization that the notion of explication appears already in *Meaning and Necessity* to Thomas Ryckman.

10. In his original expositions (1947, 1950b), Carnap refers to C. H. Langford's discussion of G. E. Moore's notion of philosophical analysis.

11. In the Schilpp volume (1963) Carnap's attitude toward ordinary language analysis comes out clearly in his exchange with P. F. Strawson.

12. For an exemplar of this style of philosophy of science see Israel Scheffler's *Anatomy of Inquiry* (1963).

13. See Friedman 1987 and 1992b and Richardson 1990, 1992a, and 1992b.

14. I learned this story from Andreas Kamlah, who knew neither of its origin nor of any evidence that it might be true.

15. This is the paper, by the way, which is reprinted in translation in Feigl and Brodbeck 1953.

16. This story too I owe to Andreas Kamlah, but this one seems to be well established.

17. Thomas Ryckman suggested to me the following elaboration of this point based on a letter of Reichenbach's to Einstein in 1936 (see Ryckman 1996). Reichenbach had clashed with Hermann Weyl over relativity theory in the 1920s. By 1936 Weil was a powerful figure in the American academic establishment, powerful enough to block any possibility of Reichenbach's obtaining a position at Princeton. Moreover, as became clear in Einstein's response to Reichenbach's contribution to the Schilpp volume (1949), Einstein himself was critical of Reichenbach's understanding of the epistemological significance of relativity theory. It is a reasonable inference that Reichenbach refused to write on space, time, and relativity for the *Encyclopedia* because he did not want these disagreements to become well known in America.

18. The foreword is dated "Istanbul, August 1934."

19. Another version, with proper Kantian overtones, appears in Karl Popper's *Logik der Forschung* (1935). Popper's version, however, seems to have had little influence on philosophical thought before publication of the revised English edition of the *Logik* in 1959 (Popper 1959).

20. One must here recall how vicious and personal the attacks on Einstein by Nazi sympathizers were in 1932–33 (Clark 1971). Einstein himself was in the United States when Hitler came to power and shortly thereafter resigned his posts from the safety of his temporary residence in Belgium. He did not return to Germany. Reichenbach, by contrast, remained in Berlin long enough to experience firsthand the attacks on his patron.

21. For a good contemporary survey of the state of American philosophy in 1930, see Adams and Montague (1930). For a recent history of American philosophy up to 1930, see Wilson 1990.

22. This is not to deny that there continued to be individual scholars of considerable renown drawing inspiration from the pragmatists, particularly Peirce, whose collected works began appearing in 1931 (Peirce 1931–58). These included Arthur Burks at Michigan, Ernest Nagel at Columbia, and both Nicholas Rescher and Wilfrid Sellars at Pittsburgh. Both Burks and Rescher published major works in the 1970s that exhibit the continuing influence of Peirce (Burks 1977; Rescher 1973a, 1973b, 1977).

23. The heart of Reichenbach's "pragmatic justification" of induction is the purely mathematical proof, utilizing the mathematical definition of the limit of an infinite se-

quence, that the "straight rule" must eventually lead to ever more accurate posits of the true limit — provided only that such exists. But if no limit exists, no rule could lead us to it. So we have everything to gain and nothing to lose by following the rule (Reichenbach 1949b, 469–82).

24. Reichenbach himself sounded this theme in the final paragraph of his contribution (1939, 192) to Dewey's Schilpp volume. There he wrote: "The early period of empiricism in which an all-round philosopher could dominate at the same time the fields of scientific method, of history of philosophy, of education and social philosophy, has passed. We enter into the second phase in which highly technical investigations form the indispensable instrument of research, splitting the philosophical campus into specialists of its various branches." In other words, Dewey's (and pragmatism's) time has passed; we enter the time of Reichenbach (and logical empiricism).

25. How different this was from the situation in Germany following World War I, where the natural sciences were often identified with the militarism that had led to Germany's destruction. I wonder whether those who experienced both periods, including Carnap and Reichenbach, realized the irony of the difference.

26. Here I simply follow the biography in Schilpp and Hahn 1939.

27. Dewey's commitment to socialism, and his involvements in the social movements of the 1930s and 1940s, are examined in the final chapters of Robert Westbrook's lucid biography (1991).

28. In a forthcoming essay, John McCumber (1996) explores connections between McCarthyism and the development of analytic philosophy following World War II. His analysis supports my speculations regarding the decline of pragmatism. I thank Professor McCumber for providing me with a copy of this manuscript.

29. McCumber (1996) reports that Carnap initially refused an appointment at UCLA because of the loyalty oath in force in California in the early 1950s. Arthur Fine told me that letters in the Einstein papers show that Carnap similarly was reluctant to accept an appointment at Princeton because of discrimination there against Jews. (Reichenbach reportedly had no such compunctions.) These private acts of conscience, however laudable, do not alter the fact that the public face of logical empiricism exhibited mostly the logical analysis of scientific theories and methods.

30. Reichenbach's popular book *The Rise of Scientific Philosophy* (1951) may be an exception to the general lack of involvement by logical empiricists in social issues of the day. David Hollinger (1995) has recently placed this work within in a genre of works of the 1940s and 1950s extolling the value of science for liberal democratic societies and thus opposing the nationalistic anticommunism exemplified by McCarthyism. I recall Wesley Salmon once telling me that Reichenbach had written the book because he had promised his second wife, Maria, that he would write her a "best-seller" after they got to America. This latter motivation is compatible with Hollinger's analysis of the role of the book in the culture of the time. It is also possible that Reichenbach himself had political motivations in writing the book but sought to conceal them from his American students.

31. In conversations, the historian of science Mitchell Ash has suggested a striking contrast between the scientific philosophers and another German immigrant group, the Gestalt psychologists. The latter were generally unsuccessful in achieving academic positions that provided access to graduate students.

32. Nagel is especially worth studying in this regard since he was identified both with pragmatism and later with logical empiricism. It is instructive to compare Cohen and Nagel 1934 with Nagel 1961. Nagel's best-known students are identified more with logical empiricism than with pragmatism, although one can find pragmatist sympathies in the work of both Isaac Levi and Patrick Suppes. Nagel himself recommended the study of

Peirce's philosophy at the Fifth International Congress for the Unity of Science at Harvard in 1939 (Nagel 1940), and he defended naturalism as late as his 1954 presidential address for the Eastern Division of the American Philosophical Association (Nagel 1956).

33. Frank had been part of the first Vienna Circle in the years 1908–12. On Einstein's recommendation, he was appointed in 1913 to fill the position vacated by Einstein as professor of physics at the German University in Prague. He was a member of the later Vienna Circle and instrumental in bringing Carnap to Prague in 1931. He remained a part-time professor of physics and philosophy at Harvard until retirement in the mid-1950s. He died in 1966.

34. In an unpublished talk for a plenary session at the 1991 meeting of the Society for Social Studies of Science, Gerald Holton suggests that the founding of the program called Science, Technology, and Society at MIT in 1977 was also partly a result of Frank's endeavors. The then president of MIT, Jerome Wiesner, and his provost, Walter Rosenblith, had both attended some meetings of the Institute for the Unity of Science and shared some of Frank's visions for science. I thank Professor Holton for providing me with a copy of this manuscript.

35. Consider arguably the most influential and prolific philosopher of science in the "third" generation, Bas van Fraassen. He was born during the war in the Netherlands and emigrated with his family to Canada while a teenager. He completed his Ph.D. with Grünbaum at Pittsburgh, thus establishing a lineage going directly back to Reichenbach by way of Hempel. Like Reichenbach, van Fraassen has combined logical and methodological works (1980, 1989) with philosophical studies of both relativity theory (1970) and quantum mechanics (1991).

Bibliography

Adams, G. P., and W. P. Montague, eds. 1930. *Contemporary American Philosophy.* 2 vols. New York: Russell and Russell.

Adler, M. 1910. Marxismus und Materialismus. *Der Kampf* 3:68–75.

———. 1930–32. *Lehrbuch der materialistischen Geschichtsauffassung.* 2 vols. Berlin: Laub. Rev. ed. *Grundgelegung der Materialistischen Geschichtsauffassung: Soziologie des Marxismus.* Vienna: Europa, 1964.

———. 1931. Wozu schreibt man Bücher? *Der Kampf* 24:125–31.

Adler, R., M. Bazin, and M. Schiffer. 1975. *Introduction to General Relativity.* 2d ed. New York: McGraw-Hill.

Adorno, T., and M. Horkheimer. 1955. *Dialektik der Aufklärung.* Amsterdam. Translated as *Dialectics of Enlightenment.* London: Verso, 1985.

Archives for Scientific Philosophy (ASP). University of Pittsburgh Libraries. References are to file folder number and page number (where applicable). All rights reserved.

Aufbau Almanac: The Immigrant's Handbook. 1941 [1934]. New York.

Aufbau und Erbe: Das Künstlerische Schaffen im Arbeitsdank. 1938. Berlin and Leipzig: Oscar Brandstetter.

Avenarius, R. 1888. *Kritik der reinen Erfahrung.* Vol. 1. Leipzig: Fues's Verlag (O.R. Reisland).

———. 1890. *Kritik der reinen Erfahrung.* Vol. 2. Leipzig: O. R. Reisland.

Ayer, A. J. 1936. *Language, Truth and Logic.* London: Gollancz.

———. 1946. *Language, Truth and Logic.* 2d ed. New York: Dover.

———, ed. 1959. *Logical Positivism.* New York: Free Press.

Ballod, K. 1919. *Der Zukunftsstaat.* Stuttgart: Dietz. 2d rev. ed. of Atlanticus [pseud.], *Produktion und Konsum im Sozialstaat.* Stuttgart: Dietz, 1898.

Bargmann, V. 1960. Relativity. In *Theoretical Physics in the Twentieth Century: A Memorial Volume to Wolfgang Pauli,* edited by M. Fierz and V. Weisskopf, 187–98. New York: Interscience.

Bar-Hillel, Y. 1963. Remarks on Carnap's *Logical Syntax of Language.* In Schilpp 1963, 519–43.

Bauer, O. 1924. Das Weltbild des Kapitalismus. In *Der lebendige Marxismus: Festgabe zum 70, Geburtstag von Karl Kautsky,* edited by O. Jenssen, 407–64. Jena: Thüringer Verlagsanstalt. Excerpts translated as The World View of Organised Capitalism, in *Austro-Marxism,* edited by T. Bottomore and P. Goode, 208–17. Oxford: Clarendon, 1978.

Bauhaus Archiv. Argon: Berlin.

Bell, D., and W. Vossenkuhl. 1992. *Science and Subjectivity: The Vienna Circle and Twentieth Century Philosophy.* Berlin: Akademie.

Bergmann, P. 1976 [1943]. *Introduction to the Theory of Relativity.* New York: Dover.

Bernstein, E. 1897. *Die Voraussetzungen des Sozialismus und die Aufgaben der Sozialdemokratie.* Stuttgart: Dietz. Partly translated as *Evolutionary Socialism.* New York: Schoken, 1961.

Boltzmann, L. 1905. Über eine These Schopenhauers. In *Populäre Schriften,* 385–402. Leipzig: Barth. Translated as On a Thesis of Schopenhauer's, in *Theoretical Physics and Philosophical Problems,* edited by B. McGuinness, 185–98. Dordrecht: Reidel, 1974.

Born, M. 1922. Hilbert und die Physik. *Die Naturwissenschaften* 10:88–93.

Bourdieu, P. 1988. *l'Ontologie politique de Martin Heidegger.* Paris: Editions de Minuit. Translated as *The Political Ontology of Martin Heidegger.* Stanford, Calif.: Stanford University Press, 1991.

Bratt, R. G., C. Hartman, and A. Meyerson, eds. 1986. *Critical Perspectives on Housing.* Philadelphia: Temple University Press.

Bridgman, P. 1927. *The Logic of Modern Physics.* New York: Macmillan.

Burks, A. W. 1977. *Cause, Chance, and Reason.* Chicago: University of Chicago Press.

Carnap, R. 1921. "Der Raum: Ein Beitrag zur Wissenschaftslehre." Ph.D. diss., University of Jena. Göttingen: Dieterich'schen Univ.-Buchdruckerei, W. Fr. Kaestner.

———. 1922. *Der Raum: Ein Beitrag zur Wissenschaftlehre.* Kant-Studien Ergängzunghefte, 56. Berlin: Reuther and Reichard.

———. 1923. Über die Aufgabe der Physik und die Anwendung des Grundsatzes der Einfachstheit. *Kant-Studien* 28:90–107.

———. 1924. Dreidimensionalität des Raumes und Kausalität. *Annalen der Philosophie und philosophischen Kritik* 4:105–30.

———. 1925. Über die Abhängigkeit der Eigenschaften des Raumes von denen der Zeit. *Kant-Studien* 30:331–45.

———. 1926. *Physikalische Begriffsbildung.* Wissen und Wirken. Einzelschriften zu den Grundfragen des Erkennens und Schaffens. Emil Ungerer, ed. Vol. 39. Karlsruhe: Braun.

———. 1927. Eigentliche und uneigentliche Begriffe. *Symposion. Philosophische Zeitschrift für Forschung und Ausprache* 1:355–74.

———. 1928a. *Der logische Aufbau der Welt.* Berlin: Weltkreis. Translated as *The Logical Structure of the World,* in Carnap 1967.

———. 1928b. *Scheinproblem in der Philosophie.* Berlin: Weltkreis. Translated as *Pseudoproblems in Philosophy,* in Carnap 1967.

———. 1929. *Abriß der Logistik. Mit besonderer Berücksichtigung der Relationstheorie und ihrer Anwendungen.* Schriften zur Wissenschaftlichen Weltauffassung. Philipp Frank and Moritz Schlick, eds. Vol. 2. Vienna: Springer.

———. 1930a. Die alte und die neue Logik. *Erkenntnis* 1:12–26.

———. 1930b. Die Mathematik als Zweig der Logik. *Blätter für deutsche Philosophie* 4:298–310.

———. 1931. Die logizistiche Grundlegung der Mathematik. *Erkenntnis* 2:91–105. Translated as The Logicist Foundations of Mathematics, in Carnap 1983, 41–52.

———. 1932a. Die physikalische Sprache als Universalsprache der Wissenschaft. *Erkenntnis* 2:432–65. Translated as The Physical Language as a Universal Language, in *The Unity of Science,* edited by M. Black. London: Kegan Paul, 1934.

———. 1932b. Erwiderung auf die vorstehenden Aufsätze von E. Zilsel und K. Duncker. *Erkenntnis* 3 (1932–33): 177–83.

———. 1932c. Psychologie in physikalischer Sprache. *Erkenntnis* 3:107–142. Translated as Psychology in Physical Language, in Ayer 1959, 165–98.

———. 1932d. Über Protokollsätze. *Erkenntnis* 2:215–28. Translated as On Protocol Sentences, in Carnap 1987a, 457–70.
———. 1932e. Überwindung der Metaphysik durch logische Analyse der Sprache. *Erkenntnis* 2:219–41. Translated as The Elimination of Metaphysics through Logical Analysis of Language, in Ayer 1959, 60–81.
———. 1934a. *Die Aufgabe der Wissenschaftslogik.* Einheitswissenschaft, 3. Vienna: Gerold.
———. 1934b. *Die logische Syntax der Sprache.* Vienna: Springer. Translated as *The Logical Syntax of Language* (Carnap 1937).
———. 1935. *Philosophy and Logical Syntax.* London: Kegan Paul.
———. 1936a. Von Erkenntnistheorie zur Wissenschaftslogik. In *Actes du Congrès Internationale de Philosophie Scientifique, Sorbonne, Paris, 1935.* Vol. 1, *Philosophie scientifique et empirisme logique,* 36–41. Paris: Hermann and Cie.
———. 1936b. Wahrheit und Bewährung. *Actes du Congrès Internationale de Philosophie Scientifique, Sorbonne, Paris, 1935.* Vol. 4, *Induction et Probabilité,* 18–23. Paris: Hermann and Cie. Translated as Truth and Confirmation, in Feigl and Sellars 1949, 119–27.
———. 1936–37. Testability and Meaning. *Philosophy of Science* 3:419–71; 4:1–40.
———. 1937. *The Logical Syntax of Language.* Translated by A. Smeaton. London: Routledge and Kegan Paul.
———. 1939. *Foundations of Logic and Mathematics.* Chicago: University of Chicago Press.
———. 1942. *Introduction to Semantics.* Cambridge, Mass.: Harvard University Press.
———. 1945. On Inductive Logic. *Philosophy of Science* 12:72–97.
———. 1947. *Meaning and Necessity.* Chicago: University of Chicago Press.
———. 1949. Truth and Confirmation. In Feigl and Sellars 1949, 119–27.
———. 1950a. Empiricism, Semantics, and Ontology. *Revue Internationale de Philosophie* 4:20–40. Reprinted in Carnap 1956, 205–21.
———. 1950b. *Logical Foundations of Probability.* Chicago: University of Chicago Press. 2d ed. 1962.
———. 1952. Meaning Postulates. *Philosophical Studies* 3:65–73. Reprinted in Carnap 1956.
———. 1954. *Einführung in die symbolische Logik. Mit besonderer Berücksichtigung ihrer Anwendungen.* Vienna: Springer.
———. 1956. *Meaning and Necessity.* 2d ed. Chicago: University of Chicago Press.
———. 1958. *Introduction to Symbolic Logic and Its Applications.* New York: Dover.
———. 1959. The Old and the New Logic. In Ayer 1959, 133–46.
———. 1963a. Intellectual Autobiography. In Schilpp 1963, 3–84.
———. 1963b. My Conception of Semantics. In Schilpp 1963, 900–905.
———. 1963c. Replies and Systematic Expositions. In Schilpp 1963, 859–1013.
———. 1963d. W. V. Quine on Logical Truth. In Schilpp 1963, 915–22.
———. 1967. *The Logical Structure of the World and Pseudoproblems in Philosophy.* Translated by R. A. George. Berkeley: University of California Press.
———. 1983. The Logicist Foundations of Mathematics. In *Philosophy of Mathematics: Selected Readings,* edited by P. Benacerraf and H. Putnam, 41–52. Cambridge: Cambridge University Press.
———. 1987a. On Protocol Sentences. Translated by R. Creath and R. Nollan. *Nous* 21:457–70.

———. 1987b. The Task of the Logic of Science. In *Unified Science: The Vienna Circle Monograph Series Originally Edited by Otto Neurath,* edited by B. McGuiness, 46–66. Dordrecht: Reidel.

Carnap, R., H. Hahn, and O. Neurath. 1929. "Wissenschaftliche Weltauffassung: Der Wiener Kreis." Vienna: Wolf. Translated as The Scientific Conception of the World: The Vienna Circle, in Neurath 1973, 299–319.

Carrier, M. 1990. Constructing or Completing Physical Geometry? On the Relation between Theory and Evidence in Accounts of Space-Time Structure. *Philosophy of Science* 57:369–94.

Cartwright, N., et al. 1996. *Otto Neurath: Philosophy between Science and Politics.* Cambridge: Cambridge University Press.

Cassirer, E. 1902. *Leibniz' System in seinen wissenschaftlichen Grundlagen.* Marburg: Elwert.

———. 1910. *Substanzbegriff und Funktionsbegriff. Untersuchungen über die Grundfragen der Erkenntniskritik.* Berlin: Bruno Cassirer. Translated in Cassirer 1923.

———. 1913. Erkenntnistheorie nebst den Grenzfragen der Logik. *Jahrbücher der Philosophie* 1:1–59.

———. 1921. *Zur Einsteinschen Relativitätstheorie. Erkenntnistheoretische Betrachtungen.* Berlin: Cassirer. Translated in Cassirer 1923.

———. 1923. *Substance and Function and Einstein's Theory of Relativity.* Chicago: Open Court. Reprint, New York: Dover, 1953.

———. 1929. *Philosophie der symbolischen Formen.* Vol. 3, *Phenomenologie der Erkenntnis.* Berlin: Cassirer. Translated in *The Philosophy of Symbolic Forms.* Vol. 3, *The Phenomenology of Knowledge.* New Haven: Yale University Press, 1957.

Cassirer, T. 1981. *Mein Leben mit Ernst Cassirer.* Hildesheim: Gerstenberger.

Cat, J., H. Chang, and N. Cartwright. 1996. Political Philosophy of Science: Otto Neurath, Unity of Science, and Socialism. In *The Disunity of Science: Boundaries, Contexts, and Power,* edited by P. Galison and D. Stump. Stanford, Calif.: Stanford University Press.

Church, A. 1943. Review of Rudolf Carnap, "Introduction to Semantics." *The Philosophical Review* 52:298–304.

Churchland, P. S. 1986. *Neurophilosophy: Toward a Unified Science of the Mind-Brain.* Cambridge, Mass.: MIT Press.

Clark, R. W. 1971. *Einstein: The Life and Times.* New York: World.

Coffa, J. A. 1979. Elective Affinities: Weyl and Reichenbach. In *Hans Reichenbach: Logical Empiricist,* edited by W. Salmon, 267–304. Dordrecht: Reidel.

———. 1985. Idealism and the *Aufbau.* In *The Heritage of Logical Empiricism,* edited by N. Rescher. Lanham, Md.: University Press of America.

———. 1991. *The Semantic Tradition from Kant to Carnap: To the Vienna Station.* Edited by L. Wessels. Cambridge: Cambridge University Press.

Cohen, H. 1902. *Logik der reinen Erkenntnis.* Berlin: Cassirer.

Cohen, M. R., and E. Nagel. 1934. *An Introduction to Logic and Scientific Method.* New York: Harcourt.

Creath, R. 1990. The Unimportance of Semantics. In *PSA 1990,* vol. 2, edited by A. Fine, M. Forbes, and L. Wessels, 405–15. East Lansing, Mich.: Philosophy of Science Association.

———. 1991. *Dear Carnap, Dear Van.* Berkeley: University of California Press.

———. 1996. Languages without Logic. In Giere and Richardson 1996.

Daston, L. 1991. The Ideal and Reality of the Republic of Letters in the Enlightenment. *Science in Context* 2:367–86.

Davidson, D. 1986. Empirical Content. In *Truth and Interpretations,* edited by E. LePore, 320–32. Oxford: Blackwell.

Dedekind, R. 1893. *Was sind und was sollen die Zahlen?* 2d ed. Braunschweig: Friedrich Vieweg und Sohn. Translated as The Nature and Meaning of Numbers, in *Essays on the Theory of Numbers.* Translated by W. W. Beman. Chicago: Open Court, 1901. Reprint, New York: Dover, 1963, 29–115.

Der Aufbau: Lagerzeitung für Camp Perry u. Nebenlager 2 (6 October 1945). (Ohio; Source: U.S. Army Military History Institute).

Der Aufbau: Österreichische Monatshefte für Siedlung und Städtebau (Vienna) 1/1 (1926): 1.

Der Aufbau: Österreichische Monatshefte für Siedlung und Städtebau (Vienna) 1/6 (1926): 1.

Der Aufbau: Sozialistische Wochenzeitung (Zurich) 1 (December 1919): 1.

Dewey, J. 1910. *The Influence of Darwin on Philosophy and Other Essays in Contemporary Thought.* New York: Holt.

———. 1938. *Logic: The Theory of Inquiry.* New York: Holt.

———. 1939a. No Matter What Happens—Stay Out! *Common Sense* 8/3:11.

———. 1939b. *Theory of Valuation.* International Encyclopedia of Unified Science 2/4. Chicago: University of Chicago Press.

Dreben, B. 1990. Quine. In *Perspectives on Quine,* edited by R. Barrett and R. Gibson, 81–95. Oxford: Basil Blackwell.

Drosdowski, D., et al., eds. 1976. *Das grosse Wörterbuch der deutsche Sprache.* 6 vols. Mannheim: Bibliographisches Institut.

Duhem, P. 1906. *La Theorie physique: Son object sa structure.* 2d ed., 1914. Translated as *The Aim and Structure of Physical Theory.* Princeton, N.J.: Princeton University Press, 1954. 2d ed., New York: Atheneum, 1962.

Eddington, A. 1921. A Generalization of Weyl's Theory of the Electromagnetic and Gravitational Fields. *Proceedings of the Royal Society of London* A99, pp. 104–27.

———. 1923. *The Mathematical Theory of Relativity.* Cambridge: Cambridge University Press.

———. 1925. *Relativitätstheorie in mathematischer Behandlung.* Berlin: Springer.

Ehlers, J. 1973. Survey of General Relativity Theory. In *Relativity, Astrophysics and Cosmology,* edited by W. Israel, 1–125. Dordrecht and Boston: Reidel.

———. 1985. Hermann Weyl's Contributions to the General Theory of Relativity. In *Exact Sciences and Their Philosophical Foundations: Proceedings of the International Hermann Weyl Congress Kiel 1985,* edited by W. Deppert et al., 83–105. Frankfurt: Peter Lang.

———. 1988. Einführung in die Raum-Zeit-Struktur mittels Lichtstrahlen und Teilchen. In *Philosophie und Physik der Raum-Zeit,* edited by J. Audretsch and K. Mainzer, 145–62. Mannheim: Bibliographische Institut.

Ehlers, J., F. A. E. Pirani, and A. Schild. 1972. The Geometry of Free Fall and Light Propagation. In *General Relativity; Papers in Honour of J. L. Synge,* edited by L. O'Raifeartaigh, 63–84. Oxford: Clarendon Press.

Einstein, A. 1914. Die formale Grundlage der allgemeinen Relativitätstheorie. *Königlich Preussische Akademie der Wissenschaften* (Berlin). *Sitzungsberichte,* 1030–85.

———. 1915. Erklärung der Perhelbewegung des Merkur aus der allgemeinen Relativitätstheorie. *Königlich Preussische Akademie der Wissenschaften* (Berlin), *Sitzungsberichte,* 831–39.

———. 1916a. Die Grundlage der allgemeinen Relativitätstheorie. *Annalen der Physik* 49:769–822. Translated as The Foundation of the General Theory of Relativity, in Lorentz et al. 1923, 111–64.

———. 1916b. Hamiltonsches Prinzip und allgemeine Relativitätstheorie. *Königlich Preussische Akademie der Wissenschaften. Physikalisch-mathematische Klasse. Sitzungsberichte,* 1111–16. Translated as Hamilton's Principle and the General Theory of Relativity, in Lorentz et al. 1923, 165–73.

———. 1919. Spielen Gravitationsfelder im Aufbau der materiellen Elementarteilchen eine wesentliche Rolle? *Königlich Preussische Akademie der Wissenschaften. Physikalisch-mathematische Klasse. Sitzungsberichte,* 349–56. Translated as Do Gravitational Fields Play an Essential Part in the Structure of Elementary Particles of Matter? in Lorentz et al. 1923, 189–98.

———. 1920. Bemerkungen. *Physikalische Zeitschrift* 21:662.

———. 1921a. Geometrie und Erfahrung. *Königlich Preussische Akademie der Wissenschaften. Physikalisch-mathematische Klasse. Sitzungsberichte,* 123–30. Reprinted as *Geometrie und Erfahrung: Erweiterte Fassung des Fest-vortrages gehalten an der Preussischen Akademie der Wissenschaften zu Berlin am 27. Januar 1921.* Berlin: Springer. Expanded and translated as Geometry and Experience, in Einstein 1922, 27–56.

———. 1921b. Über eine naheliegende Ergänzung des Fundamentes der allgemeinen Relativitätstheorie. *Königlich Preussische Akademie der Wissenschaften. Physikalisch-mathematische Klasse. Sitzungsberichte,* 261–64.

———. 1922. *Sidelights on Relativity.* London: Methuen. Reprint, New York: Dover, 1983.

———. 1923a. The Theory of the Affine Field. *Nature* 112:448–49.

———. 1923b. Zur allgemeinen Relativitätstheorie. *Königlich Preussische Akademie der Wissenschaften. Physikalisch-mathematische Klasse. Sitzungsberichte,* 32–38.

———. 1925a. Eddingtons Theorie und Hamiltonisches Prinzip. In Eddington 1925, 367–71.

———. 1925b. Einheitliche Feldtheorie von Gravitation und Elektrizität. *Königlich Preussische Akademie der Wissenschaften. Physikalisch-mathematische Klasse. Sitzungsberichte,* 414–19.

———. 1925c. Nichteuklidische Geometrie und Physik. *Die Neue Rundschau* 36:16–20.

———. 1933. On the Method of Theoretical Physics. In Einstein 1954, 270–76.

———. 1952. Relativity and the Problem of Space. In Einstein 1961, 135–57.

———. 1954. *Ideas and Opinions.* New York: Crown.

———. 1961. *Relativity: The Special and the General Theory.* New York: Crown.

———. 1983. Review of E. Meyerson, *The Relativistic Deduction.* In *The Relativistic Deduction,* by E. Meyerson, 252–56. Dordrecht: Reidel. Translation of A propos de "La Déduction Relativiste" de M. Émile Meyerson, *Revue Philosophique* 105 (1928): 161–66.

———. 1987. *The Collected Papers of Albert Einstein.* Vol. 1, *The Early Years, 1879–1902.* Edited by J. Stachel et al. Princeton, N.J.: Princeton University Press.

Einstein, A., and M. Grossmann. 1913. Entwurf einer verallgemeinerten Relativitätstheorie und einer Theorie der Gravitation. *Zeitschrift für Mathematik und Physik* 62:225–61.

Eisenstaedt, J., and A. J. Kox, eds. 1992. *Studies in the History of General Relativity.* Einstein Studies, vol. 3. Boston: Birkhäuser.

Elkana, Y. 1981. A Programmatic Attempt at an Anthropology of Knowledge. In *Sciences and Cultures,* edited by E. Mendelsohn and Y. Elkana, 1–76. Vol. 5 of *Sociology of Science Yearbook.* Dordrecht: Reidel.

Engelbrecht, L. C. 1973. The Association of Arts and Industries: Background and Origins of the Bauhaus Movement in Chicago. Ph.D. diss., University of Chicago.

Farias, V. 1987. *Heidegger et le nazisme.* Paris: Verdier. Translated as *Heidegger and Nazism.* Philadelphia: Temple University Press, 1989.

Feigl, H. 1950. De Principiis Non Disputandum . . . ? Reprinted in Feigl 1981.

———. 1969. The Wiener Kreis in America. In Fleming and Bailyn 1969, 630–73.

———. 1981. *Inquiries and Provocations.* Edited by R. S. Cohen. Dordrecht: Reidel.

Feigl, H., and M. Brodbeck, eds. 1953. *Readings in the Philosophy of Science.* New York: Appleton-Century-Crofts.

Feigl, H., and G. Maxwell, eds. 1961. *Current Issues in the Philosophy of Science.* New York: Holt, Rinehart and Winston.

Feigl, H., and M. Scriven, eds. 1956. *The Foundations of Science and the Concepts of Psychology and Psychoanalysis.* Minnesota Studies in the Philosophy of Science, vol. 1. Minneapolis: University of Minnesota Press.

Feigl, H., and W. Sellars, eds. 1949. *Readings in Philosophical Analysis.* New York: Appleton-Century-Crofts.

Fine, A. 1984a. The Natural Ontological Attitude. In *Scientific Realism,* edited by J. Leplin, 83–107. Berkeley: University of California Press. Reprinted in Fine 1986b, 112–35.

———. 1984b. And Not Antirealism Either. *Nous* 18:51–65. Reprinted in Fine 1986b, 136–50.

———. 1986a. Einstein's Realism. In Fine 1986b, 86–111.

———. 1986b. *The Shaky Game: Einstein, Realism, and the Quantum Theory.* Chicago: University of Chicago Press.

Fleck, L. 1935. *Die Entstehung und Entwicklung einer wissenschaftlichen Tatsache.* Translated as *Genesis and Development of a Scientific Fact.* Edited by T. J. Trenn and R. K. Merton. Chicago: University of Chicago Press, 1979.

Fleming, D., and B. Bailyn, eds. 1969. *Intellectual Migration: Europe and America, 1930–1960.* Cambridge, Mass.: Harvard University Press.

Fock, V. 1964. *The Theory of Space, Time and Gravitation.* 2d rev. ed. New York: Pergamon.

Forman, P. 1971. Weimar Culture, Causality, and Quantum Theory, 1918–1927: Adaptation by German Physicists and Mathematicians to a Hostile Intellectual Environment. *Historical Studies in the Physical Sciences* 3:1–115.

Frank, M. 1983. Zwei Jahrhunderte Rationalitätskritik und die Sehnsucht nach einer neuen Mythologie. *Manuskripte* 23. Reprinted in M. Frank, *Conditio Moderna.* Leipzig: Reclam, 1993, 30–50.

Frank, P. 1917. Die Bedeutung der physikalischen Erkenntnistheorie Ernst Machs für das Geisteslebens unserer Zeit. *Die Naturwissenschaften* 5:65–72. Translated as The Importance for Our Times of Ernst Mach's Philosophy of Science, in P. Frank 1949c, 61–79.

———. 1938. Ernst Mach — The Centenary of His Birth. *Erkenntnis* 7:247ff. Reprinted as Ernst Mach and the Unity of Science, in P. Frank 1949c, 79–89.

———. 1941. Introduction: Historical Background. In *Between Physics and Philosophy,* 3–16. Cambridge, Mass.: Harvard University Press.

———. 1949a. Einstein, Mach, and Logical Positivism. In Schilpp 1949, 271–86.

———. 1949b. Historical Introduction. In P. Frank 1949c, 1–51.

———. 1949c. *Modern Science and Its Philosophy.* Cambridge, Mass: Harvard University Press.
Frank, W. 1931. Ist die marxistische Geschichtsauffassung materialische? *Der Kampf* 24.
Frege, G. 1879. *Begriffsschrift.* In *From Frege to Gödel,* edited by J. van Heijenoort. Cambridge, Mass.: Harvard University Press, 1967.
———. 1884. *Die Grundlagen der Arithmetik.* Breslau: Koebner.
———. 1984. On the Foundations of Geometry: Second Series. In *Collected Papers on Mathematics, Logic, and Philosophy,* edited by B. McGuiness, 292–340. Oxford and New York: Blackwell.
French, A. P. 1968. *Special Relativity.* New York: Norton.
———, ed. 1979. *Einstein: A Centenary Volume.* Cambridge, Mass.: Harvard University Press.
Friedman, M. 1983a. Critical Notice: Moritz Schlick. *Philosophical Papers. Philosophy of Science* 50:498–514.
———. 1983b. *Foundations of Space-Time Theories.* Princeton, N.J.: Princeton University Press.
———. 1987. Carnap's *Aufbau* Reconsidered. *Noûs* 21:521–45.
———. 1988. Logical Truth and Analyticity in *Logical Syntax of Language.* In *Essays in the History and Philosophy of Modern Mathematics,* edited by W. Aspray and P. Kitcher. Minneapolis: University of Minnesota Press.
———. 1991. The Re-evaluation of Logical Empiricism. *The Journal of Philosophy* 88:505–19.
———. 1992a. Carnap and A Priori Truth. In Bell and Vossenkuhl 1992, 47–60.
———. 1992b. Epistemology in the *Aufbau. Synthese* 93:15–57.
———. 1996. Overcoming Metaphysics: Carnap and Heidegger. In Giere and Richardson 1996.
Galison, P. 1986. Otto Neurath et la communication visuelle. In Sebestik and Soulez 1986, 271–88.
———. 1990. Aufbau/Bauhaus: Logical Positivism and Architectural Modernism. *Critical Inquiry* 16:709–52.
Giere, R. N. 1988. *Explaining Science.* Chicago: University of Chicago Press.
Giere, R. N., and A. W. Richardson, eds. 1996. *Origins of Logical Empiricism.* Minnesota Studies in the Philosophy of Science, vol. 16. Minneapolis: University of Minnesota Press.
Gödel, K. 1995. Is Mathematics Syntax of Language? In *Collected Works,* vol. 3, edited by S. Feferman et al. Oxford: Oxford University Press.
Goldfarb, W. 1996. The Philosophy of Mathematics in Early Positivism. In Giere and Richardson 1996.
Goldfarb, W., and T. Ricketts. 1992. Carnap's Philosophy of Mathematics. In Bell and Vossenkuhl 1992, 61–78.
Grünbaum, A. 1973. *Philosophical Problems of Space and Time.* Dordrecht: Reidel.
Haack, S. 1977. Carnap's *Aufbau:* Some Kantian Reflections. *Ratio* 19:170–75.
Haas, A. 1919. Die Axiomatik der modernen Physik. *Die Naturwissenschaften* 6:744–50.
———. 1920. Die Physik als geometrische Notwendigkeit. *Die Naturwissenschaften* 8: 121–27.
Hacker, P. M. 1986. *Insight and Illusion.* 2d rev. ed. Oxford: Clarendon Press.
Hahn, H. 1929. Empirismus, Mathematik, Logik. *Forschungen und Fortschritte* 5.
———. 1930–31. Die Bedeutung der wissenschaftlichen Weltauffassung, insbesondere für Mathematik und Physik. *Erkenntnis* 1:96–105.
———. 1931. Diskussion zur Grundlegung der Mathematik. *Erkenntnis* 2:135–41.

———. 1933. *Logik, Mathematik, und Naturerkennen.* Einheitswissenschaft, 2. Vienna: Gerold. Translated as Hahn 1987.

———. 1980. *Empiricism, Logic, and Mathematics: Philosophical Papers.* Edited by B. McGuiness. Dordrecht: Reidel.

———. 1987. Logic, Mathematics, and Knowledge of Nature. In *Unified Science,* edited by B. McGuiness, 24–45. Dordrecht: Reidel.

Hahn, H., O. Neurath, and R. Carnap. 1929. "Wissenschaftliche Weltauffassung: Der Wiener Kreis." Veröffentlichungen des Vereines Ernst Mach. Wien: Artur Wolf.

Haller, R. 1979. Geschichte und wissenschaftliches System bei Otto Neurath. In *Wittgenstein, Der Wiener Kreis und der kritische Rationalismus: Akten des 3. internat. Wittgenstein Kongresses,* edited by H. Berghel, A. Hübner, E. Köhler, 302–7. Vienna: Hölder-Pichler-Tempsky. Translated as History and the System of Science in Otto Neurath, in Uebel, ed., 1991, 33–40.

———. 1982. Das Neurath-Prinzip. In *Arbeiterbildung in der Zwischenkriegszeit,* edited by F. Stadler, 79–87. Vienna: Löcker. Translated as The Neurath Principle: Its Grounds and Consequences, in Uebel, ed., 1991, 117–30.

———. 1985. Der erste Wiener Kreis. *Erkenntnis* 22:341–58. Translated as The First Vienna Circle, in Uebel, ed., 1991, 95–108.

Haller, R., and F. Stadler. 1993. *Wien-Berlin-Prag: Der Aufstieg der wissenschaftlicher Philosophie.* Vienna: Hölder-Pichler-Tempsky.

Hanson, N. R. 1958. The Logic of Discovery. *Journal of Philosophy* 55:1073–89.

———. 1961. Is There a Logic of Discovery? In Feigl and Maxwell 1961, 20–35.

Hatfield, G. 1990. *The Natural and the Normative: Theories of Perception from Kant to Helmholtz.* Cambridge, Mass.: MIT Press.

Havas, P. 1989. The Early History of the 'Problem of Motion' in General Relativity. In Howard and Stachel 1989, 234–76.

Hecht, H., and D. Hoffmann. 1982. Die Berufung Hans Reichenbachs an die Berliner Universitaet. *Deutsche Zeitschrift für Philosophie* 30:651–62.

Hegselmann, R. 1979. Otto Neurath: Empiristischer Aufklärer und Sozialreformer. In Hegselmann, ed., *Otto Neurath, Wissenschaftliche Weltauffassung: Sozialismus und Logischer Empirismus.* Frankfurt am Main: Suhrkamp, 7–78.

———. 1983. Empiristischer Antifaschismus: Das Beispiel Otto Neurath. *Dialektik* 7:67–74.

Heidegger, M. 1927. *Sein und Zeit.* Tübingen: Max Niemeyer. Translated as *Being and Time.* New York: Harper and Row, 1962.

———. 1929a. *Kant und das Problem der Metaphysik.* Bonn: Friedrich Cohen. Translated as *Kant and the Problem of Metaphysics.* Bloomington: Indiana University Press, 1990.

———. 1929b. *Was ist Metaphysik?* Bonn: Friedrich Cohen. Translated as What Is Metaphysics? in *Basic Writings,* edited by D. Krell. New York: Harper and Row, 1977.

———. 1943. "Nachwort" to the 4th ed. of *Was ist Metaphysik?* Frankfurt: Klostermann. Translated as "Postscript" to What Is Metaphysics? in *Existence and Being,* edited by W. Brock, 349–61. Chicago: Henry Regnery, 1949.

———. 1953. *Einführung in die Metaphysik.* Tübingen: Niemeyer. Translated as *Introduction to Metaphysics.* New York: Doubleday, 1961.

———. 1976. *Logik: Die Frage nach der Wahrheit: Gesamtausgabe.* Vol. 21. Frankfurt: Klostermann.

———. 1978. *Frühe Schriften: Gesamtausgabe.* Vol. 1. Frankfurt: Klostermann.

———. 1979. *Prolegomena zur Geschichte des Zeitbegriffs: Gesamtausgabe.* Vol. 20. Frankfurt: Klostermann.

———. 1983. *Einführung in die Metaphysik: Gesamtausgabe.* Vol. 40. Frankfurt: Klostermann.
———. 1991. *Kant und das Problem der Metaphysik: Gesamtausgabe.* Vol. 3. Frankfurt: Klostermann.
Helmholtz, H. von. 1869. The Aim and Progress of Physical Science. In Kahl 1971, 223–45.
———. 1878a. Die Thatsachen in der Wahrnehmung: Rede gehalten zur Stiftungsfeier der Friedrich-Wilhelms-Universtität zu Berlin. In *Vorträge und Reden,* 4th ed., 2:215–47, 387–406. Braunschweig: Friedrich Vieweg und Sohn, 1896.
———. 1878b. The Facts of Perception. In Kahl 1971, 366–408.
———. 1894. Introduction to the Lectures on Theoretical Physics. In Kahl 1971, 513–29.
———. 1921. *Schriften zur Erkenntnistheorie.* Edited by Paul Hertz and Moritz Schlick. Berlin: Julius Springer. Translated by M. F. Lowe as *Epistemological Writings,* edited by R. S. Cohen and Y. Elkana. Dordrecht: Reidel, 1977.
Hempel, C. 1935a. On the Logical Positivists' Theory of Truth. *Analysis* 49–59.
———. 1935b. Some Remarks on "Facts" and Propositions. *Analysis* 93–96.
———. 1945. Studies in the Logic of Confirmation. *Mind* 54:1–26, 97–121.
Hempel, C. G., and P. Oppenheim. 1948. Studies in the Logic of Explanation. *Philosophy of Science* 15:135–75.
Hentschel, K. 1990. *Interpretationen und Fehlinterpretationen der speziellen und der allgemeinen Relativitätstheorie durch Zeitgenossen Albert Einsteins.* Basel: Birkhäuser.
Herf, Jeffrey. 1984. *Reactionary Modernism: Technology, Culture, and Politics in Weimar and the Third Reich.* Cambridge: Cambridge University Press.
Hermann, A. 1979. *Wolfgang Pauli Wissenschaftlicher Briefwechsel.* Vol. 1. New York: Springer.
———, ed. 1969. *Albert Einstein/Arnold Sommerfeld Briefwechsel.* Basel/Stuttgart: Schwabe.
Herneck, F. 1976. *Einstein und sein Weltbild: Aufsätze und Vorträge.* Berlin: Buchverlag Der Morgen.
Hilbert, D. 1899. Grundlagen der Geometrie. In *Festschrift zur Feier der Enthüllung des Gauss-Weber-Denkmals,* 3–92. Leipzig: Teubner.
———. 1915. Die Grundlagen der Physik: Erste Mitteilung. *Königliche Gesellschaft der Wissenschaften zu Göttingen, Mathematische-Physikalische Klasse. Nachrichten,* 395–407.
———. 1916. Die Grundlagen der Physik: Zweite Mitteilung. *Königliche Gesellschaft der Wissenschaften zu Göttingen, Mathematische-Physikalische Klasse. Nachrichten,* 53–76.
———. 1918. Axiomatisches Denken. *Mathematische Annalen* 78:405–15.
Hoffman, B. 1972. *Albert Einstein: Creator and Rebel.* New York: Viking.
Hoffmann, R. 1982. Proletarisches Siedeln: Otto Neuraths Engagement für die Wiener Siedlungsbewegung und den Geldensozialismus von 1920 bis 1925. In Stadler 1982b, 140–48.
Hollinger, D. A. 1995. Science as a Weapon in *Kulturkämpfe* in the United States during and After World War II. *Isis* 86:440–54.
Holton, G. 1992. Ernst Mach and the Fortunes of Positivism in America. *Isis* 83:27–60.
———. Forthcoming. On the Vienna Circle in Exile: An Eyewitness Report. *Yearbook: Vienna Circle Lecture Series,* edited by F. Stadler. Boston: Kluwer.
Howard, D. 1984. Realism and Conventionalism in Einstein's Philosophy of Science: The Einstein-Schlick Correspondence. *Philosophia Naturalis* 21:616–29.
———. 1990. Einstein and Duhem. *Synthese* 83:363–84.

———. 1992. Einstein and *Eindeutigkeit:* A Neglected Theme in the Philosophical Background to General Relativity. In Eisenstaedt and Kox 1992, 154–243.
———. 1993. Was Einstein Really a Realist? *Perspectives on Science* 1:204–51.
———. 1994. Einstein, Kant, and the Origins of Logical Empiricism. In *Language, Logic, and the Structure of Scientific Theories: The Carnap-Reichenbach Centennial,* edited by W. Salmon and G. Wolters, 145–205. Pittsburgh: University of Pittsburgh Press; Konstanz: Universität Konstanz, 1994.
Howard, D., and J. D. Norton. 1993. Out of the Labyrinth? Einstein, Hertz, and the Göttingen Answer to the Hole Argument. In *The Attraction of Gravitation: New Studies in the History of General Relativity,* edited by J. Earman, M. Janssen, and J. D. Norton, 30–62. Einstein Studies, vol. 3. Boston: Birkhäuser.
Howard, D., and J. Stachel, eds. 1989. *Einstein and the History of General Relativity: Based on the Proceedings of the Osgood Hill Conference, North Andover, Massachusetts, 8–11 May 1986.* Einstein Studies, vol. 1. Boston: Birkhäuser.
Huntington, E. V. 1902. A Complete Set of Postulates for the Theory of Absolute Continuous Magnitude. *American Mathematical Society: Transactions* 3:264–79.
Husserl, E. 1900–1901. *Logische Untersuchungen.* Halle: Max Niemeyer. Translated as *Logical Investigations.* London: Routledge and Kegan Paul, 1970.
———. 1911. Philosophie als strenge Wissenschaft. *Logos* 1:289–341. Translated as Philosophy as Rigorous Science, in *Phenomenology and the Crisis of Philosophy,* edited by Q. Lauer. New York: Harper and Row, 1965.
———. 1913. *Ideen zu einer reinen Phänomenologie und phänomenologischen Philosophie.* Vol. 1. Halle: Niemeyer. Translated as *Ideas: General Introduction to Pure Phenomenology.* London: Allen and Unwin, 1931.
Huyssen, A. 1986. Mapping the Postmodern. In *After the Great Divide: Modernism, Mass Culture, Postmodernism,* 179–221. Bloomington: Indiana University Press.
Jones, C. 1994. Postmodernism. In *Dictionary of Art.* London: Macmillan.
Jordy, W. H. 1969. The Aftermath of the Bauhaus in America: Gropius, Mies, and Breuer. In Fleming and Bailyn 1969.
Kahl, R., ed. 1971. *Selected Writings of H. von Helmholtz.* Middleton, Conn.: Wesleyan University Press.
Kamlah, A. 1979. *Erlauterungen zur "Axiomatik der relativistischen Raum-Zeit-Lehre" bis §19.* In *Hans Reichenbach: Gesammelte Werke,* edited by A. Kamlah and M. Reichenbach, 3:431–64. Braunschweig/Wiesbaden: F. Vieweg and Sohn.
Kant, I. 1968. *Kant: Selected Pre-critical Writings and Correspondence with Beck.* Manchester: Manchester University Press; New York: Barnes and Noble.
Kasner, E. 1921. Einstein's Theory of Gravitation: Determination of the Field by Light Signals. *American Journal of Mathematics* 43:20–28.
Kautsky, K., 1909. Ein Brief über Marx und Mach. *Der Kampf* 2:451–52.
Keynes, J. M. 1921. *A Treatise on Probability.* London: Macmillan.
Kolakowski, L. 1978. *Main Currents of Marxism.* Vol. 2. Oxford: Oxford University Press.
Koppelberg, D. 1987. *Die Aufhebung der Analytischen Philosophie.* Frankfurt am Main: Suhrkamp.
Körner, T. 1946. Ziel und Aufgabe. *Der Aufbau* (Vienna) 1:2.
Kraft, V. 1953. *The Vienna Circle.* New York: Philosophical Library.
Kretschmann, E. 1917. Über den physikalischen Sinn der Relativitätstheorie. *Annalen der Physik* 53:575–614.
Kuhn, T. S. 1962. *The Structure of Scientific Revolutions.* Chicago: University of Chicago Press. 2d ed., 1970.

Labriola, A. 1908. *Essays on the Materialist Conception of History.* Chicago: Kerr. 1st ed., 1897.

Landauer, G. 1907. *Die Revolution.* Stuttgart. Reprint, Berlin: Kramer, 1974.

Lane, B. M. 1968. *Architecture and Politics in Germany, 1918–1945.* Cambridge, Mass.: Harvard University Press. 2d. ed., 1985.

Lask, E. 1912. *Die Lehre vom Urteil.* Tübingen: Mohr.

Latour, B. 1987. *Science in Action.* Cambridge, Mass.: Harvard University Press.

Lewin, K. 1923. Die zeitliche Geneseordnung. *Zeitschrift für Physik* 13:62–81.

Lewis, J. 1988. Schlick's Critique of Positivism. *PSA 1988,* edited by A. Fine and J. Leplin, 1:110–17. East Lansing, Mich.: Philosophy of Science Association.

———. 1989. Antimetaphysical Positivism and Empirical Realism. *Southwest Philosophy Review* 5/2:51–67.

———. 1991. Hidden Agendas: Knowledge and Verification. In *PSA 1990,* edited by L. Wessels, A. Fine, and M. Forbes, 2:159–68. East Lansing, Mich.: Philosophy of Science Association.

Lorentz, H. A., et al. 1923. *The Principle of Relativity: A Collection of Original Memoirs on the Special and General Theory of Relativity.* Notes by A. Sommerfeld. Translated by W. Perrett and G. B. Jeffery. London: Methuen. Reprint, New York: Dover, 1952.

Lotze, H. 1874. *Logik.* Leipzig: Hirzel. Translated as *Logic.* Oxford: Oxford University Press, 1884.

Mach, E. 1872. *Die Geschichte und die Wurzel des Satzes von der Erhaltung der Arbeit.* Prague: Calve.

———. 1883. *Die Mechanik in ihrer Entwicklung historisch-kritisch dargestellt.* Leipzig: Brockhaus. 4th ed. translated as *The Science of Mechanics.* Chicago: Open Court, 1902.

———. 1888. *Die Mechanik in ihrer Entwickelung historisch-kritisch dargestellt.* 2d ed. Leipzig: Brockhaus.

———. 1896. *Die Principien der Wärmelehre.* Leipzig: Barth.

———. 1897. *Die Mechanik in ihrer Entwickelung historisch-kritisch dargestellt.* 3d ed. Leipzig: Brockhaus.

———. 1900. *Die Analyse der Empfindungen und das Verhältniss des Physischen zum Psychischen.* 2d ed. Jena: Fischer.

———. 1905. *Erkenntnis und Irrtum. Skizzen zur Psychologie der Forschung.* Leipzig: Barth.

———. 1921. *Die Mechanik in ihrer Entwickelung historisch-kritisch dargestellt.* 8th ed. Edited by Joseph Petzoldt. Leipzig: Barth.

MacLean, P. 1990. *The Triune Brain in Evolution.* New York: Plenum.

Marcuse, P. 1986. A Useful Installment of Socialist Work: Housing in Red Vienna in the 1920s. In Bratt, Hartman, and Meyerson 1986, 558–85.

Marx, K. 1845. Thesen zu Feuerbach. Translated as Theses on Feuerbach. In *The Marx-Engels Reader,* edited by R. C. Tucker, 143–45. 2d ed. New York: Norton, 1978.

———. 1852. Der achzehnte Brumaire des Louis Bonaparte. Translated as The Eighteenth Brumaire of Louis Bonaparte. In *The Marx-Engels Reader,* edited by R. C. Tucker, 594–617. 2d ed. New York: Norton, 1978.

Marx, K., and F. Engels. 1935. *Correspondence, 1846–1895.* Edited by Dona Torr. London: Martin Lawrence.

McCumber, J. 1996. Time in the Ditch: American Philosophy and the McCarthy Era. *Diacritics.*

Mehra, J. 1974. *Einstein, Hilbert, and the Theory of Gravitation.* Dordrecht: Reidel.

Mehra, J., and H. Rechenberg. 1982. *The Historical Development of Quantum Theory.* Vol. 2. New York: Springer.

Messer, A. 1901. *Einführung in die Erkenntnistheories.*

Meyer, H. 1926. The New World. Translated typescript of *Das Werk,* in the Hannes Meyer Nachlass, Deutsches Architekturmuseum, Frankfurt am Main, III-1/1:82/1–20.

Mie, G. 1917. Die Einsteinsche Gravitationstheorie und das Problem der Materie. *Physikalische Zeitschrift* 18:551–80, 596–602.

Moore, G. E. 1954–55. Wittgenstein's Lectures in 1930–1933. *Mind* 63:1–15, 289–316; 64:1–27.

Morris, C. 1937. *Logical Positivism, Pragmatism, and Scientific Empiricism.* Paris: Hermann et Cie.

Moulines, C. 1985. Hintergründe der Erkenntnistheorie des frühen Carnap. *Grazer Philosophische Studien* 23:1–18.

Müller, A. 1923. Der Sinn der physikalischen Axiomatik. *Physikalische Zeitschrift* 24: 444–50.

Münsterberg, H. 1900. *Prinzipien der Psychologie.* Vol. 1.

Nagel, E. 1939. *Principles of the Theory of Probability.* Chicago: University of Chicago Press.

———. 1940. Charles S. Peirce, Pioneer of Modern Empiricism. *Philosophy of Science* 7:69–80. Reprinted in *Sovereign Reason,* 89–100. Glencoe, Ill.: Free Press, 1954.

———. 1956. Naturalism Reconsidered. In *Logic without Metaphysics,* 3–18. Glencoe, Ill.: Free Press.

———. 1961. *The Structure of Science.* New York: Harcourt, Brace, and World.

Natorp, P. 1910. *Die logischen Grundlagen der exakten Wissenschaften.* Wissenschaft und Hypothese, vol. 12. Leipzig: Teubner.

———. 1912. Kant und die Marburger Schule. *Kantstudien* 17:193–221.

Nemeth, E. 1981. *Otto Neurath und der Wiener Kreis: Wissenschaftlichkeit als revolutionärer politischer Anspruch.* Frankfurt: Campus.

———. 1982a. Die Einheit der Planwirtschaft und die Einheit der Wissenschaft. *Grazer Philosophische Studien* 16/17. Translated as The Unity of Planned Economy and the Unity of Science, in Uebel, ed., 1991, 275–83.

———. 1982b. Otto Neurath's Utopias — Der Wille zur Hoffnung. In Stadler 1982b, 94–99. Translated as Otto Neurath's Utopias — The Will to Hope, in Uebel, ed., 1991, 285–92.

———. 1994. Empiricism and the Norms of Scientific Knowledge: Some Reflections on Otto Neurath and Pierre Bourdieu. In *Norms, Values and Society,* edited by H. Pauer-Studer, 23–32. Dordrecht: Kluwer.

Neurath, O. 1910. Über die Produktivität der Volkswirtschaft: Verhandlungen des Vereins für Sozialpolitik, Wien 1909. *Schriften des Vereins für Sozialpolitik* 132:599–603.

———. 1911. Nationalökonomie und Wertlehre. *Zeitschrift für Volkswirtschaft, Sozialpolitik und Verwaltung* 20:52–114.

———. 1913a. Die Verirrten des Cartesius und das Auxiliarmotiv (Zur Psychologie des Entschlusses). *Jahrbuch der Philosophischen Gesellschaft an der Universität zu Wien 1913,* 45–59. Reprinted in Neurath 1981, 57–67. Translated as The Lost Wanderers of Descartes and the Auxiliary Motive (On the Psychology of Decision). In Neurath 1983, 1–12.

———. 1913b. Probleme der Kriegswirtschaftslehre. *Zeitschrift für die gesammte Staatswissenschaft* 69:438–501.

———. 1915. Prinzipielles zur Geschichte der Optik. *Archiv für Geschichte der Naturwissenschaften und der Technik* 5:371–89. Translated as On the Foundations of the History of Optics. In Neurath 1973, 101–12.

———. 1916. Zur Klassification von Hypothesensystemen. *Jahrbuch der philosophischen Gesellschaft an der Universität Wien 1916.* Translated as On the Classification of Systems of Hypotheses. In Neurath 1981, 13–31.

———. 1919a. Die Utopie als gesellschaftstechnische Konstruktion. In *Durch die Kriegswirtschaft zur Naturalwirtschaft.* Munich: Callwey, 228–31. Translated as Utopia as a Social Engineer's Construction. In Neurath 1973, 150–55.

———. 1919b. Utopien. *Wirtschaft und Lebensordnung* 1/1:5–6. Reprinted in Neurath 1981, 137–38.

———. 1920a. Ein System der Sozialisierung. *Archiv für Sozialwissenschaft und Sozialpolitik* 48:44–73.

———. 1920b. *Vollsozialisierung.* Jena: Diederichs.

———. 1920c. Wirtschaftsplan, Planwirtschaft, Landesverfassung und Völkerordnung. *Der Kampf* 13:224–27.

———. 1921. *Anti-Spengler.* Munich: Callwey. Translated in Neurath 1973.

———. 1922. Vollsozialisierung und gemeinwirtschaftliche Anstalten. *Der Kampf* 15:54–60.

———. 1926a. Das Neue Bauhaus in Dessau. *Der Aufbau: Österreichische Monatshefte für Siedlung und Städtebau* 1/11–12:210–11.

———. 1926b. Rationalismus, Arbeiterschaft und Baugestaltung. *Der Aufbau: Österreichische Monatshefte für Siedlung und Städtebau* 1/4:49–54.

———. 1931a. *Empirische Soziologie.* Vienna: Springer. Reprinted in Neurath 1981, 423–527. Translated (in part) as Empirical Sociology, in Neurath 1973, 391–421.

———. 1931b. Physicalism: The Philosophy of the Vienna Circle. *The Monist* 41:618–23. Reprinted in Neurath 1983, 48–51.

———. 1931c. Physikalismus. *Scientia* 50:297–303. Translated as Physicalism in Neurath 1983, 52–57.

———. 1931d. Weltanschauung und Marxismus. *Der Kampf* 24:447, 451. Reprinted in Neurath 1981, 407–12.

———. 1932a. Die "Philosophie" im Kampf gegen den Fortschritt der Wissenschaft. *Der Kampf* 25:385–89. Reprinted in Neurath 1981, 571–76.

———. 1932b. Protokollsätze. *Erkenntnis* 3:204–14. Reprinted in Neurath 1981, 577–86. Translated as Protocol Statements, in Neurath 1983, 91–99.

———. 1932c. Soziologie im Physikalismus. *Erkenntnis* 2:393–431. Reprinted in Neurath 1981, 533–62. Translated as Sociology in the Framework of Physicalism, in Neurath 1983, 58–90.

———. 1934. Radikaler Physikalismus und "Wirkliche Welt." *Erkenntnis* 4:346–62. Translated as Radical Physicalism and the Real World, in Neurath 1983, 100–114.

———. 1935. Einheit der Wissenschaft als Aufgabe. *Erkenntnis* 5:16–22. Translated as The Unity of Science as a Task, in Neurath 1983, 115–20.

———. 1939. *Modern Man in the Making.* London: Secker and Warburg; New York: Knopf.

———. 1941. Universal Jargon and Terminology. *Proceedings of the Aristotelian Society,* n.s., 41:127–48. Reprinted in Neurath 1983, 213–29.

———. 1944. *Foundations of the Social Sciences. International Encyclopedia of Unified Science.* Vol. 2, no.1. Chicago: University of Chicago Press.

———. 1973. *Empiricism and Sociology.* Edited by M. Neurath and R. S. Cohen. Dordrecht: Reidel.

———. 1981. *Gesammelte philosophische und methodologische Schriften.* Edited by R. Haller and H. Rutte. Vienna: Hölder-Pichler-Tempsky.

———. 1983. *Philosophical Papers: 1913–1946.* Edited by R. S. Cohen and M. Neurath. Dordrecht: Reidel.

Neurath, O., R. Carnap, and C. Morris, eds. 1955. *Foundations of the Unity of Science.* Vol 1. Chicago: University of Chicago Press.

Neurath, P. 1982. Souvenirs des débuts des statistiques illustrées et de l'isotype. In Sebestik and Soulez 1986, 289–97.

Norton, J. 1984. How Einstein Found His Field Equations. *Historical Studies in the Physical Sciences* 14:253–316. Reprinted in Howard and Stachel 1989, 101–59.

———. 1992. Einstein, Nordström and the Early Demise of Scalar, Lorentz-Covariant Theories of Gravitation. *Archive for History of Exact Science* 45:17–94.

———. 1993. General Covariance and the Foundations of General Relativity: Eight Decades of Dispute. *Reports on Progress in Physics* 56:791–858.

Oberdan, T. 1990. Positivism and the Pragmatic Theory of Observation. In *PSA 1990,* edited by A. Fine, M. Forbes, and L. Wessels, 1:25–37. East Lansing, Mich.: Philosophy of Science Association.

———. 1992. The Concept of Truth in Carnap's *Logical Syntax of Language. Synthese* 93:239–60.

———. 1993. *Protocols, Truth and Convention.* Studien zur Oesterreichischen Philosophie, vol. 19. Amsterdam and Atlanta: Rodopi.

Ott, H. 1988. *Martin Heidegger: Unterwegs zu seiner Biographie.* Frankfurt: Campus. Translated as *Martin Heidegger: A Political Life.* London: HarperCollins, 1993.

Pais, A. 1982. *"Subtle Is the Lord..." : The Science and Life of Albert Einstein.* New York: Oxford University Press.

Pauli, W. 1919a. Merkurperihelbewegung und Stralenablenkung in Weyls Gravitationstheorie. *Verhandlungen der Deutschen Physikalischen Gesellschaft* 21/22:742–50.

———. 1919b. Zur Theorie der Gravitation und der Elektrizitätat von Hermann Weyl. *Physikalische Zeitschrift* 20:457–67.

———. 1921. Relativitätstheorie. *Encyklopädie der mathematischen Wissenschaften.* Vol. 19. Leipzig: Teubner. Translated as *Theory of Relativity.* New York: Pergamon, 1958.

———. 1926. Review of Eddington 1925. *Die Naturwissenschaften* 13:273–74.

Pedoe, D. 1988. *Geometry: A Comprehensive Course.* New York: Dover.

Peirce Edition Project. Indiana University–Purdue University at Indianapolis.

Peirce, C. S. 1931–58. *Collected Works.* Edited by C. Hartshorne, P. Weiss, and A. Burks. 8 vols. Cambridge, Mass.: Harvard University Press.

Petzoldt, J. 1890. Maxima, Minima und Ökonomie. *Vierteljahrsschrift für wissenschaftliche Philosophie und Soziologie* 14:206–39, 354–66, 417–42. Reprinted as *Maxima, Minima und Ökonomie.* Inaugural-Dissertation zur Erlangung der Doctorwürde der philosophischen Facultät der Georgs-August-Universität zu Göttingen. Altenburg: Schnuphase (Max Lippold), 1891.

———. 1895. Das Gesetz der Eindeutigkeit. *Vierteljahrsschrift für wissenschaftliche Philosophie und Soziologie* 19:146–203.

———. 1900. *Einführung in die Philosophie der reinen Erfahrung.* Vol. 1, *Die Bestimmtheit der Seele.* Leipzig: Teubner.

———. 1904. *Einführung in die Philosophie der reinen Erfahrung.* Vol. 2, *Auf dem Wege zum Dauernden.* Leipzig: Teubner.

———. 1906. *Das Weltproblem von positivistischem Standpunkte aus.* Aus Natur und Geisteswelt, no. 133. Leipzig: Teubner.

———. 1912a. *Das Weltproblem vom Standpunkte des relativistischen Positivismus aus, historisch-kritisch dargestellt.* Wissenschaft und Hypothese, 14. Leipzig: Teubner. 2d ed. of Petzoldt 1906.

———. 1912b. Die Relativitätstheorie im erkenntnistheoretischer Zusammenhange des relativistischen Positivismus. *Deutsche Physikalische Gesellschaft. Verhandlungen* 14: 1055–64.

———. 1914. Die Relativitätstheorie der Physik. *Zeitschrift für positivistische Philosophie* 2:1–53.

———. 1921a. Das Verhältnis der Machschen Gedankenwelt zur Relativitätstheorie. Postscript to Mach 1921, 490–517.

———. 1921b. *Die Stellung der Relativitätstheorie in der geistigen Entwicklung der Menschheit.* Dresden: Sibyllen.

Pevsner, N. 1975. *Pioneers of Modern Design: From William Morris to Walter Gropius.* New York: Penguin.

Planck, M. 1909. The Unity of the Physical Universe. In *A Survey of Physical Theory,* 1–26. Translated by R. Jones and D. Williams. New York: Dover, 1960.

———. 1913. New Paths of Physical Knowledge. In *A Survey of Physical Theory,* translated by R. Jones and D. Williams, 45–55, 1960. New York: Dover Publications.

Plekhanov, G. V. 1897. The Materialist Conception of History. Reprinted in Plekhanov 1908.

———. 1908. *Fundamental Problems of Marxism.* London. Reprint, New York: International Publishers, 1969.

Poincaré, H. 1902. *Science et Hypothese.* Paris. Translated as *Science and Hypothesis.* Translated by W. J. Greenstreet. London: Scott, 1905. Reprint, New York: Dover, 1952. Also translated by G. B. Halsted in Poincaré, *The Foundations of Science.* Lancaster, Pa.: Science Press, 1946.

———. 1905a. La Valeur de science, Paris. Translated as *The Value of Science.* Translated by G. B. Halsted in Poincaré, *The Foundations of Science.* Lancaster, Pa.: Science Press, 1946.

———. 1905b. Les mathématiques et la logique. *Revue de métaphysique et de morale* 13:815–35.

Popper, K. R. 1935. *Logik der Forschung: Zur Erkenntnistheorie der Modernen Naturwissenschaft.* Vienna: Springer.

———. 1959. *The Logic of Scientific Discovery.* London: Hutchinson.

Pos, H. 1949. Recollections of Ernst Cassirer. In *The Philosophy of Ernst Cassirer,* edited by P. Schilpp, 63–72. La Salle, Ill.: Open Court.

Proust, J. 1986. *Questions de forme: Logique et proposition analytique de Kant à Carnap.* Paris: Librairie Arthème Fayard. Translated as *Questions of Form: Logic and the Analytic Proposition from Kant to Carnap.* Minneapolis: University of Minnesota Press, 1989.

Putnam, H. 1963. An Examination of Grünbaum's Philosophy of Geometry. Reprinted in *Philosophical Papers,* vol. 1. 2d ed., 93–129. 2d ed. New York: Cambridge University Press, 1979.

———. 1976. Realism and Reason. Reprinted in *Meaning and the Moral Sciences,* 123–38. Boston: Routledge and Kegan Paul, 1978.

Quine, W. V. O. 1936. Truth by Convention. In *Philosophical Essays for A. N. Whitehead,* edited by O. H. Lee, 90–124. New York: Longman. Reprinted in Quine 1976, 107–32.

———. 1948. On What There Is. *Review of Metaphysics* 2:21–38. Reprinted in Quine 1953, 1–19.

———. 1950. *Methods of Logic.* New York: Holt.

———. 1951a. *Mathematical Logic.* Cambridge, Mass.: Harvard University Press.
———. 1951b. Two Dogmas of Empiricism. *Philosophical Review* 60:20–43. Reprinted in Quine 1953, 20–47.
———. 1953. *From a Logical Point of View.* Cambridge, Mass.: Harvard University Press.
———. 1960. Carnap and Logical Truth. *Synthese* 12:350–74. Reprinted in Quine 1976, 107–32.
———. 1969. Epistemology Naturalized. In *Ontological Relativity and Other Essays,* 69–90. New York: Columbia University Press.
———. 1976. *Ways of Paradox.* Cambridge, Mass.: Harvard University Press.
Reichenbach, H. 1916–17. Der Begriff der Wahrscheinlichkeit für die mathematische Darstellung der Wirklichkeit. *Zeitschrift für Philosophie und Philosophische Kritik,* vols. 161–63.
———. 1919. Student und Sozialismus. *Der Aufbau: Flugblätter an Jugend* (Berlin) 5:3–9.
———. 1920. *Relativitätstheorie und Erkenntnis Apriori.* Berlin: Springer. Translated as Reichenbach 1965.
———. 1921. Bericht über den Axiomatik der relativistischen Raum-Zeit Lehre. *Physikalische Zeitschrift* 22:683–86.
———. 1922a. Der gegenwärtige Stand der Relativitätsdiskussion. *Logos* 10:316–78. Translated (with omissions) as The Present State of the Discussion on Relativity: A Critical Investigation, in Reichenbach 1987, 3–47.
———. 1922b. La Signification philosophique de la théorie de la relativité. *Revue Philosophique* 94:5–61.
———. 1922c. Review of Helmholtz 1921. *Zeitschrift für angewandte psychologie* 20:421–22.
———. 1924. *Axiomatik der Relativistischen Raum-Zeit-Lehre.* Braunschweig: Vieweg. Translated as Reichenbach 1969.
———. 1925. Über die physikalischen Konsequenzen der relativistischen Axiomatik. *Zeitschrift für Physik* 34:32–48.
———. 1927. Lichtgeschwindigkeit und Gleichzeitigkeit. *Annalen der Philosophie* 6: 128–44.
———. 1928. *Philosophie der Raum-Zeit-Lehre.* Berlin: de Gruyter. Translated as Reichenbach 1958.
———. 1933. Die logischen Grundlagen des Wahrscheinlichkeits-begriffs. *Erkenntnis* 3:401–25.
———. 1935. *Wahrscheinlichkeitslehre.* Leiden: Sijthoff.
———. 1936. Logistic Empiricism in Germany and the Present State of Its Problems. *The Journal of Philosophy* 33:141–60.
———. 1938. *Experience and Prediction.* Chicago: University of Chicago Press.
. 1939. Dewey's Theory of Science. In Schilpp and Hahn 1939, 157–92.
———. 1944. *Philosophic Foundations of Quantum Mechanics.* Berkeley: University of California Press.
———. 1947. *Elements of Symbolic Logic.* New York: Macmillan.
———. 1949a. The Philosophical Significance of the Theory of Relativity. In Schilpp 1949, 287–311.
———. 1949b. *The Theory of Probability.* Berkeley: University of California Press.
———. 1951. *The Rise of Scientific Philosophy.* Berkeley: University of California Press.
———. 1954. *Nomological Statements and Admissible Operations.* Amsterdam: North-Holland.

———. 1956. *The Direction of Time.* Edited by M. Reichenbach. Berkeley: University of California Press.

———. 1958. *The Philosophy of Space and Time.* Translated by M. Reichenbach and J. Freund. New York: Dover.

———. 1965. *The Theory of Relativity and A Priori Knowledge.* Berkeley: University of California Press.

———. 1969. *Axiomatization of the Theory of Relativity.* Berkeley: University of California Press.

———. 1978. *Hans Reichenbach: Selected Writings, 1909–1953.* Vol. 2. Edited by M. Reichenbach and R. Cohen. Dordrecht: Reidel.

Reisch, G. A. 1994. Planning Science: Otto Neurath and the "International Encyclopedia of Unified Science." *British Journal for the History of Science* 27:153–75.

Rescher, N. 1973a. *The Primacy of Practice.* Oxford: Blackwell.

———. 1973b. *Conceptual Idealism.* Oxford: Blackwell.

———. 1977. *Methodological Pragmatism.* New York: New York University Press.

———, ed. 1985. *The Heritage of Logical Empiricism.* Lanham, Md.: University Press of America.

Richardson, A. 1990. How Not to Russell Carnap's *Aufbau.* In *PSA 1990,* edited by A. Fine, M. Forbes, and L. Wessels, 1:3–14. East Lansing, Mich.: Philosophy of Science Association.

———. 1992a. Idealisms in the "Aufbau." *Grazer Philosophische Studien* 43:45–72.

———. 1992b. Logical Idealism and Carnap's Construction of the World. *Synthese* 93: 59–92.

———. 1994. The Limits of Tolerance. *Proceedings of the Aristotelian Society,* supplementary volume.

Rickert, H. 1892. *Der Gegenstand der Erkenntnis.* Tübingen: Mohr.

———. 1909. Zwei Wege der Erkenntnistheorie. *Kantstudien* 14:169–228.

———. 1911. Das Eine, die Einheit und die Eins. *Logos* 2:26ff.

Ricketts, T. 1982. Rationality, Translation, and Epistemology Naturalized. *Journal of Philosophy* 79:117–36.

———. 1994. Carnap's Principle of Tolerance, Empiricism, and Conventionalism. In *Reading Putnam,* edited by P. Clark and B. Hale, 176–200. Oxford: Blackwell.

———. 1996. Carnap: From Logical Syntax to Semantics. In Giere and Richardson 1996.

Robb, A. A. 1914. *A Theory of Time and Space.* Cambridge: Cambridge University Press.

Russell, B. 1903. *The Principles of Mathematics.* London: Allen and Unwin

———. 1912. *The Problems of Philosophy.* Oxford: Oxford University Press.

———. 1914. On Scientific Method in Philosophy. In *Mysticism and Logic,* 75–93. London: George Allen and Unwin, 1917.

———. 1939. Dewey's New *Logic.* In Schilpp and Hahn 1939, 135–56.

———. 1940. *An Inquiry into Meaning and Truth.* London: George Allen and Unwin.

Ryckman, T. 1991. Conditio Sine Qua Non? Zuordnung in the Early Epistemologies of Cassirer and Schlick. *Synthese* 88:57–95.

———. 1992. "P(oint) C(oincidence) Thinking": The Ironical Attachment of Logical Empiricism to General Relativity (and Some Lingering Consequences). *Studies in History and Philosophy of Science* 23:471–93.

———. 1994. Weyl, Reichenbach and the Epistemology of Geometry. *Studies in the History and Philosophy of Science* 25:831–70.

———. 1996. Einstein *Agonists:* Weyl and Reichenbach on Geometry and the General Theory of Relativity. In Giere and Richardson 1996.

Salmon, W. C. 1989. *Four Decades of Scientific Explanation.* Minneapolis: University of Minnesota Press.

Sarkar, S. 1992. "The Boundless Ocean of Unlimited Possibilities": Logic in Carnap's "Logical Syntax of Language." *Synthese* 93:191–237.

Sauer, W. 1985. Carnaps *Aufbau* in kantianischer Sicht. *Grazer philosophische Studien* 23:19–35.

———. 1989. On the Kantian Background of Neopositivism. *Topoi* 8:111–19.

Scheffler, I. 1963. *The Anatomy of Inquiry: Philosophical Studies in the Theory of Science.* New York: Knopf.

———. 1982. *Science and Subjectivity.* 2d ed. Indianapolis: Hackett.

Schilpp, P. A., ed. 1949. *Albert Einstein: Philosopher-Scientist.* New York: Tudor.

———. 1963. *The Philosophy of Rudolf Carnap.* The Library of Living Philosophers, 11. La Salle, Ill.: Open Court.

Schilpp, P. A., and L. E. Hahn, eds. 1939. *The Philosophy of John Dewey.* La Salle, Ill.: Open Court.

Schlick, M. 1910. Das Wesen der Wahrheit nach der modernen Logik. *Vierteljahrsschrift für wissenschaftliche Philosophie und Soziologie* 34:386–477.

———. 1915. Die philosophische Bedeutung des Relativitätsprinzips. *Zeitschrift für Philosophie und philosophische Kritik* 159:129–75. Translated as The Philosophical Significance of the Principle of Relativity, in Schlick 1979a, 153–89.

———. 1917a. *Raum und Zeit in den gegenwärtigen Physik. Zur Einführung in das Verständnis der allgemeinen Relativitätstheorie.* Berlin: Springer. Translated as Space and Time in Contemporary Physics, in Schlick 1979a, 207–69.

———. 1917b. Raum und Zeit in den gegenwärtigen Physik. Zur Einführung in das Verständnis der allgemeinen Relativitätstheorie. *Die Naturwissenschaften* 5:161–67, 177–86.

———. 1918. *Allgemeine Erkenntnislehre.* Berlin: Springer.

———. 1919. *Raum und Zeit in den gegenwärtigen Physik. Zur Einführung in das Verständnis der Relativitäts- und Gravitationstheorie.* 2d ed. Berlin: Springer.

———. 1920. *Raum und Zeit in den gegenwärtigen Physik: Zur Einführung in das Verständnis der Relativitäts- und Gravitationstheorie.* 3d ed. Berlin: Springer.

———. 1921. Kritizistische oder empiristische Deutung der neuen Physik. *Kant-Studien* 26:96–111.

———. 1922a. *Raum und Zeit in den gegenwärtigen Physik. Zur Einführung in das Verständnis der Relativitäts- und Gravitationstheorie.* 4th ed. Berlin: Springer.

———. 1922b. Die Relativitätstheorie in der Philosophie. In *Verlhandlungen der Gesellschaft Deutscher Natuforscher und Ärzte. 87. Versammlung. Hundertjahrfeier,* 58–69. Leipzig.

———. 1925. *Allgemeine Erkenntnislehre.* 2d ed. Berlin: Springer. Translated as *General Theory of Knowledge,* in Schlick 1975.

———. 1926. Erleben, Erkennen, Metaphysik. *Kant Studien* 31:146–58. Translated as Experience, Cognition, and Metaphysics, in Schlick 1979b, 99–111.

———. 1928. Preface to Friedrich Waismann, *Logik, Sprache, Philosophie.* In Schlick 1979b, 130–38.

———. 1929. Review of Reichenbach, *Philosophie der Raum-Zeit-Lehre* (1928). *Die Naturwissenschaften* 17:549.

———. 1931. Die Kausalitätat in der gegenwärtigen Physik. *Die Naturwissenschaften* 19:145–62. Translated as Causality in Contemporary Physics, in Schlick 1979b, 176–209.

———. 1932. Form and Content: An Introduction to Philosophical Thinking. In Schlick 1979b, 285–369.

———. 1934. Über das Fundament der Erkenntnis. *Erkenntnis* 4:79–99. Translated as On the Foundation of Knowledge, in Schlick 1979b, 370–87.

———. 1934–35. *Logik und Erkenntnistheorie.* Unpublished typescript held at the Archives of the Vienna Circle Foundation, Rijksarchief in Noord-Holland. All rights reserved.

———. 1935a. De la Relation entre les notions psychologiques et les notions physiques. *Revue de Synthese* 10:5–26. Translated as On the Relation between Psychological and Physical Concepts, in Schlick 1979b, 420–36.

———. 1935b. Facts and Propositions. *Analysis* 2:65–70. Reprinted in Schlick 1979b, 400–404.

———. 1935c. Sur les "Constatations": Sur le Fondement de la connaissance. *Actualites Scientifiques et Industrielles* 289:44–54. Paris: Hermmann et Cie. Translated as Introduction and On "Affirmations," in Schlick 1979b, 405–13.

———. 1936a. Meaning and Verification. *Philosophical Review* 45:339–69. Reprinted in Schlick 1979b, 456–81.

———. 1936b. Sind die Naturgesetze Conventionen? *Induction et Probabilite.* Actes du Congres International de Philosophie Scientifique (Paris), 8–17. Translated as Are Natural Laws Conventions? in Schlick 1979b, 437–45.

———. 1938. Form and Content: An Introduction to Philosophical Thinking. In Moritz Schlick, *Gesammelte Aufsätze.* Wien: Gerold, 151–249. Reprinted in Schlick 1979b, 285–369.

———. 1975. *General Theory of Knowledge.* Translated by H. Feigl and A. Blumberg. New York: Springer.

———. 1979a. *Philosophical Papers.* Vol. 1, edited by H. L. Mulder and B. F. B. van de Velde-Schlick. Vienna Circle Collection. Vol. 11, edited by Henk Mulder. Dordrecht: Reidel.

———. 1979b. *Philosophical Papers.* Vol. 2, edited by H. L. Mulder and B. F. B. van de Velde-Schlick. Vienna Circle Collection.

———. 1985. *General Theory of Knowledge.* La Salle, Ill.: Open Court. Translation of Schlick 1925.

———. 1987. *The Problems of Philosophy in Their Interconnection.* Edited by H. Mulder, A. Kox, and R. Hegselmann. Dordrecht: Reidel.

Schnädelbach, H. 1979. Über historistische Aufklärung. *Allgemeine Zeitschrift für Philosophie.* Reprinted in *Vernunft und Geschichte* 1987. Frankfurt am Main: Suhrkamp.

Schneeberger, G. 1962. *Nachlese zu Heidegger.* Bern: Suhr.

Schorske, C. 1981. *Fin de Siècle Vienna: Politics and Culture.* New York: Vintage.

Schouten, J. A., and D. J. Struik. 1921. On Some Properties of General Manifolds Relating to Einstein's Theory of Gravitation. *American Journal of Mathematics* 43:213–16.

Schüller, H. 1921. Revolution-Aufbau. *Der Aufbau: Flugblätter an Jugend* (Berlin) 1:2–7.

Schuster, F. 1926. Gedanken über den Hausbau der Zukunft. *Der Aufbau: Österreichische Monatshefte für Siedlung und Städtebau* (Vienna) 1/1:7–11.

Sebestik, J., and A. Soulez, eds. 1986. *Le Cercle de Vienne: Doctrines et controverses.* Paris.

Seelig, C. 1960. *Albert Einstein.* Zurich: Europa.

Shapin, S., and S. Schaffer. 1985. *Leviathan and the Air-Pump: Hobbes, Boyle, and the Experimental Life.* Princeton, N.J.: Princeton University Press.

Sigurdsson, S. 1991. *Hermann Weyl, Mathematics, and Physics, 1900–1927.* Ph.D. diss., Harvard University.

Simmel, G. 1895. Über eine Beziehung der Selektionslehre zur Erkenntnistheorie. *Archiv für systematische Philosophie* 1:34–45. Translated as On a Relationship between the Theory of Selection and Epistemology, in *Learning, Development and Culture,* edited by H. C. Plotkin. New York: Wiley and Sons, 1982.

Sklar, L. 1985. Facts, Conventions, and Assumptions in the Theory of Spacetime. Reprinted in Sklar, *Philosophy and Spacetime Physics.* Berkeley: University of California Press, 73–147.

Sluga, H. 1980. *Gottlob Frege.* London: Routledge.

———. 1993. *Heidegger's Crisis: Philosophy and Politics in Nazi Germany.* Cambridge, Mass.: Harvard University Press.

Smith, C., and M. N. Wise. 1989. *Energy and Empire: A Biographical Study of Lord Kelvin.* New York: Cambridge University Press.

Speziali, P., ed. 1972. *Albert Einstein-Michele Besso: Correspondance 1903–1955.* Paris: Hermann.

Stachel, J. 1989a. Einstein's Search for General Covariance, 1912–1915. In Howard and Stachel 1989, 63–100.

———. 1989b. The Rigidly Rotating Disk as the "Missing Link" in the History of General Relativity. In Howard and Stachel 1989, 48–62.

———. 1993. The Meaning of General Covariance. In *Philosophical Problems of the Internal and External Worlds: Essays on the Philosophy of Adolf Grünbaum,* edited by J. Earman et al., 129–60. Pittsburgh: University of Pittsburgh Press.

Stadler, F. 1982a. *Vom Positivismus zur wissenschaftlichen Weltauffassung.* Vienna: Löcker.

———, ed. 1982b. *Arbeiterbildung in der Zwischenkriegszeit Otto Neurath–Gerd Arntz.* Vienna: Löcker.

———, ed. 1993. *Scientific Philosophy: Origins and Developments.* Dordrecht: Kluwer.

Straumann N. 1987. *"Zum Ursprung der Eichtheorien bei Hermann Weyl." Physikalischen Blätter* 43:414–21.

Struik, D. J. 1989. Schouten, Levi-Cività, and the Emergence of Tensor Calculus. In *The History of Modern Mathematics,* edited by D. Rowe and J. McCleary, 2:99–105. Boston: Academic Press.

Study, E. 1914. *Die realistische Weltansicht und die Lehre vom Raume.* Die Wissenschaft, 54. Braunschweig: Friedrich Vieweg and Sohn.

Suppe, F. 1989. *The Semantic View of Theories and Scientific Realism.* Chicago: University of Illinois Press.

Tarski, A. 1936. On the Concept of Logical Consequence. In *Logic, Semantics, Metamathematics,* 409–20. Oxford: Clarendon Press, 1956.

Tönnies, F. 1887. *Gemeinschaft und Gesellschaft.* 8th ed. Leipzig, 1935. Translated as *Fundamental Concepts of Sociology.* New York: American Book Company, 1940. Reprinted as *Community and Association.* London: Routledge and Kegan Paul, 1955.

Torretti, R. 1983. *Relativity and Geometry.* Oxford: Pergamon.

Uebel, T. E. 1991. Neurath's Programme for Naturalistic Epistemology. *Studies in the History and Philosophy of Science* 22:623–46.

———. 1992a. Neurath vs. Carnap: Naturalism vs. Rational Reconstructionism before Quine. *History of Philosophy Quarterly* 9:445–70.

———. 1992b. *Overcoming Logical Positivism from Within: The Emergence of Neurath's Naturalism in the Vienna Circle's Protocol Sentence Debate.* Amsterdam-Atlanta: Rodopi.

———. 1992c. Rational Reconstruction as Elucidation? Carnap in the Early Protocol Sentence Debate. *Synthese* 93:107–40.

———. 1993a. Neurath's Protocol Statements: A Naturalistic Theory of Data and Pragmatic Theory of Theory Acceptance. *Philosophy of Science* 60:587–607.

———. 1993b. Wilhelm Neurath's Argument against "Materialist" Darwinism. In Stadler 1993, 209–28.

———. 1995. Otto Neurath's Idealist Inheritance: The Socio-economic Theories of Wilhelm Neurath. *Synthese* 103:87–121.

———. 1996. The Enlightenment Ambition of Epistemic Utopianism: Notes on the History of Otto Neurath's Theory of Science. In Giere and Richardson 1996.

———, ed. 1991. *Rediscovering the Forgotten Vienna Circle: Austrian Studies on Otto Neurath and the Vienna Circle.* Dordrecht: Kluwer.

Umsturz und Aufbau (Berlin). 1919. 3:22–23. Reprinted in 1974.

Unity of Science Movement Papers. Department of Special Collections, University of Chicago Library.

van Fraassen, B. C. 1970. *An Introduction to the Philosophy of Time and Space.* New York: Random House.

———. 1980. *The Scientific Image.* Oxford: Oxford University Press.

———. 1989. *Laws and Symmetry.* Oxford: Oxford University Press.

———. 1991. *Quantum Mechanics: An Empiricist View.* Oxford: Oxford University Press.

Veblen, O. 1904. A System of Axioms for Geometry. *American Mathematical Society: Transactions* 5:343–84.

Volkmann, P. 1896. *Erkenntnistheoretische Grundzüge der Naturwissenschaften und ihre Beziehungen zum Geistesleben der Gegenwart. Allgemein wissenschaftliche Vorträge.* Leipzig: Teubner.

Waismann, F. 1958. *The Blue and Brown Books.* Oxford: Blackwell.

———. 1967. *Wittgenstein und der Wiener Kreis.* Edited by B. F. McGuinness. Frankfurt: Suhrkamp.

Wald, R. 1984. *General Relativity.* Chicago: University of Chicago Press.

Wartofsky, M. W. 1982. Positivism and Politics: The Vienna Circle as a Social Movement. *Grazer Philosophische Studien* 16/17:79–101.

Webb, J. 1992. Reconstruction from Recollection and the Refutation of Idealism: A Kantian Theme in the "Aufbau." *Synthese* 93:93–106.

Wellmer, A. 1985. *Zur Dialektik von Moderne und Postmoderne: Vernunftkritik nach Adorno.* Frankfurt: Suhrkamp.

Westbrook, R. B. 1991. *John Dewey and American Democracy.* Ithaca, N.Y.: Cornell University Press.

Weyl, H. 1917. Zur Gravitationstheorie. *Annalen der Physik* 54:117–45.

———. 1918a. Gravitation und Elektrizität. *Sitzungsberichte, Preussischen Akademie der Wissenschaften, Physikalische-Mathematische Klasse,* 465–80.

———. 1918b. *Raum-Zeit-Materie.* Berlin: Springer.

———. 1918c. Reine Infinitesimalgeometrie. *Mathematische Zeitschrift* 2:384–411.

———. 1919a. Eine Neue Erweiterung der Relativitätstheorie. *Annalen der Physik* 59: 101–33.

———. 1919b. *Raum-Zeit-Materie.* 3d ed. Berlin: Springer.

———. 1920. Elektrizität und Gravitation. *Physikalische Zeitschrift* 21:649–51.

———. 1921a. Electricity and Gravitation. *Nature* 106:800–802.

———. 1921b. Feld und Materie. *Annalen der Physik* 65:541–63.

———. 1921c. *Raum-Zeit-Materie.* 4th ed. Berlin: Springer. Translated as *Space-Time-Matter* (Weyl 1922c).

———. 1921d. Über die physikalischen Grundlagen der erweiterte Relativitätstheorie. *Physikalische Zeitschrift* 22:473–80.
———. 1921e. Zur Infinitesimalgeometrie: Einordnung der projektiven und der konformen Auffassung. *Nachrichten: Königlichen Gesellschaft der Wissenschaften zu Göttingen, Mathematische-Physikalische Klasse,* 99–112.
———. 1922a. Die Einzigartigkeit der Pythagoreischen Massbestimmung. *Mathematische Zeitschrift* 12:114–46.
———. 1922b. Die Relativitätstheorie auf der Naturforscherversammlung in Bad Nauheim. *Jahresbericht der Deutschen Mathematische-Vereinigung,* 51–63.
———. 1922c. *Space-Time-Matter.* London: Methuen. Reprint, New York: Dover, 1952.
———. 1923a. *Mathematische Analyse der Raumproblems: Vorlesungen gehalten in Barcelona und Madrid.* Berlin: Springer.
———. 1923b. *Raum-Zeit-Materie.* 5th ed. Berlin: Springer.
———. 1924. Rezension von Hans Reichenbach, Axiomatik der relativistischen Raum-Zeit-Lehre. *Deutsche Literaturzeitung* 30, col. 2122–27.
———. 1927. *Philosophie der Mathematik und Naturwissenschaft.* Munich: Oldenbourg. Translated as Weyl 1949.
———. 1930. Axiomatik. Vorlesungen der Winter Semester 1930. Göttingen Universität.
———. 1949. *Philosophy of Mathematics and Natural Science.* Princeton, N.J.: Princeton University Press.
———. 1951. 50 Jahre Relativitätstheorie. *Die Naturwissenschaften* 38:73–83.
Whitehead, A. N., and B. Russell. 1910. *Principia Mathematica.* Vol. 1. Cambridge: Cambridge University Press.
Willett, J. 1978. *The New Sobriety: Art and Politics in the Weimar Period, 1917–1933.* London: Thames and Hudson.
Williams, R. 1983. *Keywords: A Vocabulary of Culture and Society.* New York: Oxford University Press.
Wilson, Daniel J. 1990. *Science, Community, and the Transformation of American Philosophy, 1860–1930.* Chicago: University of Chicago Press.
Wingler, H. M. 1978. *The Bauhaus: Weimar, Dessau, Berlin, Chicago.* Edited by J. Stein. Cambridge, Mass.: MIT Press.
Wittgenstein, L. 1922. *Tractatus Logico-Philosophicus.* London: Routledge and Kegan Paul.
———. 1933. "The Big Typescript." Item 213 of the Cornell University Microfilms. Unpublished.
Wolin, R. 1991. *The Heidegger Controversy: A Critical Reader.* New York: Columbia University Press.
Worrall, J. 1989. Structural Realism: The Best of Both Worlds? *Dialectica* 43:99–124.
Wundt, W. 1894. Über psychische Causalität und das Princip des psycho-physischen Parallelismus. *Philosophische Studien* 10:1–124.
Yang C. N. 1985. Hermann Weyl's Contributions to Physics. In *Hermann Weyl 1885–1985: Centenary Lectures at the ETH Zürich,* edited by K. Chandrasekharan, 7–21. Berlin: Springer.
Zilsel, E. 1931a. Materialismus und marxistische Geschichtsauffassung. *Der Kampf* 24: 66–75.
———. 1931b. Partei, Marxismus, Materialismus, NeoKantianismus. *Der Kampf* 24: 213–20.

Contributors

Nancy Cartwright is professor of philosophy, logic, and scientific method at the London School of Economics and Political Science. In 1993 she became director of the newly formed London School of Economics Center for the Philosophy of the Natural and Social Sciences. Her two major publications are *Nature's Capacities and Their Measurement* and *How the Laws of Physics Lie.* She, along with Thomas Uebel, Jordi Cat, and Lola Fleck, has recently completed a book on Otto Neurath entitled *Otto Neurath: Philosophy between Science and Politics* (forthcoming).

Jordi Cat has written on Maxwell's philosophy of science and coauthored (with Nancy Cartwright, Lola Fleck, and Thomas Uebel) *Otto Neurath: Philosophy between Science and Politics.* He is currently a research associate in the physics department at Harvard.

Richard Creath is professor of philosophy at Arizona State University. His works include numerous essays on Carnap and Quine and in philosophy of science more generally. He is the editor of *Dear Carnap, Dear Van: The Quine-Carnap Correspondence and Related Work.*

Michael Friedman received his Ph.D. from Princeton University. He has taught at Harvard University, the University of Pennsylvania, the University of Illinois at Chicago, the University of California–Berkeley, and the University of Konstanz. He is presently Ruth N. Halls Professor of Arts and Humanities at Indiana University. He is the author of *Foundations of Space-Time Theories: Relativistic Physics and Philosophy of Science* and *Kant and the Exact Sciences.*

Peter Galison is the Mallinckrodt Professor of the History of Science and of Physics at Harvard. His primary interest centers on the history and philosophy of experimentation, the subject of his works *How Experiments End; Big Science: The Growth of Large-Scale Research,* edited with Bruce Hevly; and his current project, entitled *Image and Logic: The Material Culture of Modern Physics.* His work on the intersection of history of science and philosophy of science includes *The Disunity of Science,* edited with David Stump (forthcoming), and "Aufbau/Bauhaus: Logical Positivism and Architectural Modernism," in *Critical Inquiry.*

Ronald N. Giere is professor of philosophy and director of the Center for Philosophy of Science at the University of Minnesota. In addition to many essays on the philosophy of science, he is the author of an elementary textbook, *Understanding Scientific Reasoning,* and of *Explaining Science: A Cognitive Approach,* and editor of *Cognitive Models of Science.*

Warren Goldfarb is Walter Beverly Professor of Modern Mathematics and Mathematical Logic and professor philosophy at Harvard University. He is the coauthor (with Burton Dreben) of *The Decision Problem: Solvable Classes of Quantificational Formulas;* editor of Jacques Herbrand's *Logical Writings;* coeditor of Kurt Gödel's *Collected Works,* volume 3; and author of articles on mathematical logic, on the development of analytical philosophy, and on Wittgenstein. He is currently writing a book on Wittgenstein's *Philosophical Investigations.*

Don Howard is a philosopher of science specializing in the philosophy of physics and the history of the philosophy of science. Author of numerous articles on Einstein, Bohr, the interpretation of quantum mechanics, and other topics, Howard is now preparing a book on Einstein as a philosopher of science. He is also the cofounder (with Alan Richardson) of HOPOS, the History of Philosophy of Science Working Group.

Thomas Oberdan studied philosophy, as well as history and philosophy of science, at Indiana University. His work focuses on topics in the history of early logical positivism, including issues in the philosophy of logic and mathematics, as well as the analysis of scientific observation. He is the author of a monograph on the protocol-sentence controversy in the Vienna Circle, entitled *Protocols, Truth and Convention.* Currently he is associate professor of philosophy at Clemson University in South Carolina.

Alan W. Richardson is assistant professor of philosophy at the University of British Columbia. He is currently finishing a book on the early philosophy of Rudolf Carnap, entitled *Carnap's Construction of the World.*

Thomas Ricketts is associate professor of philosophy at the University of Pennsylvania. He has published essays on Frege, Wittgenstein, Carnap, and Quine.

T. A. Ryckman is assistant professor of philosophy at Northwestern University. His current work explores early philosophical interpretations of the general theory of relativity and their consequences.

Joia Lewis Turner received her Ph.D. in philosophy of science from Indiana University. Her published works comprise analyses of the interplay between the positivist and realist assumptions underlying Vienna Circle philosophy. She is currently on the faculty at the University of San Diego. Her current research interests are in the development of a philosophy of neuroscience encompassing the full range of non-cognitive models as well as the standard cognitive models of interest to philosophers of mind.

Thomas E. Uebel, currently lecturer in philosophy at the London School of Economics, has held visiting appointments and fellowships at Northwestern, Pittsburgh, Berlin, and Vienna. He is the editor of *Rediscovering the Forgotten Vienna Circle: Austrian Studies on Otto Neurath and the Vienna Circle;* the author of *Overcoming Logical Positivism from Within: The Emergence of Neurath's Naturalism in the Vienna Circle's Protocol Sentence Debate;* and coauthor (with Nancy Cartwright, Jordi Cat, and Lola Fleck) of *Otto Neurath: Philosophy between Science and Politics* (forthcoming).

Index of Authors

Index of Subjects